BRITISH FISHES.

LONDON:
PRINTED BY SAMUEL BENTLEY,
Dorset Street, Fleet Street.

A

HISTORY

OF

BRITISH FISHES.

BY

WILLIAM YARRELL, V.P.Z.S. F.L.S.

ILLUSTRATED BY NEARLY 400 WOODCUTS.

IN TWO VOLUMES.

VOL. II.

ABSQUE DEO NIHIL

LONDON:

JOHN VAN VOORST, 3, PATERNOSTER ROW.

M.DCCC.XXXVI.

BRITISH FISHES.

*ABDOMINAL
MALACOPTERYGII.*   *SALMONIDÆ.**

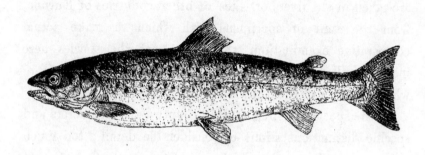

THE SALMON.

SMOLT, *young.*   GRILSE, *first year.*

Salmo salar, LINNÆUS.
    ,,   ,,   BLOCH, pt. i. pl. 20, female.
    ,,   ,,   ,, pt. iii. pl. 98, male in autumn.
    ,,   ,,   *Salmon,* PENN. Brit. Zool. vol. iii. p. 382.
    ,,   ,,   ,, FLEM. Brit. An. p. 179, sp. 40.

Generic Characters.—Head smooth; body covered with scales; two dorsal fins, the first supported by rays, the second fleshy, without rays; teeth on the vomer, both palatine bones, and all the maxillary bones; branchiostegous rays varying in number, generally from ten to twelve, but sometimes unequal on the two sides of the head of the same fish.

THE SALMON is so well known for its quality as an article of food, as well as for the immense quantities in which it is taken, that it requires no other claims to recommend it strongly to our notice; and probably, in no country

* The family of the Salmon and Trout.

VOL. II.   B of the world, in proportion to its size, are the fisheries so extensive, or the value of so much importance, as in the United Kingdom.

The history of the Salmon, and of the species of the genus *Salmo*, in this work, will extend to a considerable length ; and some doubts existing as to the extent of their identity with the species of the *Salmonidæ* generally which are taken in the rivers or lakes of other countries of Europe, from the want of specimens with which to make actual comparative examination, the account of the species here inserted will be confined more particularly to a detail of what is known of them in this country only.

Of the species existing in this country, the characters and specific distinctions admit of considerable detail : too much reliance has been placed upon colour, without resorting sufficiently to those external indications, founded on organic structure, which may with greater certainty be depended upon.

In the scale of the relative value of parts affording characters for distinction, the organs of digestion, respiration, and motion are admitted by systematic authors to hold high rank ; and in the hope to induce sportsmen to become zoologists—so far at least as to enable them to determine the various species they may meet with by a reference to those external characters which are the most important,— the specific distinctions in the genus *Salmo* will be illustrated by referring to the number and situation of the teeth, the form of the different parts of the gill-covers, and the size, form, and relative situation of the fins.

The outlines here introduced represent a front view of the mouth, and a side view of the head, of a common Trout. Of the first figure on the left hand, No. 1 marks

 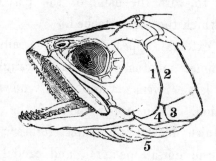

the situation of the row of teeth that are fixed on the central bone of the roof of the mouth, called the vomer: Nos. 2, 2, refer to the teeth on the right and left palatine bones; and the row of teeth outside each palatine bone on the upper jaw are those of the superior maxillary bones: No. 3, refers to the row of hooked teeth on each side of the tongue, outside of which are those of the lower jaw-bones. The Trout is chosen as showing the most complete series of teeth among the *Salmonidæ;* and the value of the arrangement, as instruments for seizure and prehension, arises from the interposition of the different rows, the four lines of teeth on the lower surface alternating when the mouth is closed with the five rows on the upper surface, those on the vomer shutting in between the two rows on the tongue, &c.

The second figure represents, in outline, a side view of the head, of which No. 1 is the preoperculum; No. 2, the operculum; No. 3, the suboperculum; No. 4, the interoperculum; No. 5, the branchiostegous rays: the four last parts together forming the moveable gill-cover. The different fins are sufficiently indicated by being coupled, when referred to, with the name of the part of the body of the fish to which they are attached.

The external appearance of the adult Salmon during the summer months, when it is caught in the estuaries of our large rivers, is too well known to require much description. The upper part of the head and back is dark bluish black; the sides lighter; the belly silvery white; the dorsal, pectoral, and caudal fins dusky black; the ventral fins white on the outer side, tinged more or less with dusky on the inner surface; the anal fin white; the small, soft, fleshy fin on the back, without rays, called the adipose, fat fin, or the second dorsal fin, is of the same colour nearly as the part of the back from which it emanates. There are mostly a few dark spots dispersed over that part of the body which is above the lateral line, and the females usually exhibit a greater number of these spots than the males.

These colours, differing but little, are, however, in a great degree common at the same period of the year to the three species that are the most numerous, as well as the most valuable; namely, the true Salmon, the Bull-Trout, and the Sea or Salmon Trout; which are also further distinguished from the other species of the genus *Salmo* by their seasonal habit of moving from the pure fresh water to the brackish water, and thence to the sea, and back to the fresh water again, at particular periods of the year. Further specific distinctions are therefore necessary; and those that will be pointed out as existing constantly in these species will, it is hoped, enable observers to identify not only each of these, but also the other species of the genus, at any age or season.

The vignette above represents the form of the different parts of the gill-cover in the three species just named; of which the figure on the left hand is that of the Salmon, the middle one is the gill-cover of the Bull-Trout, and that on the right hand is the gill-cover of the Sea or Salmon Trout: the differences are immediately apparent when thus brought into comparison.

In the Salmon, the posterior free edge of the gill-cover, as shown in the left-hand figure, forms part of a circle; the lower margin of the suboperculum is a line directed obliquely upwards and backwards: the line of the union of the suboperculum with the operculum is also oblique, and parallel with the lower margin of the suboperculum; the interoperculum is narrow vertically, and its union with the operculum is considerably above the line of the junction between the suboperculum and the operculum. The teeth of the Salmon are short, stout, pointed, and recurved: as stated in the generic characters, they occupy five situations at the top of the mouth; that is, a line of teeth on each side of the upper jaw, a line on each palatine

bone, with a few only on the vomer between the palatine
bones : the teeth on the vomer seldom exceeding two in
number, sometimes only one, and that placed at the most
anterior part; no other teeth extending along the vomer as
in the Salmon-Trout, and more particularly so in some
of those Trout that do not migrate.

The inner surface of the pectoral fin is in part dusky :
the tail very much forked when young ; the central caudal
rays growing up, the tail is much less forked the second
year, and by the fourth year it is become nearly or quite
square at the end.

The descriptions of the gill-covers of the other species
will be given in the account of the fish to which they
belong ; but it may be remarked here, that looking at
the form of the three gill-covers, it will be obvious that a line
drawn from the front teeth of the upper jaw to the longest
backward projecting portion of the gill-cover, in either species, will occupy a different situation in respect to the
eye ; that the line will fall nearest the centre of the eye
in the first, that of the Salmon, and farthest below it in the
second, that of the Bull-Trout.

As further specific distinctions in the Salmon, I may
add that, according to Dr. Richardson, the cæcal appendages are in number from sixty-three to sixty-eight; and
several observers have stated the number of vertebræ to be
sixty, which I have repeatedly found to be correct.

Commencing, then, with the true Salmon, which ascend
the rivers, in the state as to colour before mentioned, sooner
or later in the spring or summer months, it is observed
that some rivers are much earlier than others, the fish
in them coming into breeding condition and beginning
to spawn at an earlier period.

Rivers issuing from large lakes afford early Salmon, the

waters having been purified by deposition in the lakes: on the other hand, rivers swollen by melting snows in the spring months are later in their season of producing fish, and yield their supply when the lake rivers are beginning to fail. " The causes influencing this," says Sir William Jardine, to whom I am indebted for much valuable information on the *Salmonidæ*, as well as many specimens, " seem yet undecided ; and where the time varies much in the neighbouring rivers of the same district, they are of less easy solution. The Northern rivers, with little exception, are, however, the earliest,—a fact well known in the London markets ; and going still farther north, the range of the season and of spawning may be influenced by the latitude." Artedi says, " in Sweden the Salmon spawn in the middle of summer."

" It has been suggested that this variation in the season depended on the warmth of the waters ; and that those Highland rivers which arose from large lochs were all early, owing to the great mass and warmer temperature of their sources,—that the spawn there was sooner hatched. There are two rivers in Sutherlandshire which show this late and early running under peculiar circumstances. One, the Oikel, borders the county, and springs from a small alpine lake, perhaps about half a mile in breadth ; the other, the Shin, is a tributary to the Oikel, joins it about five miles from the mouth, but takes its rise from Loch Shin, a large and deep extent of water, and connected to a chain of other deep lochs. Early in the spring, all the Salmon entering the common mouth diverge at the junction, turn up the Shin, and return as it were to their own and warmer stream, while very few keep the main course of the Oikel until a much later period."

Dr. Heysham, in his Catalogue of Cumberland Animals,

has supplied similar evidence. "The Salmon," it is there observed, "is plentiful in most of our rivers, in all of which they spawn; but they evidently prefer, during the winter and spring, the Eden to the Esk, the Caldew, or the Peteril. Although the Esk and the Eden pour out their waters into the same estuary, and are only separated at the mouths by a sharp point of land, yet there is scarcely an instance of a new Salmon ever entering the former until the middle of April or beginning of May. The fishermen account for this curious fact from the different temperature of these two rivers; the water of the Eden, they allege, being considerably warmer than the water of the Esk; which is not altogether improbable, for the bed of the Esk is not only more stony and rocky than the Eden, but is likewise broader, and the stream more shallow; consequently its waters must be somewhat colder in the winter season. It is an undoubted fact, that snow water prevents the Salmon from running up even the Eden: it is probable this circumstance may have considerable effect in preventing them from entering the Esk till the beginning of summer, when the temperature of the two rivers will be nearly the same. The Peteril joins the Eden a little above, and the Caldew at Carlisle; yet up these rivers the Salmon never run unless in the spawning season, and even then in no great numbers."

The number of fish obtained in the spring in a proper state for food is small compared with the quantity procured as the summer advances. During the early part of the season, the Salmon appear to ascend only as far as the river is influenced by the tide, advancing with the flood, and generally retiring with the ebb, if their progress be not stopped by any of the various means employed to catch them, which will be explained hereafter. It is observed that the female fish ascend before the males ; and the young fish of the year, called Grilse till they have spawned once, ascend earlier than those of more adult age. As the season advances, the Salmon ascend higher up the river beyond the influence of the tide : they are observed to be getting full of roe, and are more or less out of condition according to their forward state as breeding fish. Their progress forwards is not easily stopped ; they shoot up rapids with the velocity of arrows, and make wonderful efforts to surmount cascades and other impediments by leaping, frequently clearing an elevation of eight or ten feet, and gaining the water above, pursue their course. If they fail in their attempt and fall back into the stream, it is only to remain a short time quiescent, and thus recruit their strength to enable them to make new efforts.

These feats of the Salmon are frequently watched with all the curiosity such proceedings are likely to excite. Mr. Mudie, in the British Naturalist, describes from personal observation some of the situations from which these extraordinary efforts can be witnessed. Of the fall of Kilmorac, on the Beauly, in Invernesshire, it is said, " The pool below that fall is very large; and as it is the head of the run in one of the finest Salmon rivers in the North, and only a few miles distant from the sea, it is literally thronged with Salmon, which are continually attemping to pass the fall, but without success, as the limit of their perpendicular spring does not appear to exceed twelve or fourteen feet : at least, if they leap higher than that they are aimless and exhausted, and the force of the current dashes them down again before they have recovered their energy. They often kill themselves by the violence of their exertions to ascend ; and sometimes they fall upon the rocks and are captured. It is indeed said that one of the wonders which the Frasers of

Lovat, who are lords of the manor, used to show their guests, was a voluntarily cooked Salmon at the falls of Kilmorac. For this purpose a kettle was placed upon the flat rock on the south side of the fall, close by the edge of the water, and kept full and boiling. There is a considerable extent of the rock where tents were erected, and the whole was under a canopy of overshadowing trees. There the company are said to have waited until a Salmon fell into the kettle and was boiled in their presence. We have seen as many as eighty taken in a pool lower down the river at one haul of the seine, and one of the number weighed more than sixty pounds."

The fish having at length gained the upper and shallow pools of the river, preparatory to the important operation of depositing the spawn in the gravelly beds, its colour will be found to have undergone considerable alteration during the residence in fresh water. The male becomes marked on the cheeks with orange-coloured stripes, which give it the appearance of the cheek of a *Labrus ;* the lower jaw elongates, and a cartilaginous projection turns upwards from the point, which, when the jaws are closed, occupies a deep cavity between the intermaxillary bones of the upper jaw ; the body partakes of the golden orange tinge, and the Salmon in this state is called a red-fish. The females are dark in colour, and are as commonly called black-fish ; and by these terms both are designated in those local and precautionary regulations intended for the protection and preservation of the breeding fish.

The process of spawning has been described by various observers. " A pair of fish are seen to make a furrow, by working up the gravel with their noses, rather against the stream, as a Salmon cannot work with his head down stream, for the water then going into his gills the wrong way, drowns him. When the furrow is made, the male and female retire to a little distance, one to the one side and the other to the other side of the furrow : they then throw themselves on their sides, again come together, and rubbing against each other, both shed their spawn into the furrow at the same time. This process is not completed at once ; it requires from eight to twelve days for them to lay all their spawn, and when they have done they betake themselves to the pools to recruit themselves. Three pairs have been seen on the spawning-bed at one time, and were closely watched while making the furrow and laying the spawn."*

The following extracts are made from a valuable paper by Dr. Knox, published in the Transactions of the Royal Society of Edinburgh.

" November 2.—Salmon are observed to be spawning in the various tributary streams of the Tweed which join that river from the north, and a pair are watched. The ova observed to be deposited near the sources of the stream on the 2nd of November, and covered up with gravel in the usual way."

" February 25, or a hundred and sixteen days after being deposited, the ova, on being dug up, are found to be unchanged. If removed at this time, and preserved in bottles filled with water, the developement of the egg may be hastened almost immediately by being put into warm rooms : it is not necessary to change the water. The fry so hatched, i. e. artificially, cannot be preserved alive in bottles longer than ten days ; they eat nothing during their confinement."

" March 23.—The ova now changing ; the outer shell cast ; the fry are lying imbedded in the gravel, as fishes

* Ellis on the Natural History of the Salmon.

c 2

somewhat less than an inch in length, being now twenty
weeks from the period of their deposition."

" April 1. — On reopening the spawning-bed, most of
the fry had quitted it by ascending through the gravel.
During a former series of observations I have found the
ova imbedded in the gravel unchanged on the 10th of
April, and as fry or fishes, but still imbedded in the gravel,
on the 17th : they were taken that year, with fly, as Smolts,
on the 22nd of April, about the size of the little finger."

Some specimens of Salmon fry now before me, with a
portion of the ovum still attached to the abdomen of each
fish, measure one inch in length : the head and eyes are
large ; the colour of the body pale brown, with nine or
ten dusky grey marks across the sides.   These dusky
patches, longer vertically than wide, are common, I have
reason to believe, to the young of all the species of the
genus *Salmo*.   I have seen them in the young of the
Salmon, Bull-Trout, Salmon-Trout, Parr, Common Trout,
and Welsh Charr.   I have never had an opportunity of
examining the young of the Northern Charr, or the Great
Lake-Trout ; yet I have no doubt but that, when only two
or three inches long, they also are marked in the same
manner.   In a specimen of the young of the Salmon six
inches long, these transverse marks are still observable when
the fish is viewed in a particular position in reference to the
light ; and if the scales are removed, the marks are much
more obvious.   In a Parr of the same length these marks
are still more conspicuous ; they are also very distinct in
the Common Trout and in the Welsh Charr for a considerable time ; and as far as my own examination has gone,
these lateral markings observable in the fry of the species of
Salmo are lost, or become indistinct, sooner or later, depending on the ultimate natural size attained by the particular

species : thus, they are soonest lost in the Salmon and in the Bull-Trout, and are borne the longest in the Common Trout and Parr ; indeed, I have never seen the Parr, at any age or size, without some trace of the remains of these markings. It is this similarity in markings and appearance of the fry which has caused the difficulty in distinguishing between the various species when so young; and experimenters, believing they had marked young Parr only, have been surprised to find some of their marked fish return as Grilse, young Bull-Trout or Whitling, Salmon-Trout, River-Trout, and true Parr.

There are striking examples in other animals of this similarity in the markings, or family likeness, in the young of the various species of the same genus, however different may be the colours of the parent animals. The young of the lion and the puma are as much marked for a time as the young of the tiger and leopard, or, indeed, of any of the other cats, whether striped or spotted ; and the young of all deer are said, and many are known to be, spotted, though it is also known that the greater number of the adult animals are perfectly plain.

To return to the Salmon. The adult fish having spawned, being out of condition and unfit for food, are considered as unclean fish. They are usually called Kelts ; the male fish is also called a Kipper, the female a Baggit. With the floods of the end of winter and the commencement of spring they descend the river from pool to pool, and ultimately gain the sea, where they quickly recover their condition, to ascend again in autumn for the same purpose as before ; but always remaining for a time in the brackish water of the tide-way before making either decided change ; obtaining, it has been said, a release from certain parasitic animals, either external or internal, by each seasonal change ; those of the salt

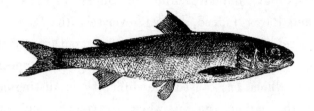

water being destroyed by contact with the fresh, and *vice versâ*.

The fry are observed to collect in small pools and mill-dam heads preparatory to quitting the river. The specimen from which the figure on the page was taken was obtained in the Thames, in which river they are occasionally caught in the season, with other fry of *Salmonidæ*, by fishermen who work at night with a casting-net on the gravelly shallows for Gudgeons to supply the London fishmongers.

My own specimens of the young of the Salmon having been preserved in spirits, and the colours thereby affected, the following description is from Dr. Heysham's Catalogue before referred to, premising that some differences in colour may be expected in specimens from different rivers.

" Length seven inches and a half; circumference three inches and one-eighth : head dark green; gill-covers fine silvery white, marked with a dark-coloured spot ; belly and sides up to the lateral line of the same silvery colour; back and sides down to the lateral line dusky, inclining to green ; sides above the lateral line marked with numerous blackish spots ; along the lateral line, and both a little above and beneath it, several dull obscure red spots : dorsal fin has twelve rays, marked with several blackish spots ; pectoral fin has twelve rays, of a dusky olive colour; ventral fin eight rays of a silvery white ; anal fin ten rays of the same colour. When the scales were carefully taken off with a knife, the obscure red spots became of a fine vermilion, and were nineteen in number; and ten obscure oval bars of a dusky bluish colour appeared, which crossed the lateral line. In a young fry which has not acquired scales, these bars are very distinct."

Whether the river be considered an early or a late river, the descent of the fry is said to take place much about the same time in all. It begins in March, and continues through April and part of May. It rarely happens that any Salmon fry are observed in the rivers late in June. The Smolt, or young Salmon, is by the fishermen of some rivers called a Laspring; and various couplets refer to the fish, as well as to the time and circumstances under which the descent is made.

> The *last spring* floods that happen in May,
> Carry the Salmon fry down to the sea.

And again,

> The floods of May
> Take the Smolts away.

But the uncertainty of popular or provincial names is a source of great perplexity to the naturalist. The Laspring of some rivers is the young of the true Salmon; but in others, as I know from having had specimens sent me, the Laspring is really only a Parr; and it must also be recollected that the fry of two other species at least descend to the sea about the same time as those of the Salmon.

The Salmon fry at first keep in the slack water by the sides of the river; after a time, as they become stronger, they go more towards the mid-stream; and when the water is increased by rain, they move gradually down the river. On meeting the tide, they remain for two or three days in that part where the water becomes a little brackish from the mix-

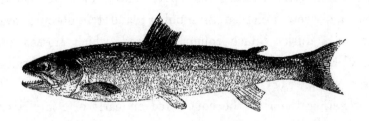

ture of salt water, till they are inured to the change, when they go off to the sea all at once. There, their growth appears to be very rapid, and many return to the brackish water, increased in size in proportion to the time they have been absent. Fry marked in April or early in May have return-ed by the end of June weighing from two to three pounds and upwards. The London markets during the latter part of June, and the months of July and August, exhibit fish of the year varying in weight from two to six pounds. I have one, here figured, that weighed only fifteen ounces, which, judging from its appearance when I bought it, that it had been to sea, is the smallest specimen I have ever seen that had been once to salt water.

These small-sized fish, when under two pounds' weight, are called by some of the London fishmongers Salmon-Peal; when larger, Grilse. These fish of the year breed during the first winter; they return from the sea with the roe enlarged; the ova in a Grilse being of nearly the same comparative size as those observed in a Salmon, but they mature only a much smaller number. The Grilse visit the estuary, remain-ing for a considerable time in the brackish water, afterwards in the tide-way above, ultimately pushing up to the sources of the tributary streams, and, as before observed, rather ear-lier in the season, in the same river, than the more adult fish.

It has been a constantly received opinion, that all the young fish after their first visit to the sea return to the rivers in which they had been bred ; and numbers of marked fish are stated to have been retaken in their native rivers : but it is equally certain that some have been taken in other rivers not far off. The difficulty of supposing that they could find and return to the same spot after roving for miles along the coast remains to be solved. That they do thus rove for miles is proved by the thousands that are taken in nets placed in the bays along the coast. Very many Tweed Salmon have been caught opposite Hopetoun House on the Forth ; and a very successful fishing there is generally followed by a scarce one in the Tweed. It is therefore very probable, from the remarks of Dr. Heysham and Sir William Jardine, that if the fish happen to have roved far from the estuary of their native river, they run at the proper season up any stream, even the first they encounter, the temperature and condition of which are congenial to them.

The growth of the Salmon from the state of fry to that of Grilse has been shown to be very rapid ; and the increase in weight attained during the second and each subsequent year is believed to be equal, if not to exceed, the weight gained within the first. The increase in size is principally gained during that part of the year in which the fish may be said to be almost a constant resident in the sea. That the food sought for and obtained to produce and sustain so rapid an increase of size must be very considerable in quantity, as well as most nutritious in quality, cannot be doubted. That the Salmon is a voracious feeder, may be safely inferred from the degree of perfection in the arrangement of the teeth, and from its own habits, of which proof will be adduced, as well as from the well-known habits of the species most closely allied to it ; yet of the many observers who have examined the stomach of the Salmon to ascertain the exact nature of that food which must constitute their principal support, few have been able to satisfy themselves. Dr. Knox states, " that the food of the Salmon, and that on which all its estimable qualities, and, in his opinion, its very existence, depend, and which the fish can obtain only in the ocean, he has found to be the ova or eggs of various kinds of echinodermata, and some of the crustacea. From the richness of the food on which the true Salmon solely subsists, arises, at least to a certain extent, the excellent qualities of the fish as an article of food. Something, however, must be ascribed to a specific distinction in the fish itself : for though he has ascertained that the Salmon-Trout lives very much in some localities on the same kind of food as the true Salmon, yet under no circumstances does this fish acquire the same exquisite flavour as the true Salmon."

That they occasionally, however, take other food, is also well known. Faber, in his Natural History of the Fishes of Iceland, remarks, " The common Salmon feeds on small fishes, and various small marine animals." Dr. Fleming says, " Their favourite food in the sea is the Sand Eel ;" and I have myself taken the remains of Sandlaunce from the stomach. Sir William Jardine says, " In the north of Sutherland a mode of fishing for Salmon is sometimes successfully practised in the firths, where Sand Eels are used as bait : a line is attached to a buoy or bladder, and allowed to float with the tide up the narrow estuaries. The Salmon are also said to be occasionally taken at the lines set for Haddocks, baited with Sand Eels. At the mouths of rivers they rise freely at the artificial fly within fifty yards of the sea ; and the common earth-worm is a deadly bait for the clean Salmon. All the other marine Salmon are known to be very voracious ; and there is nothing in the structure of the mouth or strong teeth of the common Salmon, to warrant us to suppose that there is any material difference in their food." The following is an extract from a letter sent me by Sir William Jardine, dated St. Boswell's, 15 April 1835 :—" The fisherman who rents this part of the Tweed, fishing with worm one day last week, had his hooks and tackle taken away by a fish. He put on a new set, and again with worm in ten minutes hooked and killed a Salmon with his former hooks and bait in his mouth. This will either prove extreme voracity, or little sensibility in the parts of the mouth. I have often heard fishermen mention a similar fact, but never before knew an instance on which I could depend."

Several observers have borne testimony to the partiality of the Salmon to the Sandlaunce as food ; and I have a record by an angler of Salmon caught in the Wye with a Minnow.

The present London season, 1835, has been more than of usually remarkable for large Salmon. I have seen ten different fish varying from thirty-eight to forty pounds each. A notice appeared in the public papers of one that weighed fifty-five pounds ; and, from the inquiries made, there is reason to believe most of these large-sized Salmon were sent from the Tay. Salmon, however, of much larger size have been occasionally taken. Mr. Mudie has recorded one of sixty pounds. In a note to the history of the Salmon in several editions of Walton, one is mentioned that weighed seventy pounds ; Pennant has noticed one of seventy-four pounds : the largest known, as far as I am aware, came into the possession of Mr. Groves, the fishmonger of Bond-street, about the season of 1821. This Salmon, a female, weighed eighty-three pounds ; was a short fish for the weight, but of very unusual thickness and depth. When cut up, the flesh was fine in colour, and proved of excellent quality.

The Salmon of the largest size killed by angling of which I have been able to collect particulars, are, in the Thames, October 3, 1812, at Shepperton Deeps, Mr. G. Marshall, of Brewer-street, London, caught and killed a Salmon with a single gut, without a landing-net, that weighed twenty-one pounds four ounces.

Sir H. Davy used occasionally to visit the Tweed for the sake of angling for Salmon. This river is famed for affording amusement to the Salmon fisher, more especially from the middle of March to the beginning of May. " We have heard," says Mr. Stoddart, in his Art of Angling as practised in Scotland, " that on one occasion Sir H. Davy happened by good fortune to hit upon an immense fish, weighing about forty-two pounds, immediately above Yair-bridge, and captured him after a severe struggle. This feat he makes no mention of in his Salmonia, although certainly worthy of some notice."

Mr. Lascelles, in his Letters on Sporting, Part I. Angling, says at page 21, " The largest Salmon I ever knew taken with a fly was in Scotland : it weighed fifty-four pounds and a half."

It may be stated generally, that Salmon pass the summer in the sea, or near the mouth of the estuary : in autumn they push up rivers, diverging to the tributary streams ; in winter they inhabit the pure fresh water, and in spring descend again to the sea. The question has frequently arisen, Could Salmon be preserved permanently in fresh water ? and from some facts to be adduced, it appears that they might, but not without some diminution in size or quality, or both.

Mr. Lloyd, in his Field Sports of the North of Europe, vol. i. p. 301, says, " Near Katrineberg there is a valuable fishery for Salmon, ten or twelve thousand of these fish being taken annually. These Salmon are bred in a lake, and, in consequence of cataracts, cannot have access to the sea. They are small in size, and inferior in flavour. The year 1820 furnished 21,817."

A large landed proprietor in Scotland, whose name I do not know that I am at liberty to mention, wrote as follows:— " In answer to your inquiry about the Salmon fry I have put into my newly-formed pond, I must tell you, the water was first let in about the latter end of 1830, and some months afterwards, in April 1831, I put in a dozen or two of small Salmon fry, three or four inches long, taken out of a river here, thinking it would be curious to see whether they would grow without the possibility of their getting to the sea or salt water. As the pond, between three and four acres in extent, had been newly stocked with Trout, I did not allow any fishing till the summer of 1833, when we caught with the fly several of these Salmon, from two to three pounds' weight, perfectly well shaped, and filled up, of the best Salmon colour outside, the flesh well-flavoured and well-coloured, though a little paler than that of new-run fish."

It remains to describe the different modes by which the Salmon are taken; and these are as various, and the fisheries are as numerous and as extensive, as the value and quantity of the fish would lead us to expect. The rights of the proprietors, which have arisen in various ways, some by royal grants, others by possession or occupation of the soil, are generally farmed or hired at a rent depending on the extent or value of the local stations. The first attack made upon the fish is in the summer months, when the Salmon rove along the coast in quest of the mouths of the different rivers, in which they annually cast their spawn. " On these expeditions, the fish generally swim pretty close to the shore, that they may not miss their port; and the fishermen, who

are well aware of this coasting voyage of the Salmon, take
care to project their nets at such places as may be most con-
venient for intercepting them in their course."

" It so happens that Carrick-a-rede (the rock in the road),
between Ballycastle and Portrush, eastward of Ballintoy,
is the only place on this abrupt coast (the northern coast of
the county of Antrim) which is suited for the purpose. The
net is projected directly outward from the shore with a slight
bend, forming a bosom in that direction in which the Salmon
come. From the remote extremity of the net a rope is
brought obliquely to another part of the shore, by which
the net may be swept round at pleasure and drawn to the
land : a heap of small stones is then prepared for each person. All things being ready, as soon as the watchman
perceives the fish advancing to the net, he gives the watch-
word. Immediately some of the fishermen seize the oblique
rope, by which the net is bent round to enclose the Salmon,
while the rest keep up an incessant cannonade with their
ammunition of stones, to prevent the retreat of the fish till
the net has been completely pulled round them ; after which
they all join forces, and drag the net and fish quietly to the
rocks." *

Pursuing a course along the shore and arrived at an
estuary, on each side of the mouth, and for miles up on both
sides, stake-nets are used, of which the vignette represents
the form. The distance between high and low water mark
on the shore is the site occupied. The shallow extremity
of the net on the left hand in the figure, which is fixed and
supported by stakes, is placed on the shore at high-water
mark ; the deepest part of the net, at low-water mark ; the
concavity of the sweep of the net between its two ends,

---

* Letters concerning the Natural History of the Basalts on the Northern
Coast of the County of Antrim, by the Rev. William Hamilton, A.B.

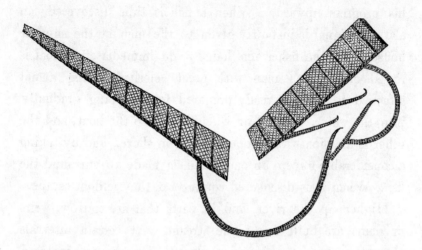

called the court, being opposed or open to the flood-tide running up the river, the Salmon which in their passage up along-shore strike against any part of the net are conducted by its form to the chambers, from whence they can find no retreat.

Many fish, in the wide part of the estuaries, ascending with each flood-tide and returning with the ebb, it is not unusual to have stake-nets placed in the reverse position, with the courts open to the ebb-tide, on purpose to meet this disposition in the Salmon; and they do actually sometimes catch as many fish in their downward as in their upward course.

The central portions of the streams, many of which are very wide, are worked incessantly by fishermen in boats called cobles, with long sweeping seine-like nets. Another mode of fishing is with a net dropped into the water from the stern of a boat, as the boat is rowed away from the shore. Men are stationed at particular places near the river, where the water is shallow, to watch the fish coming up; and so habituated are they to this, that they can discover by a ripple on the surface of the water even a solitary fish making his progress upward. When a fish is thus discovered, an alarm or signal is instantly given to the men at the shiel or house where the fishermen lodge ; and immediately a boat is rowed off by one man with great celerity, having a net attached to it, and ready prepared for dropping gradually into the water, one end of which is tied to the boat, and the other is dragged with a rope by men on shore ; and by taking a considerable sweep, an endeavour is made to surround the fish. When thus discovered coming up, they seldom escape.

Higher up the river, and in parts that are narrow, weirs or dams are built across the stream. At certain intervals along these weirs, cruives are placed. Cruives are enclosed spaces formed in the dam wall ; the fish enter these spaces, through which the water rushes, as they push up the stream, and are prevented by a grating of a peculiar contrivance from returning or getting out. All the wide and open pools of the river between these artificial, or any other natural contractions of the stream, are fished with the coble and sweep net.

In the work by the Rev. William Hamilton already quoted, and in the second series of Mr. Jesse's Gleanings in Natural History, an interesting account is related of the assistance afforded by a water-dog to some Salmon fishermen when working nets in shallow pools. The dog takes his post in a ford or on a scour where the water is not very deep, and at a distance below the net : if a Salmon escapes the net, the fish makes a shoot down the river in the direction towards the sea : the dog watches and marks his approach by the ripple on the water, and endeavours to turn the fish back towards the net, or catch him ; if he fails in both attempts, and the fish passes him, the dog then quits the water, in which the pace of the fish is too fast for him, and runs with all his speed down the bank of the river to intercept the fish at the next shallow ford, where another opportunity and a second diverting attempt occurs.—I learn also from Mr. Bicheno, that dogs are occasionally used when trying for Salmon in that part of Glamorganshire where he now resides. These dogs appear to take great pleasure in the pursuit, exhibiting by turns the most patient watchfulness, persevering exertion, or extraordinary sagacity, as either quality may best effect the wishes of the master. In some parts of Wales, where the rivers are narrow, and the Salmon are caught in a net drawn by men on each bank, dogs are trained to swim over from side to side with the head and ground lines of the net, as required.

Sir Walter Scott, in his novels of Redgauntlet and Guy Mannering, has described with his well-known skill and effect the animated scenes which occur when parties are engaged in spearing Salmon either by daylight or torchlight, as practised in the North. These works are familiar to all, and repetition would be useless. For the following description of two other modes of taking Salmon I am indebted to the kindness of Dr. Richardson.

A particular kind of fishing is peculiar to the Solway Firth, or at least can be practised with success only where the tide flows, as it does there, over extensive flats. The instrument used is termed a " halve," and consists of a funnel-shaped net ending in a pocket or bag. The mouth of this net is stretched on an oblong frame about three yards wide, to which there is attached a handle or pole. When the tide begins to flow, a number of fishermen proceed over the sands, and range themselves in a close line across the current of the flood, each with the halve resting on the bottom, and its pole against his shoulder: as the tide rises, it becomes too deep for the man farthest from the shore, who then raises his net and places himself at the other ex-

VOL. II. D

tremity of the line, where he is shortly succeeded by another and another, the whole thus changing places continually. When a fish strikes the halve, its mouth is instantly elevated above the surface by the fisherman, so as to prevent its retreat until it can be carried into shallow water and secured. During the ebb a similar plan is pursued in a reversed order; the mouths of the nets are still turned to the current, but the fishermen now move in turn to the end of the line which stands deepest in the water. Flat-fish are the principal returns of this fishing; but prime Salmon are occasionally taken both on the flood and ebb.

This kind of fishing being as yet open to all, and unfettered by parliamentary enactments, there is scarcely a cottage on the shores of the Solway Firth where the halve-net may not be seen suspended. The fishermen have all some other employment by which they maintain their families, being mostly artisans; and they generally consume the produce of the halve-net at home, unless they chance to take a fish whose value is sufficient to compensate them for the time spent in going to market, sometimes ten or twelve miles distant.

Somewhat akin to this is the Salmon fishery in the Frith of Forth. Narrow stages or platforms, supported on wooden pillars, are carried from the shore for a considerable distance into the river. Upon each of these half-a-dozen or more fishermen station themselves with bag-nets, which are dropped down from the side of the stage with the current of the tide. The owner concealed, and also sheltered by a straw hurdle, such as is used in decoys for water-fowl, watches his net, and on a fish being taken, instantly secures it. When the tide ebbs, the net is shifted to the opposite side of the stage.

" A singular method of taking Salmon is practised at

Invermoriston, in the county of Inverness, where the river flows in a narrow chasm between two projecting rocks. The fisherman seats himself on a cleft of this rock, right over the cascade, with a spear in his hand, which has a line fixed to the upper end of the shaft, similar to the practice of fishing for Whales with harpoons. Whenever the Salmon makes a spring to gain the ascent over the cataract, the spearman strikes the fish and lets the shaft go, holding only by the line until the fish has exhausted his strength ; then the spear and fish are thrown ashore by the stream, and taken out at the lower side of the pool."

The mode of fishing for Salmon in the Severn, and other rivers of Wales, with coracles and nets, requires a short and concluding notice. The coracle is a small boat constructed with willow twigs in the manner of basket-work, or with split slips of elastic wood, both the form and the material varying in different counties. In the neighbourhood of Shrewsbury, the framework is covered with canvass and painted ; in Cardiganshire it is covered with flannel, and afterwards with a coating of tar. The boat is something less than six feet long, and about four feet wide, with a seat across the middle. The form of the paddle with which this little boat is impelled and guided along is also varied : in the Severn, the blade is square, as represented in the specimen lying on the ground in the vignette ; the more elongated blade of the paddle in the hand of the fisherman is the form in use on the Dee. The boat, which in appearance is not unlike one half of a walnut-shell, is so light and portable that the fisherman carries it to and from the water on his back. These coracles,* so called, it is said, from *corium*, the hide of the beast with which they were

* This word is sometimes written coriacle, and may be derived from *coriago*, hide-bound.

D 2

formerly covered, are of great antiquity : they were known in Cæsar's time, and are described by Lucan to be very nearly the same as in our own days.

> " With twisted osiers the first boats were made,
>   O'er which the skins of slaughter'd beasts were laid ;
> With these the Britains on the oceans row,
>   And the Venetians on the swelling Po."

The custom of alternately carrying or being carried, as practised by the fisherman and his boat, is whimsically alluded to in the following lines, extracted from an old MS. history of Shropshire.

> " Some sportsmen in pursuit of prey,
>   Their horses on their shoulders lay ;
> But seizing of the booty, then
>   They sit their steeds like other men.
> Returning home when all is o'er,
>   Their steeds they carry as before."

The coracle is in frequent request with fly-fishers,* the banks of the rivers being in some places very rugged and steep, in others overgrown with wood to the water's edge.

The fishing for Salmon in coracles is performed by two men, each in his little boat, drawing between them down the stream a single-walled trammel, called there a horn-net, from its sliding by means of rings of horn, instead of corks, along the top. Through these rings runs a line, the end of which is held by one of the fishermen. By pulling upon this running line, which is distinct from the drag-line, the net is quickly closed when a fish strikes it. Various modifications of this sort of net occur in different rivers. Captain Medwin, in his Angler in Wales, says, " We stood on the bridge at Machynlleth for some time, to watch the operations of two fishermen in coracles. They were about to drag for Salmon ; and it must have been difficult to

---

* Hansard's Trout and Salmon Fishing in Wales, pages 145 and 184.

preserve the balance in such frail and fragile machines. The net was attached to the two boats, and connected them. When all was clear, the fishermen made with their paddles a considerable circle, and then reunited, drawing in cautiously the sweep. They seemed very dexterous in the management of their canoes, and perfectly unconscious of danger. The first essay was a failure; a Salmon of ten or twelve pounds' weight leaped over the corks."—Long doubly-walled trammel-nets are now in use near Shrewsbury.

The length of the head of the Salmon, as compared to the whole length of the fish, is as one to five : the eye rather small, placed nearer to the point of the nose than to the posterior edge of the gill-cover : the peculiarities of the teeth and the parts of the operculum have been already described : the origin of the last ray of the dorsal fin about half-way between the point of the nose and the end of the tail ; the first two rays simple and shorter than the third, which is the longest and branched ; all the other rays of this fin branched ; the last ray double, but arising from a single origin, is only counted as one : the posterior edge of the base of the adipose fin is half-way between the origin of the last dorsal fin-ray and the end of the tail, and over the origin of the last ray of the anal fin. The pectoral fin two-thirds of the length of the head ; ventral fin in a vertical line under the middle of the dorsal fin, with an axillary scale two-fifths of the length of the ventral fin itself; the anal fin commences about half-way between the origin of the ventral fin and the commencement of the lower caudal fin-rays, the third ray the longest, the first two rays simple, the others branched : the form of the tail has been already noticed. The body is elongated ; the dorsal and abdominal line about equally convex ; the lateral line near the middle of the body, dividing it equally ; the fleshy portion of the tail slender, and ending in the form of one half of a hexagon; the scales moderate in size, oval and thin, easily removed when young, adherent when old. The fin-rays in number are—

D. 13 : P. 12 : V. 9 : A. 9 : C. 19.   Vertebræ 60.

Salmon, and indeed all the *Salmonidæ*, like other fish that swim near the surface of the water, cannot be eaten too fresh: its fine flavour, as well as its value, diminish rapidly after capture.   In London a Thames Salmon commands the highest price : the next in point of value is that sent up either from Woodmill or Christchurch in Hampshire ; then those fish received from the Severn, which are usually brought by the mail from Gloucester.

A Thames Salmon is a prize to a fisherman, which, like other prizes, occurs but seldom.   The last Thames Salmon I have a note of was taken in June 1833.   The appearance of the Common Tern, or Sea-Swallow, which on its arrival in May wings its flight for miles up the Thames, is the signal to the fishermen to keep a good look-out for a Salmon : the occasionally coincident reappearance of a Tern and a Salmon has induced some of the Thames fishermen to apply to the former the name of the Salmon-bird.

ABDOMINAL
MALACOPTERYGII.   *SALMONIDÆ.*

THE BULL-TROUT.

THE GREY TROUT.   WHITLING.   ROUNDTAIL.

Salmo eriox,   LINNÆUS.
  ,, *cinereus aut griseus,* WILLUGHBY, p. 193.
  ,, *griseus seu cinereus,* RAY, p. 63, A. 3.
  ,, *eriox,*   *Grey,* PENN. Brit. Zool. vol. iii. p. 394.
  ,, *Cambriscus, Sewin,* DON. Brit. Fish. pl. 91.
  ,, *eriox,*   *Grey,* FLEM. Brit. An. p. 180, sp. 46.

THE BULL-TROUT is distinguished from the Salmon
and Salmon-Trout by several specific peculiarities. The
gill-cover differs decidedly in form, as examination of the
central figure of the illustration at page 5 will show.
The operculum is larger; the free vertical margin much
more straight; the inferior posterior angle more elongated
backwards; the line of union with the suboperculum not so
oblique, but nearly parallel with the axis of the body of
the fish: the inferior edge of the suboperculum parallel to
the line of union with the operculum: the interoperculum

much deeper vertically ; the vertical edge of the preoperculum more sinuous.   The teeth in the Bull-Trout are
longer and stronger than those of the Salmon ; but, like
the Salmon, the two or three teeth that may be seen on
the vomer occupy the most anterior part only. The tail
is square by the time this fish is twelve months old, as is
shown in the figure above, from a female fish in its first
winter, at which period and during its second season it is
called a Whitling in the Tweed ; it is afterwards called
a Bull-Trout : and the central rays of the tail continuing
to increase in length with age, the posterior edge becomes
convex ; a variation in form which has caused this fish
to be designated in the Annan by the name of Roundtail
when old, and Sea-Trout when young. It is to this species also that the names of Norway Trout and Norway
Salmon are believed to refer, as used occasionally on Tweed,
and some of the northern parts of Scotland. The Wark-
worth Trout and Coquet Trout of Northumberland and
Durham are the young of the Bull-Trout.

   The Bull-Trout, in all its stages of growth, is probably
better known in the Tweed than elsewhere : it is there
as abundant as the Salmon. I have had proof of the existence of this species in some of the rivers of Dorsetshire

and Cornwall: it occurs in the estuary of the Severn, and I have seen it from the rivers of South Wales. Dr. Heysham includes this fish among those of the rivers of Cumberland that run into the Solway. Mr. Low says it is found in the loch of Stenness, Orkney.

The Bull-Trout appears to be the *Salmo maculis cinereis caudæ extremo æquali* of Artedi, page 23, sp. 2; and probably also, as quoted, the *Graia Salmo cinereus seu griseus* of Willughby and Ray, whose specific names have precedence of *eriox*. This fish sometimes attains the weight of twenty pounds; but it more commonly occurs under fifteen pounds' weight. It ascends rivers for the purpose of spawning, in the same manner as the Salmon, but earlier in the season; and the fry are believed to go down to the sea sooner than the fry of the Salmon. This species affords good sport to anglers: it feeds voraciously, taking any fly or bait freely; and, from its great muscularity, it is a powerful fish when hooked, frequently leaping out of the water. It is not, however, held in the same degree of estimation as food as the Salmon or Salmon-Trout: the flesh, even when the fish is in season, is of a pale orange colour, at other times yellowish white. But few are sent to the London markets, and these produce comparatively but an inferior price.

The description is taken from an adult male of thirty-two inches in length, from which the cut at the head of this article was drawn and engraved.

The length of the head compared to that of the body only is as one to four; the teeth and the form of the parts of the gill-covers have been already described; the elongation of the under jaw is peculiar to the males only, but is not in the Bull-Trout so conspicuous as in the Salmon; the dorsal fin commences half-way between the

point of the nose and the origin of the short upper caudal
rays; the base of the dorsal fin longer than the longest
of its rays; the adipose fin large, and nearer to the end
of the tail than to the origin of the last dorsal fin-ray;
the form of the tail at different ages has been noticed;
the length of the pectoral fin very little more than half
the length of the head. The scales of the Salmon are
thin in substance, oval, with numerous concentric lines
only: the number of scales forming an oblique line from
the lateral line up to the base of the anterior part of
the dorsal fin, following the oblique arrangement of the
scales, about twenty; and the number in a row from
the axillary scale of the ventral fin up to the lateral
line about eighteen. The scales of the Bull-Trout are
rather smaller than those of the Salmon in fish of equal
size, the number forming a continuous oblique row from
the lateral line up to the base of the dorsal fin being
about twenty-six; the number of those forming a row
from the ventral axillary scale up to the lateral line,
whether taking the line that ascends obliquely forward
or backward, is about twenty-five; the axillary scale
of the ventral fin nearly half as long as the fin itself:
the anal fin nearer the tail than in the Salmon; all the fins
muscular.

The fin-rays of the Bull-Trout in number are—

D. 11 : P. 14 : V. 9 : A. 11 : C. 19 : Vertebræ 59.

In six specimens out of seven, the number of vertebræ
was fifty-nine; in the other, sixty. Fifty-nine will probably prove to be the normal number in the Bull-Trout.

The form of the body of this fish is similar to that
of the Salmon, but the nape and shoulders are thicker,

the fleshy portion of the tail and the base of each of the fins more muscular: the males are the strongest in the water, but the females are the most eager for bait, and their teeth are rather smaller. The colours of the males in the spawning season are—the head olive brown, the body reddish brown or orange brown, that of the females a blackish grey; the dorsal fin reddish brown, spotted with darker brown; the tail dark brown; the other fins dusky brown. The general colour at other times like that of the Salmon-Trout.

The *Salmo hucho* of English authors is probably the same as the Bull-Trout.

ABDOMINAL
MALACOPTERYGII.   *SALMONIDÆ.*

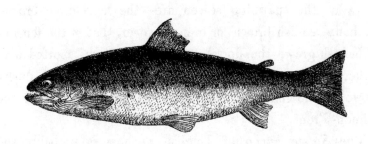

THE SALMON-TROUT.

Salmo trutta, Linnæus.
   ,,   ,,   Willughby, p. 198.
   ,,   ,,   *Sea-Trout,*   Penn. Brit. Zool. vol. iii. p. 397 ?
   ,,   ,,   ,,   Flem. Brit. An. p. 180, sp. 45.

The Salmon-Trout is, of the migrating species in this country, the next in value to the Salmon. It is most abundant in the rivers of Scotland, and its flesh is excellent. It is distinguished by the gill-cover being intermediate in its form between that of the Salmon and Bull-Trout. The representation on the right-hand of the vignette at page 5 is that of the Salmon-Trout. The posterior free margin, it will be observed, is less rounded than that of the Salmon on the left hand, but more so than that of the Bull-Trout, which is represented by the middle figure. The line of union of the operculum with the sub-operculum, and the inferior margin of the suboperculum, are oblique, forming a considerable angle with the axis of the body of the fish. The posterior edge of the preoper-culum rounded,—not sinuous, as in the Bull-Trout. The

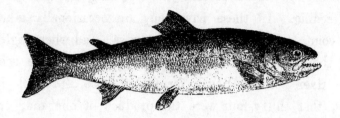

teeth are more slender as well as more numerous than in the Salmon or Bull-Trout; those on the vomer extending along a great part of its length, and indenting the tongue deeply between the two rows of teeth that are there placed, one row along each side. The tail is less forked at the same age than that of the Salmon, but becomes, like it, square at the end after the third year. The size and surface of the tail also is much smaller than that of the Salmon, from the comparative shortness of the caudal rays. The figure at the head of this article represents the Salmon-Trout in its second year; the second figure represents this species in its first year.

This fish is the White Trout of Devonshire, Wales, and Ireland; it is found in the Severn, in the rivers of Cornwall, and is plentiful in the Esk and the Eden, which communicate with the Solway, where it is called Sea-Trout.

The habits of this species are also very like those of the Salmon, and the females are said to run up the rivers before the males. Sir William Jardine says, " In approaching the entrance of rivers, or in seeking out, as it were, some one they preferred, shoals of this fish may be seen coasting the bays and headlands, leaping and sporting in great numbers, from about one pound to three or four pounds in weight; and in some of the smaller bays the shoal could be traced several times circling it, and apparently feeding. In these bays they are occasionally taken with a common hang-net stretched across; and when angled for in the estuaries, with the ordinary flies which are used in the rivers of the South for Grilse, rose and took so eagerly, that thirty-four were the produce of one rod, engaged for about an hour and a half. They enter every river and rivulet in immense numbers, and when fishing for the Salmon are annoying from their quantity. The food of those taken with the rod in the estuaries appeared very indiscriminate; occasionally the remains of some small fish, which were too much digested to be distinguished; sometimes flies, beetles, or other insects, which the wind or tide had carried out; but the most general food seemed to be the *Talitris locusta,* or common sand-hopper, with which some of their stomachs were completely crammed. It is scarcely possible to arrive with any certainty at the numbers of this fish. Two hundred are frequently taken at a single draught of a sweep-net, and three hundred have occasionally been counted." They are much more numerous in the Don, the Spey, and the Tay, than in the Tweed.

Great quantities of this Salmon-Trout are sent to the London market; those from Perth, Dundee, Montrose, and Aberdeen appear, from their comparative depth of body, to be better fed, are higher in colour, and considered to be finer in flavour than from some other localities. The Fordwich Trout of Isaac Walton is the Salmon-Trout; and its character for affording " rare good meat," besides the circumstance of its being really an excellent fish, second only to the Salmon, was greatly enhanced, no doubt, by the opportunity of eating it very fresh. Fordwich is about two miles east-north-east of Canterbury. The stream called the Stour was formerly very considerable; it communicates with the sea opposite the back of the Isle of Sheppy, and from Fordwich one branch going eastward, again enters the sea at Sandwich. The ancient right to the fishery at Fordwich was enjoyed jointly by two religious establishments: it is now vested in six or seven individuals, who receive a consideration for their several interests. It was formerly the custom to visit the nets at Fordwich every morning to purchase the fish caught during the night. I have seen specimens of the Salmon-Trout from the Sandwich river exposed for sale in the fishmongers' shops at Ramsgate, during the season for visiting that watering-place; and the Salmon-Trout is also occasionally taken in the Medway by fishermen who work long nets for Smelts during the autumn and winter. I have obtained a young fish of the year in the Thames from the men who fish for Shads above Putney-bridge in the months of June and July.

The largest adult fish of this species I have ever seen was in the possession of Mr. Groves, the fishmonger of Bond-street: this specimen, which occurred in June 1831, was a female in very fine condition, and weighed seventeen pounds.

Dr. Mac Culloch states, that the Salmon-Trout, or Sea-Trout, as it is called in Scotland, is now a permanent resident in a fresh-water lake in the island of Lismore, one of the Hebrides, and without the power of leaving it or reaching the sea. There it has been known for a long course of years, perfectly reconciled to its prison, and propagating without any apparent difficulty."[*]

The length of the head is, when compared with the length

* Journal of the Royal Institution, No. xxxiv. p. 211.

of the body alone, as one to four; the depth of the body compared to the whole length of the fish is also as one to four: the teeth small and numerous, occupying five rows on the upper surface of the mouth; those of the central row on the vomer extending some distance along it, the points turning outwards alternately to each side; one row upon each side of the under jaw, and three or four teeth on each side of the tongue, strong, sharp, and curving backwards, well calculated to assist in holding a living prey, or to convey food towards the pharynx: the middle of the eye situated half-way between the point of the nose and the posterior edge of the preoperculum: the form of the parts of the gill-cover have been already described and figured. The first ray of the dorsal fin is short; the second ray long, equal to the length of the base of the fin; the articulation at the base of the last dorsal fin-ray exactly half-way between the point of the nose and the end of the tail; the fleshy fin on the back being also half-way between the base of the last ray of the dorsal fin and the end of the tail. The body of the fish rather deep for its length; the lateral line very nearly straight, and passing along the middle of the body: the scales adhering closely; in form rather a longer oval than those of the Salmon, and having about twenty-three in the usual line up to the dorsal fin, and twenty-two below it. The fin-rays in number are—

D. 12 : P. 13 : V. 9 : A. 10 : C. 19.　Vertebræ 58.

The upper part of the head and back dark bluish black, becoming lighter on the sides, which are marked with numerous spots, somewhat resembling in form the letter X: these spots are mostly above the lateral line. The lower part of the sides and belly silvery white; cheeks and gill-covers silvery white; the dorsal fin, fleshy fin, and tail, nearly as dark as the colour of the back ; the pectoral fin rather small and bluish white ; the ventral fins white, arising in a vertical line under the middle of the dorsal fin ; the anal fin white, the base of the fin one-third shorter than the longest of its fin-rays. When the Salmon-Trout is placed by the side of a Salmon, it is, in comparison, darker in colour in the body, but lighter in the colour of the fins.

The Phinock or Hirling of the North, the *Salmo albus* of Dr. Fleming, is perhaps distinct from the Salmon-Trout just described ; but the specimens of it which I have received so exactly resemble the young of the Salmon-Trout in its first year, that I am unable to point out any sufficiently distinguishing specific character ; it is therefore omitted as a species, in the hope that it will be hereafter described and figured by some naturalist who has better opportunities, and more materials for proving its specific distinction.

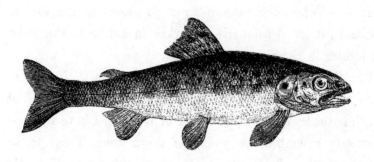

THE PARR, OR SAMLET.

Salmo salmulus, WILLUGHBY, p. 192.
    ,,   ,,   RAY, Syn. p. 63, sp. 2.
    ,,   ,,   PENN. Brit. Zool. vol. iii. p. 404.

THIS little fish, one of the smallest of the British *Sal-monidæ,* has given rise to more discussion than any other species of the genus. Abounding in our Salmon rivers, and conspicuous for those lateral marks which are now known to be borne also for a time by the young of the Trout as well as the fry of the other *Salmonidæ,* and this fish always appearing of small comparative size, it has frequently been insisted upon as the young of the Salmon, and local regulations have as generally been invoked for its preservation.

The fry, however, of the different species of migratory *Salmonidæ* are even now probably accurately known only to a few persons: their great similarity when very small has so frequently deceived even those who have lived the greater part of their lives on the Salmon river banks, that the fry marked by them, in their experiments, believing them all to be what they considered the young of the Parr, have been retaken as Grilse, Bull-Trout, Salmon-Trout, and River-Trout. That the Parr is not the young of the Salmon, or indeed of any other of the larger species of *Salmonidæ*, as still considered by some, is sufficiently obvious from the circumstance that Parrs by hundreds may be taken in the rivers all the summer, long after the fry of the year of the larger migratory species have gone down to the sea ; and the greater part of those Parrs taken even in autumn do not exceed five inches in length, when no example of the young of the Salmon can be found under sixteen or eighteen inches, and the young of the Bull-Trout and Salmon-Trout are large in proportion. As has been before stated, the transverse dusky bars from which this fish has obtained the name of Brandling and Fingerling are family marks, borne by all the species of the genus for a time, are obliterated by degrees, and at periods depending on the ultimate size attained by the individual species when adult ; the soonest probably in the Salmon, and certainly the latest in the Parr.

" Some of the rivers of Scotland being unprotected, are poached to such an extent that very few Salmon or Salmon-Trout escape the nets or spears of their relentless pursuers ; yet the Parr swarms in shoals."—*Statistics of Scotland.*

" In the Western Isles there are streams in which Parrs are common, although Salmon never visit them ; and although the Salmon and the Sea-Trout, *Salmo trutta,* frequent some of the lakes, yet the Parr has never been seen in these lakes."*

Dr. Heysham, of Carlisle, devoted particular attention to the history of this fish, which is there called Branlin and

* By the Editor of the Edinburgh Quarterly Journal of Agriculture.

E 2

Samlet; and some of his observations are here repeated, adopting only the name of Samlet, for reasons that will be hereafter explained.

" The old Samlets begin to deposit their spawn in December, and continue spawning the whole of that month, and perhaps some part of January. As this season of the year is not favourable for angling, few or no observations are made during these months. As soon as they have spawned, they retire, like the Salmon, to the sea, where they remain till the autumn, when they again return to the rivers."

" The spawn deposited by the old Samlets in the sand begins to exclude the young or fry, according to the temperature of the season, either in April or May. The young Samlets remain in the rivers where they were spawned during the whole of the spring, summer, and autumn, and do not acquire their full size till the autumn, about which time the old ones return from the sea. Hence it is evident that, although there are Samlets of various sizes in the spring and fore part of the summer, there will be no very large ones till the autumn, when the young ones have nearly acquired their full size, and the old ones have returned to associate with their offspring."

" If the weather be mild and open in January and February, Samlets are taken when retiring to the sea with empty bellies, and in a weak emaciated condition. In short, we see Samlets of various sizes—we see them with milt and roe in various stages, and we see them perfectly empty; all which circumstances clearly prove that they are a distinct species."

Sir William Jardine, during an excursion in Sutherlandshire, observed that the Parr decreased in numbers as he proceeded northward; and detailing the result of his obser-

vations made on the Parr of the Tweed, further adds, "that the difference of opinion among ichthyologists, or rather the difficulty which they appear to have in forming one, whether this fish is distinct, or only the young of some others, has rendered the solution of it interesting. The greatest uncertainty, however, has latterly resolved itself into, whether the Parr was distinct, or a variety or young of the common Trout, *S. fario ;* with the migratory Salmon it has no connexion whatever."

" Among the British *Salmonidæ,* there is no fish whose habits are so regular, or the colours and marking so constant. It frequents the clearest streams, delighting in the shallower fords or heads of the streams, having a fine gravelly bottom, and hanging there in shoals, in constant activity, apparently day and night. It takes any bait at any time with the greatest freedom ; and hundreds may be taken when no Trout, either large or small, will rise, though abundant among them. That part of its history only which is yet unknown is the breeding. Males are found so far advanced as to have the milt flow on being handled ; but at that time, and indeed all those females which I have examined, had the roe in a backward state ; and they have not been discovered spawning in any of the shallow streams or lesser rivulets, like the Trout."

" In the markings they are so distinct as to be at once separated from the Trout by any observer. The row of blue marks which is also found in the young Trout, and in the young of several *Salmonidæ*, in the Parr are narrower and more lengthened. The general spotting seldom extends below the lateral line, and two dark spots on the gill-cover are a very constant mark. On a still closer comparison between the young Trout and Parr of similar size, the following distinctions present themselves :—The Parr is altogether more delicately formed ; the nose is blunter, the tail more forked ; but the chief external distinction is in the immense comparative power of the pectoral fin : it is larger, much more muscular, and nearly one-third broader ; and we at once see the necessity for this greater power, when we consider that they serve to assist in almost constantly suspending this little fish in the most rapid streams. Scales of the Parr taken from the lateral line below the dorsal fin were altogether larger, the length greater by nearly one-third, the furrowing more delicate, and the form of the canal not so apparent or so strongly marked towards the basal end of the scale. The greater delicacy of the bones in the Parr is still kept up very distinctly. The operculum forming the posterior edge of the gill-cover is much more rounded than in the Trout, approaching in this respect to the Salmon ; in the Trout the lower part is decidedly angular. The interoperculum in the Parr is longer and narrower. The maxillary bone is broader at the posterior corner, but much shorter in the Parr ; the vomer is much weaker ; the bones or rays of the gill-covers are longer and much narrower than those of the Trout. The teeth of the Parr are smaller ; the bone of the tongue longer, weaker, and not so broad ; the under jaw much weaker, and the distance between the two sides of the under jaw in the Parr about one-third less. These are the most conspicuous distinctions, but every bone varies ; and not in one only, but in the many specimens which I have lately examined, the distinctions were the same, and at once to be perceived. In this state, therefore, I have no hesitation in considering the Parr not only distinct, but one of the best and most constantly marked species we have, and that it ought to remain in our systems as the *Salmo salmulus* of Ray." *

By the kindness of various friends, I have received Parrs from several rivers on the east, south, and west shores ; and from close comparative examination of specimens from distant localities, and these with the young of others of the *Salmonidæ*, I believe the Parr to be a distinct fish. The largest I possess measures full eight inches, and was sent to me by Dr. George Johnston, with several others not more than five inches long, from the Tweed, and taken in the month of July. The representations which illustrate this subject were taken from these specimens. The smallest Parr I have preserved measures but three inches and one quarter, and was sent me with others nearly double the size by Sir William Jardine, Bart. : these were taken in summer from the Annan. I have also received Parrs, the Skirling of Pennant, not exceeding four inches in length, from Glamorganshire and from the Ribble, in June, as well as from other localities between the Tweed and the Annan.

An opinion prevails that the Parrs are hybrids, and all of them males. Dr. Heysham, at different times and seasons, opened and examined three hundred and ninety-five Parrs, or Samlets, as they are called at Carlisle, and found one hundred and ninety-nine males, and one hundred and ninety-six females. I am indebted to J. C. Heysham, Esq. for

* Sir William Jardine, Bart. Edinburgh New Philosophical Journal for January 1835.

a specimen measuring seven inches in length, having both lobes of roe in a forward state : no such accumulation, I venture to say, will be found in the young of the Salmon, Bull-Trout, or Salmon-Trout, when only seven inches long. The specimen just referred to was taken in the middle of February. Mr. Heysham, among other communications on the subject of fishes, sent me word he had seen a female taken in March, in which the ova were very large : and the Rev. W. F. Cornish, of Totness in Devonshire, on the· Dart, where this fish is called the Heppar, preserved a specimen of a female, also taken in March, in which the ova were very large ; much larger, he said, than he could have thought it possible so small a fish could have matured. The three specimens last mentioned might be examples of late breeders, and Dr. Heysham's view of the breeding period is probably the correct one : the Parr being, as that gentleman considered, a migratory species, deposits its spawn in the depth of winter, like the other migratory species of the same genus.

The Skegger of the Thames is the Parr or Samlet. Laleham, between Staines and Chertsey, where the water is shallow, formerly afforded the greatest quantity ; forty and even fifty dozen have been taken in one day by a skilful fly-fisher ; but the numerous gas and other manufactories on the banks of the river are considered so greatly to have affected the quality of the water, that a Salmon or a Skegger in the Thames is now but rarely seen. It was customary to permit fishing for Skeggers only, before the usual period for angling in the Thames,* from the belief that these fish were migratory and their return uncertain.

* Angling in the Thames, within the conservancy of the Lord Mayor of London, which extends to Staines Bridge, is prohibited during the months of March, April, and May, under a penalty, and with loss of rod and line.

The length of the head is, as compared to the whole length of the head, body, and tail, including the caudal rays, as one to five; the body of greater girth than that of the young of the Salmon when of the same length; the pectoral fin of great breadth and length, nearly as long as the head; the base of the last ray of the dorsal fin exactly half-way between the point of the nose and the end of the upper half of the tail; the base of the dorsal fin considerably shorter than the third ray of that fin, which is the longest; the second dorsal or fleshy fin half-way between the origin of the first ray of the dorsal fin and the end of the upper half of the tail, and in a line over the origin of the last ray of the anal fin; the tail deeply forked, much more so than that of the Trout; the lower jaw shorter than the upper; the teeth small, placed in five lines on the upper inner surface of the mouth; two or three small hooked teeth on the superior lateral portion of the tongue on each side towards the tip, and a row of small teeth on each side of the under jaw: the eye large, its diameter one-fourth of the length of the head, and placed at the distance of its diameter from the point of the nose. The fin-rays in number are—

D. 13 : P. 14 : V. 9 : A. 9 : C. 19 : Vertebræ 60.

The following description of the colours of the Parr is derived from Dr. Heysham's paper already quoted, my specimens being affected by immersion in spirits.

" Head green and ash colour. Gill-covers tinged with a variable green and purple, and marked with a round dark-coloured spot: in some specimens there are two of these spots on each gill-cover. Back and sides, down to the lateral line, dusky and marked with numerous dark-coloured spots. Belly white. Along the lateral line there are from sixteen to thirty bright vermilion spots. The sides are marked with nine or ten oval bars of a dusky bluish colour. Dorsal fin with a few dusky spots; colour of the lower fins inclining to yellow." The number of scales in a row above the lateral line twenty-two, below it nineteen.

In its feeding, the Parr is voracious: the stomachs of several examined were distended with the larvæ of water-beetles of various sizes,—*Dytiscidæ*.

It would be very desirable to discontinue the use of all the names bestowed upon this fish except those of Parr and Samlet; the terms Brandling, Fingerling, Skirling, Gravelling, Laspring, Sparling, &c. not being sufficiently defined, but referring either to some quality or habit observed in other species.

The Gravelling of the river Taw, as figured in the Magazine of Natural History for January 1835, is the young of the Salmon, but with a greater number of spots than I have usually seen them.

ABDOMINAL
MALACOPTERYGII.   *SALMONIDÆ.*

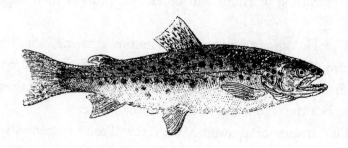

THE COMMON TROUT.

Salmo fario, Linnæus.
   ,,   ,,   *River-Trout,*   Penn. Brit. Zool. vol. iii. p. 399, pl. 70.
   ,,   ,,   *Trout,*   Don. Brit. Fish. pl. 85.
   ,,   ,,   *Common Trout,* Flem. Brit. An. p. 181, sp. 47.

The Common Trout is too widely diffused and too generally known to make any enumeration of particular localities necessary : it is an inhabitant of most of the rivers and lakes of Great Britain, and so closely identified with the pursuits and gratifications of sportsmen, that those landed proprietors who possess streams of water favourable to the production and growth of Trout preserve them with great care and at considerable expense. The Trout, though a voracious feeder, and thus affording excellent diversion to the experienced angler, is so vigilant, cautious, and active, that great skill as well as patience are required to ensure success. During the day, the larger sized fish move but little from their accustomed haunts ; but towards evening and during the night they rove in search of small fish, insects, and their various larvæ, upon which they feed with eagerness. The young Trout fry may be seen throughout the day sporting on the shallow gravelly scours of the stream, where the want of sufficient depth of water, or the greater caution of larger and older fish, prevent their appearance.

Though vigilant and cautious in the extreme, the Trout is also bold and active. A Pike and a Trout put into a confined place together had several battles for a particular spot, but the Trout was eventually the master.

The season of spawning with the Trout is generally in the month of October, at which period the adult fish make their way up the stream ; and the under jaw of the old male exhibits in a smaller degree the elongation and curvature observed to obtain in the male Salmon, of which an instance will be shown.

The Trout varies considerably in appearance in different localities ; so much so, as to have induced the belief that several species exist. It is, indeed, probable that more than one species of river Trout may exist in this country ; but when we consider geologically the various strata traversed by rivers in their course, the effect these variations of soil must produce upon the water, and the influence which the constant operation of the water is likely to produce upon the fish that inhabit it ;—when we reflect also on the great variety and quality of the food afforded by different rivers, depending also on soil and situation, and the additional effect which these combined causes in their various degrees are likely to produce ;—we shall not be much surprised at the variations both in size and colour which are found to occur. That two Trout of very different appearance and quality should be found within a limited locality in the same lake or river, is not so easily explained ;

and close examination of the various parts which afford the most permanent characters should be resorted to, with a view to determine whether the subject ought to be considered only as a variety, or entitled to rank as a species. In these examinations the character of the internal organs also, and the number of the bones forming the vertebral column, should be ascertained. The normal number of vertebræ in *Salmo fario*, our Common Trout, I believe to be fifty-six.

Sir William Jardine, Bart. in a paper on the *Salmonidæ*, published in the Edinburgh New Philosophical Journal for January 1835, has described at considerable length the variations observed in the Trout of some of the lochs of Sutherlandshire. Other lochs abound with Trout which are reddish, dark, or silvery, according to the clearness of the water. Mr. Neill, in his Tour, has noticed the black-moss Trout of Loch Knitching, and Loch Katrine is said to abound also with small black Trout; an effect considered to be produced in some waters by receiving the drainings of boggy moors. In streams that flow rapidly over gravelly or rocky bottoms, the Trout are generally remarkable for the brilliancy and beauty of their spots and colours. Trout are finest in appearance and flavour from the end of May till towards the end of September; an effect produced by the greater quantity and variety of nutritious food obtained during that period. Two specimens of the Common Trout taken early in January were unusually fine in colour for that season of the year; their stomachs on examination were distended with ova of large size, which, from circumstances attending the capture of the Trout, were known to be the roe of the Bull-Trout. The albuminous nature of this sort of food, which the Trout availed themselves of, was believed to be the cause of their colour; since other

Trout, procured at the same time from localities where no such food could be obtained, were of the usual dark colour of that season of the year.

Mr. Stoddart, in his " Art of Angling as practised in Scotland," mentions an interesting experiment made with Trout, some years ago in the south of England, in order to ascertain the value of different food. " Fish were placed in three separate tanks, one of which was supplied daily with worms, another with live minnows, and the third with those small dark-coloured water-flies which are to be found moving about on the surface under banks and sheltered places. The Trout fed with worms grew slowly, and had a lean appearance; those nourished on minnows, which, it was observed, they darted at with great voracity, became much larger; while such as were fattened upon flies only, attained in a short time prodigious dimensions, weighing twice as much as both the others together, although the quantity of food swallowed by them was in nowise so great."

Of four Trout fed in a stew together, three of them weighed fifteen pounds each, the fourth attained the weight of seventeen pounds; but neither the food nor the time consumed was recorded.

Stephen Oliver the younger, in his agreeable Scenes and Recollections of Fly-fishing, mentions a Trout " taken in the neighbourhood of Great Driffield, in September 1832, which measured thirty-one inches in length, twenty-one in girth, and weighed seventeen pounds." A few years since, a notice was sent to the Linnean Society of a Trout that was caught on the 11th of January 1822, in a little stream, ten feet wide, branching from the Avon, at the back of Castle-street, Salisbury. On being taken out of the water, its weight was found to be twenty-five pounds. Mrs. Powell, at the bottom of whose garden the fish was first discovered, placed it in a pond, where it was fed and lived four months, but had decreased in weight at the time of its death to twenty-one pounds and a quarter.

The age to which Trout may arrive has not been ascertained. Mr. Oliver mentions, that in August 1809, "a Trout died which had been for twenty-eight years an inhabitant of the well at Dumbarton Castle. It had never increased in size from the time of its being put in, when it weighed about a pound ; and had become so tame, that it would receive its food from the hands of the soldiers." In August 1826, the Westmoreland Advertiser contained a paragraph stating that a Trout had lived fifty-three years in a well in the orchard of Mr. William Mossop, of Board Hall, near Broughton-in-Furness.

The Thames at various places produces Trout of very large size. Among the best localities may be named Kingston, opposite the public-house called the Angler, Hampton-Court bridge and wear, and the wears at Shepperton and Chertsey. These large Trout are objects of great attraction to some of the best London anglers, who unite a degree of skill and patience rarely to be exceeded. The most usual mode practised to deceive these experienced fish is by trolling or spinning with a small Bleak, Gudgeon, or Minnow ; and Trout of fifteen pounds' weight are occasionally taken.

On the 21st of March in the present year, 1835, a male Trout of fifteen pounds' weight was caught in a net. The length of this fish was thirty inches. On the 14th of April following, a male Trout of eleven pounds' weight, and measuring twenty-eight inches in length, was also caught in a net. From this second fish the representation

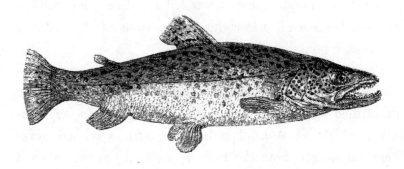

here shown was taken, by permission of Mr. Groves, who allowed a drawing to be made, which was engraved for this work.

Some deep pools in the Thames above Oxford afford excellent Trout, and some of them of very large size. I have before me a record of six, taken by minnow spinning, which weighed together fifty-four pounds, the largest of them thirteen pounds. Few persons are aware of the difficulty of taking a Trout when it has attained twelve or fourteen pounds' weight, and it is very seldom that one of this size is hooked and landed except by a first-rate fisherman : such a fish, when in good condition, is considered a present worthy a place at a royal table.

Among performances in Trout catching, the following may be mentioned, as found in the MS. of the late Colonel Montagu.

" Mr. Popham, of Littlecot, in the county of Wilts, was famous for a Trout fishery. They were confined to a certain portion of a river by grating, so that fish of a moderate size could not escape. To the preserving and fattening these fish much trouble and expense were devoted, and fish of seven and eight pounds' weight were not uncommon. A gentleman at Lackham, in the same county, had a

favourite water-spaniel that was condemned to suffer death for killing all the Carp in his master's ponds, but was reprieved at the desire of Mr. Popham, who took charge of him, in the belief that so shy and so swift a fish as a Trout was not to be caught by a dog. However, in this he was mistaken, for the dog soon convinced him that his largest Trout were not a match for him." Mr. Stoddart also, in his Scottish Angling, page 119, has recorded the propensities of a fish-catching dog.

I am indebted to William Thompson, Esq. of Belfast, for a very fine specimen of the Gillaroo Trout of Lough Neagh, measuring twenty-two inches in length, from which fish the representation on this page was taken. The internal surface of the stomach presented an indurated cuticle, but the parietes were not thicker than those of other Trout; the cavity was filled with some dozens of the *Paludina impura* of Lamarck. The fin-rays and vertebræ were—

D. 12 : P. 14 : V. 9 : A. 11. : C. 19. Vertebræ 56.

So little difference appeared to exist between this and English specimens of *S. fario*, as to induce the belief that the Gillaroo is only a variety of the Common Trout, as stated by Pennant.

VOL. II. F

Several loughs in Ireland produce this fish, which sometimes attains the weight of ten or twelve pounds. The teeth are remarkably small, but in number and situation like those of *S. fario.*

The figure of the Trout at the head of this article, and the following description, were taken from a Hampshire fish of twelve inches in length.

The length of the head compared to the length of the head and body, not including the caudal rays, was as one to four; the depth of the body rather more than the length of the head: the dorsal fin commenced half-way between the point of the nose and the commencement of the upper caudal rays; the third ray of the dorsal fin, which is the longest, longer than the base of the fin: the origin of the adipose fin half-way between the commencement of the dorsal fin and the end of the upper half of the tail; the pectoral fin two-thirds of the length of the head; the ventral fins under the middle of the dorsal fin, and half-way between the origin of the pectoral fin and the end of the base of the anal fin; the anal fin begins half-way between the origin of the ventral fin and the commencement of the inferior caudal rays. The tail but slightly forked, and growing slowly up to square in old fish, or even very slightly convex, as seen in the figure of the large Thames Trout. The fin-rays in number are—

D. 14 : P. 14 : V. 9 : A. 11 : C. 19. Vertebræ 56.

The form of the head blunt; the eye large, placed one diameter and a half from the end of the nose; the irides silvery, with a tinge of pink: the lower jaw in the *Salmonidæ* appears to be the longest when the mouth is opened, but it shuts within the upper jaw when the mouth is closed; the teeth numerous, strong, and curving inwards, extending along the whole length of the vomer ; the disposition of the teeth and the form of the gill-cover shown in outline at page 3 ; the convexity of the dorsal and ventral outline nearly similar : the colour of the back and upper part of the sides made up of numerous dark reddish brown spots on a yellow brown ground ; eleven or twelve bright red spots along the lateral line, with a few other red spots above and below the line ; the lower part of the sides golden yellow ; belly and under surface silvery white ; the spots on the sides liable to great variation in number, size, and colour ; dorsal fin and tail light brown, with numerous darker brown spots ; the adipose fin brown, frequently with one or two darker brown spots, and edged with red ; the pectoral, ventral, and anal fins uniform pale orange brown. The number of scales in a row above and underneath the lateral line about twenty-five.

Deformed Trout are not uncommon ; mention of them as occurring in some of the lakes of Wales is made by Pennant, Oliver, and Hansard. " In 1829," says the writer of the article on Angling in the seventh edition of the Encyclopædia Britannica, " we received some very singular Trouts from a small loch called Lochdow, near Pitmain, in Inverness-shire. Their heads were short and round, and their upper jaws were truncated, like that of a bull-dog. They do not occur in any of the neighbouring lochs, and have not been observed beyond the weight of half a pound." Such a Trout from Lochdow was presented to the Museum of the Zoological Society by the Honourable Twiselton Fiennes : the vignette is a representation of the head of that specimen.

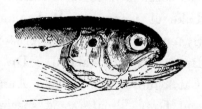

F 2

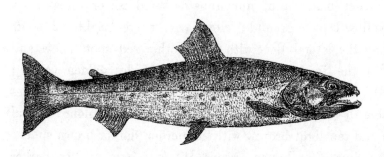

THE GREAT GREY TROUT.

THE GREAT LAKE TROUT.

Salmo ferox,　　JARDINE and SELBY.
　,,　*lacustris, Lake Trout,* BERKENHOUT's Syn. edit. 1795, vol. i. p. 79, sp. 3.

THE GREAT LAKE TROUT of Loch Awe, to which attention has lately been drawn by the various notices that have appeared in print of the fish, as well as of the beauties of the locality, was shortly noticed by Pennant, in the editions of the British Zoology, as a native of Ullswater Lake in Cumberland, and of Lough Neagh in Ireland, and was considered to be identical with the Great Trout of the Lake of Geneva. Berkenhout includes this fish in his Synopsis of the Natural History of Great Britain and Ireland, as quoted above. Dr. Heysham records it in his Catalogue of Cumberland Animals as the Ullswater Trout and Grey Trout, some specimens of which were said to weigh between fifty and sixty pounds; and the Rev. Mr.

Low, in his *Fauna Orcadensis*, mentions a Trout of thirty-six pounds' weight or more, which, besides the Common Trout, occurs both there and in Shetland. Mr. William Thompson of Belfast, when at the meeting of the British Association at Edinburgh in 1834, saw a specimen of the Great Trout of Loch Awe, and recognised it as identical with the Great Trout, or Buddagh, of Lough Neagh. Two examples of large size, about thirty-five inches each in length, were lately exhibited at the Zoological Society by Mr. Thompson.* These were obtained from Lough Neagh, where the younger and smaller sized fish of this species are called Dolachans.

According to Sir William Jardine, this fish, as far as can be traced, seems to have been first noticed about forty-five years since by the late Mr. Morrison of Glasgow, who used to exhibit them to his friends as the trophies of his expeditions. The first specimens taken in Loch Awe by Mr. Selby and Sir William Jardine were considered as a species undescribed and new to Britain ; and the name of *Salmo ferox* was given to it, from its extreme voracity and rapacious habits. M. Agassiz, who saw specimens of this fish when he was in Edinburgh, pronounced it to be different from any of the large Continental species.

" In Scotland this fish appears to be generally distributed in all the larger and deeper lochs. Loch Awe, Loch Laggan, the upper end of Loch Shin, Lochs Loyal and Assynt, they certainly inhabit, roving indiscriminately, and feeding almost entirely upon the smaller fish. By persons residing on the banks they are taken by night-lines, few rising at the artificial fly ; but they may always be taken by strong trolling tackle, baited with a small Trout. They are extremely voracious, and having seized the bait, they will allow them-

* See the Report of the Proceedings of the Society for June 9th, 1835.

selves to be dragged by the teeth for forty or fifty yards ; and when accidentally freed, will immediately again seize it."

This Great Trout is almost entirely confined to the lochs, seldom venturing far either up or down any of the streams communicating with them, and never descending to the sea. It is known to spawn in September.

The most usual mode of fishing for this Great Lake Trout is from a boat, which is rowed gently through the water ; the bait, as before mentioned, a small Trout, guarded by six or eight large hooks ; the rod and line of great strength ; for this fish is considered to be even stronger than a Salmon of the same size, but not so active. Young fish from one to two pounds' weight rise freely to the usual Trout flies.

For the opportunity of obtaining a drawing of this fish I am indebted to the kindness of Mr. Selby and Dr. Richardson.

The length of the head is to the whole length of head, body, and tail, as one to four and a half ; the depth of the body not equal to the length of the head : the teeth large, strong, and numerous, occupying five lines above and four below, thus extending along the length of the vomer : the free edge of the gill-cover rounded in the female, more angular below in the males ; the inferior edge of the suboperculum and the line of its junction with the operculum oblique. The commencement of the dorsal fin is half-way between the point of the nose and the origin of the first upper caudal ray ; the third dorsal fin-ray, which is the longest, equal to the length of the base of the dorsal fin ; the soft fleshy fin half-way between the last dorsal fin-ray and the end of the caudal rays, and in a line over the origin of the last ray of the anal fin : the ventral fins each with an elongated axillary scale ; the fins rather small in size, but very muscular.

The fin-rays in number in this specimen from Loch Awe were—

D. 13 : P. 14 : V. 9 : A. 11 : C. 19.

The specimens examined from Lough Neagh had one ray less in the dorsal, pectoral, and anal fins, and one ray more in the caudal fin. For an excellent account of this fish, see the article Angling in the seventh edition of the Encyclopædia Britannica, from which I derive the following description of the colours of this species, which was probably taken from a recent specimen.

" When in perfect season and full-grown, it is a very handsome fish, though the head is always too large and long to be in accordance with our ideas of perfect symmetry in a Trout. The colours are deep purplish brown on the upper parts, changing into reddish grey, and thence into fine orange yellow on the breast and belly. The whole body, when the fish is newly caught, appears as if glazed over with a thin tint of rich lake colour, which fades away as the fish dies, and so rapidly, that the progressive changes of colour are easily perceived by an attentive eye. The gill-covers are marked with large dark spots ; and the whole body is covered with markings of different sizes, and varying in number in different individuals. In some these markings are few, scattered, and of a large size ; in others they are thickly set, and of smaller dimensions. Each spot is surrounded by a paler ring, which sometimes assumes a reddish hue ; and the spots become more distant from each other as they descend beneath the lateral line. The lower parts of these fish are spotless. The dorsal fin is of the same colour with the upper part of the fish ; it is marked with large dark spots ; the pectoral, ventral, and anal fins are of a rich yellowish green colour, darker towards their extremities. The tail is remarkable for its breadth and consequent power.   In adults it is perfectly square, or might even be described as slightly rounded at its extremity : in the young it is slightly forked, and appears to fill up gradually as the fish advances in age."

" The flavour of this great lacustrine species is coarse and indifferent. The colour of the flesh is orange yellow, not the rich salmon-colour of a fine Common Trout in good season." Pennant states from experience that it is but an indifferent fish. The stomach is very capacious, and is almost always found gorged with fish. I have not had an opportunity of ascertaining the number of vertebræ.

The form of the scale is decidedly different from that of the Trout, and more circular than those of any of the migrating species : they are thin, flexible, and covered with a delicate membrane.

The vignette below represents three states of the angler's May-fly.

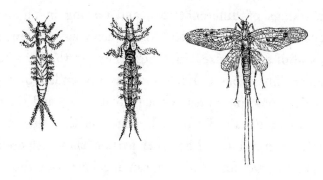

ABDOMINAL
MALACOPTERYGII.   *SALMONIDÆ.*

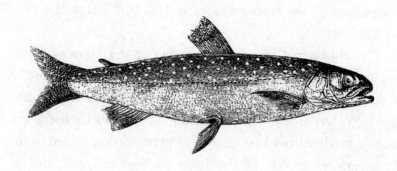

THE NORTHERN CHARR.

Salmo umbla,   LINNÆUS.   BLOCH, pt. iii. pl. 101.
,,   ,,   CUVIER, Règne An. t. ii. p. 305.
,,   *alpinus, Charr,*   PENN. Brit. Zool. vol. iii. p. 411, pl. 71.
,,   ,,   *Alpine Salmon,* DON. Brit. Zool. pl. 61.
,,   ,,   *Case Charr,*   FLEM. Brit. An. p. 180, sp. 43.

M. AGASSIZ, when in the North of England in the autumn of 1834, had several opportunities of examining the Charr of the Northern lakes, which he declared to be identical with the *Ombre Chevalier* of the Lake of Geneva; and in his " Remarks on the different species of the genus *Salmo* which frequent the various rivers and lakes of Europe," read at the meeting of the British Association at Edinburgh, that gentleman considered the *S. umbla, alpinus, salvelinus,* and *salmarinus* of Linnæus as different states only of the same fish.*

Examples of the Charr of the Northern lakes of England

* See the Reports of the Fourth Meeting of the British Association, p. 617 to 623.

agree exactly with the description and figure of the *Ombre Chevalier* in M. Jurine's paper on the fishes of Lake Leman, Geneva. The Charr of the lakes of Wales is, however, distinct from that of the Northern lakes, as will be seen by a comparison of the description of it that will follow the present subject.

The Northern Charr inhabits many of the lakes of Cumberland, Westmoreland, and Lancashire, which are annually visited by the admirers of fine scenery. Keswick, Crummock Water, Buttermere, Winandermere, and Coniston are among the localities best known to produce this delicate fish. It occurs in several of the lochs of Scotland, and also in Lough Esk, Egish, Neagh, Dan, Luggelaw, and probably many other loughs in various counties of Ireland.

The Charr generally inhabit the deepest parts of those lakes in which they are found, and afford but little amusement to the angler. The most successful mode of fishing for them is to trail a very long line after a boat, using a minnow for a bait, with a large bullet of lead two or three feet above the bait, to sink it deep in the water. By this mode a few may be caught in the beginning of summer, at which time they are in the height of perfection, both as to colour and flavour. The flyfisher when whipping for Trout, which frequently abound in the same lakes, occasionally takes a Charr; but this does not happen often: they are believed to feed principally during the night. The stomachs of those I have examined were empty; but Sir William Jardine has found abundance of minute *Entomostraca* in the stomachs of some of those examined by him. The forms of two species of these very minute animals will be represented when describing the Vendace.

The Charr are very seldom known to wander into any of the streams by which these lakes are either supplied or drained, except at the season of spawning, and their decided partiality for clear water and a hard bottom is then very conspicuous. Winandermere has two principal feeders, the rivers Rothay and Brathay: the Rothay has a sandy bottom, but the channel of the Brathay is rocky. These streams unite at the western corner of the head of the lake, below Clappers-gate, at a place called the Three-foot-brander, and after a short course boldly enter the lake together. The spawning season is in November and December; about which time the Charr in shoals make their way up both these rivers, but invariably, before depositing their spawn, those fish which have ascended over the sandy bed of the Rothay return and pass up the rocky channel of the Brathay.* A few Charr also spawn in the lake; and it is observed that they frequent the stony parts only which resemble the bottom of the Brathay.

Charr, even at the same season of the year, exhibit considerable difference in colour, which has been attributed to different causes. M. Jurine, when describing the *Salmo umbla* of the Lake of Geneva, which fish there is little or no doubt is identical with our Northern Charr, says the females are the finest in colour. Mr. Mascall, in a communication to the Magazine of Natural History for April 1835, states that he found the males of the Charr of Ennerdale Lake in Cumberland superior to the females in colour. It is not improbable that the degree of colour may depend, not so much upon the sex, as upon the constitutional vigour of the individual fish ; a circumstance observed in the periodical assumption of peculiar tints in other animals. In reference to these variations in the intensity of the colours, several distinctions have been supposed to exist in the Charr of the Northern lakes, and the names of Case Charr, Gilt Charr, Red Charr, and Silver

* The Trout, in their spawning season, prefer the Rothay.

Charr have been applied to them : Pennant, however, states that, after the closest examination, he was unable to discover any specific difference.

The most common size of our Northern Charr is from nine to twelve inches in length ; they are said occasionally to attain the length of two feet ; the largest specimen in my possession measures eighteen inches. The finest coloured specimen I ever saw was brought me by my friends T. B. Giles and W. C. Hewitson, from Coniston Water, in the month of May. They are considered to be in the greatest perfection as food from July to October.

The length of the head compared to the length of the head and body is as one to five ; the depth of the body greater than the length of the head : the commencement of the dorsal fin is half-way between the point of the nose and the adipose fin ; the posterior edge of the base of the adipose fin half-way between the origin of the last dorsal fin-ray and the end of the longest caudal ray ; the longest dorsal fin-ray but one-fourth longer than the base of that fin : the pectoral fin small ; the ventral fins originate half-way between the point of the nose and the commencement of the under caudal rays ; the ventral axillary scale nearly half as long as the fin ; the anal fin small, the longest ray but little longer than the base of the fin ; the tail deeply forked, the longest rays more than as long again as those in the centre ; all the fins of small comparative size. The fin-rays in number are—

D. 13 : P. 12 : V. 9 : A. 11 : C. 19. Vertebræ 59.

The diameter of the eye is less than one-fourth of the length of the whole head ; it is placed at the distance of one diameter from the point of the nose : the teeth small ; a few on the anterior part only of the vomer ; the other four rows above and four rows below, as usual in the fishes of this genus. The peculiarities in the form and size of the parts of the gill-cover are shown in the representation of the heads of the two Charr introduced in the description of the next fish: the branchiostegous rays vary from ten to twelve, and frequently differ in number on the two sides of the head of the same fish.

The top of the head and all the upper parts of the back umber brown; the sides lighter; the whole of the belly, pectoral, ventral, and anal fins, deep reddish orange; the first ray of the ventral and anal fins white; the sides above and below the lateral line marked with numerous red spots; the irides orange; gill-covers yellowish olive; dorsal and caudal fins dark brown, tinged with purple brown; the lateral line straight, but rising gradually from the parallel of the point of the pectoral fin to the top of the operculum; the scales very small, more than thirty in an oblique line from the base of the dorsal fin to the lateral line, and as many from the ventral axillary scale upwards to the lateral line. In this state as to colour, this fish is considered to be the *S. salvelinus* of Continental authors.

When not in fine condition as to colour, the top of the head and along the back are pale purplish brown, becoming lighter lower down; the sides silvery; the belly tinged with pale orange; above the lateral line are numerous small round white spots; irides and gill-covers silvery, with a tinge of yellow; pectoral, ventral, and anal fins, brownish red; dorsal fin brown; caudal fin purple brown.

In this state, as to colour, this fish is considered to be the *S. alpinus* of Continental authors, Bloch excepted, whose *S. alpinus*, part iii. plate 104, appears to be a Trout.

THE WELSH CHARR.

Salmo salvelinus,   *Salvelian Charr,* Don. Brit. Fish. pl. 112.
   „   „   *Torgoch,* Flem. Brit. An. p. 183, sp. 43.

The Charr of Wales was described and figured by Mr. Donovan, in his Natural History of British Fishes, under the name of Salvelian Charr, *Salmo salvelinus;* and he appears to have considered it distinct from the Charr of the Northern lakes of England, which is also described and figured under the name of *S. alpinus.* Pennant included his notice of the Welsh Charr under that of the Northern Charr; but examination of a specimen of each fish side by side will convince the observer that they are perfectly distinct. The Northern Charr is an elegantly-shaped slender-bodied fish, with fins of small comparative size. The Welsh Charr is a short fish, considerably deeper for its length, with very large fins; it has also in its form much of the character of an adult Parr of the Tweed, and carries for a long time the same sort of dusky lateral markings, but is immediately

distinguished from that species by having only a few teeth
on the most anterior part of the vomer :* but the teeth, the
gape, and the eye are much larger in the Welsh Charr than
in that from the North. The outlines of the heads introduced
as a vignette show these distinctions : the great difference in
the size of the eye of the two species is an obvious character,
and the form and relative size of the operculum and suboperculum of the one compared with the same parts in the other
are additional distinguishing features. Dr. Richardson, in
the third part of the *Fauna Boreali-Americana*, devoted to
fishes, has figured the forms of the gill-covers and teeth of
several species of *Salmo*. Plate 92, fig. 5, A, B, represent
the head and teeth of a Welsh Charr from Llyn Cawellyn,
which is evidently of the same species as the fish here described, derived from another locality.

The Welsh Charr is the Torgoch or Red-belly of Wales,
and was formerly to be taken either in Llanberris Lake, or in
Llyn Cawellyn, two deep lakes situated on the east and west
sides of Snowden. The waters from a neighbouring copper-
mine are said to have destroyed or driven out the Charr from
Llanberris, where they were formerly very numerous ; and
it was remarked that some of these fish were caught in the
sea, at the mouths of rivers on this coast, after they disap-
peared from the lake.

" Llyn Cawellyn," says Mr. Donovan, " is a vast lake of
unknown depth, sheltered on one side by an abrupt moun-
tain, which rises immediately out of the water, and in the
deep recesses at the base of which the Torgoch is supposed
to pass the milder seasons of the year in perfect security.
These fish approach the shallower parts of the lake in winter,
about the middle of December, appearing in small troops at

---

* The teeth in the Parr extend along the whole length of the vomer.

a short distance from the shores, and are at this season taken in some plenty by a poor cottager who resides in the vicinity of the lake, and derives a small annual profit from the fishery ; this delicious fish being in much request for the tables of the neighbouring gentry."

I am indebted to the kindness of the Rev. F. W. Hope for specimens of the Welsh Charr from a locality near Barmouth in Merionethshire, unnoticed by Mr. Donovan, but recorded by Willughby. The piece of water is called Coss-y-gedawl—the lake of the fruitful marsh.* This fish is certainly identical with that of Llyn Cawellyn, and distinct from the Charr of the Northern lakes of England. I have not seen specimens of the Charr of the various lakes of Scotland or Ireland, but have here endeavoured to furnish the means of identifying them. The name bestowed by Mr. Donovan is retained, till the Continental name, if it has one, can be ascertained : I have, however, some reason to believe that this species is unknown on the Continent.

The length of the head compared to the length of the head and body is as one to four ; the depth of the body equal to the length of the head : the commencement of the dorsal fin is half-way between the point of the nose and the origin of the upper caudal rays ; the posterior edge of the adipose fin is half-way between the commencement of the dorsal fin and the end of the longest upper caudal ray ; the longest dorsal fin-ray as long again as the base of that fin : the pectoral fin large ; the ventral fin originates half-way between the posterior edge of the orbit of the eye and the end of the fleshy portion of the tail ; ventral axillary scale small, not one-third the length of the fin ; the longest anal

* The meaning of this term is said to be questionable : the primitive from which it is derived means gift, relief, or profit. By Willughby it is spelt Casa-geddor ; by others, Cors-y-gedol.

fin-ray one-third longer than the base of the anal fin ; the tail not very deeply forked ; the longest rays but one-third longer than those in the middle of the tail : all the fins of large comparative size. The fin-rays in number are—

D. 12 : P. 11 : V. 8 : A. 11 : C. 19.   Vertebræ 62.

The eye large, placed less than the breadth of its diameter from the point of the nose ; the breadth of the orbit more than one-fourth of the whole length of the head : the teeth large ; those on the vomer occupying the most anterior part only ; the other eight rows as usual : the form of the different parts of the gill-cover are shown in the figures of the head.

The top of the head, and the whole of the upper part of the back, dark brown or blackish green ; the sides lighter, approaching to olive, and passing by a golden tinge below the lateral line, to a deep reddish orange, which pervades the whole of the belly : the lateral line straight, but ascending rather abruptly to the upper edge of the operculum when within half an inch of that part ; above the lateral line a few small round yellowish white spots ; upon and below the line are numerous red spots ; the sides exhibit traces across the lateral line of the dusky patches common to young fishes of the genus *Salmo :* the irides hazel : cheeks and gill-covers light olive ; pectoral, dorsal, and caudal fins brown ; the ventral and anal fins dark reddish orange, like the belly : the first ray of each of these fins yellowish white. This is the only state in which I have seen the Charr from Wales, but it most likely does not retain these brilliant colours throughout the year. The largest specimen did not exceed seven inches in length.

VOL. II.   G

This species has been called a Red-bellied Trout ; but the latter name is far from applicable, as the Trout have a formidable row of teeth along the whole line of the vomer, and the tail at all ages much less forked.

The vignette represents the heads of two Charr ; that on the left is from Keswick, the other from Wales. The Northern fish has the mouth, teeth, and eyes smaller comparatively than those of the Welsh Charr : there is also a difference in the form and proportions of the gill-cover.

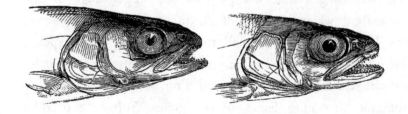

*ABDOMINAL
MALACOPTERYGII.*   *SALMONIDÆ.*

THE SMELT.

SPIRLING AND SPARLING. *Scotland.*

Osmerus eperlanus,	*Smelt,*	FLEM. Brit. An. p. 181, sp. 48.
,, ,,		CUVIER, Règne An. t. ii. p. 305.
Eperlanus Rondeletii,		WILLUGHBY, p. 202.
,, *Schonfeldii,*	,,	WILLUGHBY, tab. N. 6, fig. 4.
Salmo eperlanus,		LINNÆUS. BLOCH, pt. i. pl. 28. 2.
,, ,,	,,	PENN. Brit. Zool. vol. iii. p. 416, pl. 72.
,, ,,	,,	DON. Brit. Fish. pl. 48.

Generic Characters.—Body elongated, covered with small scales : two dorsal fins, the first with rays, the second fleshy, without rays ; ventral fins in a vertical line under the commencement of the first dorsal fin : teeth on the jaws and tongue very long, two distinct rows on each palatine bone, none on the vomer except at the most anterior part ; branchiostegous rays 8.

THE SMELT, as a British fish, appears to be almost exclusively confined to the eastern and western coasts of Great Britain. I am not aware of any good authority for the appearance of the true Smelt between Dover and the Land's End.* The fish called Smelt and Sandsmelt along the

* Mr. Salter, in his Angler's Guide, page 169, says he has caught very fine Smelts by angling in Portsmouth harbour ; but there is very little doubt that the Sandsmelt, or Atherine, which is there abundant, is the fish alluded to.

G 2

extended line of our southern coast is in reality the Atherine, as stated in the account of that fish, volume i. page
214 : but the Atherine, though furnished with two dorsal
fins, and otherwise something like the Smelt, is immediately
distinguished from it by the numerous rays supporting the
second dorsal fin ; which fin in the true Smelt is without any
rays whatever, like the adipose fin of the species of the genus
Salmo last described.

On the eastern side of our island, the Smelt occurs in the
Tay, in the Frith of Forth, in the Ure on the Yorkshire
coast ; it is taken in abundance in the Humber, and on
the Lincolnshire coast ; in the Thames, and the Medway.
On the western side, the Smelt is taken in the Solway
Firth, and may be traced as far south as the parallel line
formed by the Mersey, the Dee, the Conway, and Dublin
Bay.

The Smelt inhabits fresh water from August to May.
After spawning in March or the beginning of April, they
return to the sea. The ova are small and yellowish in
colour. The fry are found about three inches long, swimming near the surface in shoals in the rivers in the month of
August, ascending and descending with the tide, when the
adult fish are again visiting the fresh water. Their food is
small fish, with crustaceous and testaceous animals : Dr.
Fleming says, the principal food of the Smelt in the Tay is
the shrimp.

Two modes of fishing for Smelts are in practice ; one
on the sandy shallow shores of the sea, on the eastern coast,
particularly Lincolnshire, where large quantities are taken
in spring ; the other is the river-fishing within the tide-way.
The excellence of the Smelts of the Medway is well known.
The Thames and Medway fishing with small-meshed nets
for Smelts is permitted, under the jurisdiction of the Lord

Mayor of London, from the 28th of August (St. Augustine) till Good-Friday. Formerly, the Thames from Wandsworth to Putney-bridge, and from thence upwards to the situation of the present suspension-bridge at Hammersmith, produced abundance of Smelts, and from thirty to forty boats might then be seen working together; but very few are now to be taken, the state of the water, it is believed, preventing the fish advancing so high up. The particular cucumber-like smell of this fish is well known; and it is very considerably more powerful when they are first taken out of the water.

The Smelt is generally in great request from its delicate and peculiar flavour. This quality, coupled with the circumstance of the fish passing six or seven months of the year in fresh water, has induced two or three experiments to retain it in ponds, one of which was attended with complete success, and the attempts might be multiplied with advantage. Colonel Meynell, of Yarm in Yorkshire, kept Smelts for four years in a fresh-water pond having no communication with the sea: they continued to thrive, and propagated abundantly. They were not affected by freezing, as the whole pond, which covered about three acres, was so frozen over as to admit of skating. When the pond was drawn, the fishermen of the Tees considered that they had never seen a finer lot of Smelts. There was no loss of flavour or quality.

From the point of the lower jaw to the end of the gillcover, the length is, as compared to the body alone, as one to three; the depth of the body not equal to the length of the head: the dorsal fin commences half-way between the point of the nose and the end of the fleshy portion of the tail; the first ray of this fin less than half the length of the second, which is as long as the third; the second and third are the longest rays in the fin, nearly as high as the body of the fish is deep, and as long again as the base of the fin ; the two first rays simple, all the others branched : the anterior edge of the adipose fin is half-way between the base of the last ray of the dorsal fin and the end of the fleshy portion of the tail, and in a vertical line over the middle of the anal fin ; pectoral fins long and narrow ; the ventral fins commence on the same plane as the dorsal fin : the base of the anal fin long, commencing half-way between the origin of the ventral fins and the end of the fleshy portion of the tail ; the third ray the longest, but not so long as the base of the fin ; the other rays diminish in length gradually : the tail slender and deeply forked. The fin-rays are—

$$\text{D. 11 : P. 11 : V. 8 : A. 15 : C. 19.}$$

The lower jaw much longer than the upper; the gape deeper than wide : the teeth long, and curving inwards ; those on the anterior parts of the tongue and palatine bones are the longest : the breadth of the eye about one-fifth of the whole length of the head, the irides silvery white ; the gill-cover triangular ; the upper part of the head flat ; the nape and back rising ; the form of the body elongated and slender ; the dorsal and abdominal lines slightly convex : the colour of the upper part of the body pale ash green ; all the lower parts, cheeks, and gill-covers, brilliant silvery white : the scales oval, small, and deciduous : all the fins pale yellowish white ; the ends of the caudal rays tipped with black.

The specimen described measured seven inches in length. Occasionally Smelts may be seen in the London markets ten and eleven inches long, but this is an unusually large size. Pennant mentions having seen one that was thirteen inches long, and weighed eight ounces.

ABDOMINAL
MALACOPTERYGII.   SALMONIDÆ.

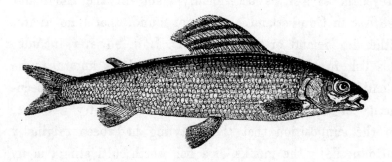

THE GRAYLING.

Thymallus vulgaris, Cuvier, Règne An. t. ii. p. 306.
  ,,   ,, Willughby, p. 187, N. 8.
Salmo thymallus, Linnæus. Bloch, pt. i. pl. 24.
  ,,   ,, *Grayling,* Penn. Brit. Zool. vol. iii. p. 414, pl. 72.
  ,,   ,,   ,, Don. Brit. Fish. pl. 88.
Coregonus ,,   ,, Flem. Brit. An. p. 181, sp. 49.

Generic Characters.—Head and body elongated ; the sides marked with longitudinal bands ; two dorsal fins, the first much longer than high, with numerous
rays ; the second small, adipose, without rays : the mouth small, the orifice
square ; the teeth very small ; branchiostegous rays 7 or 8.

The Grayling, though abundant in some streams,
is yet a very local fish. Similar in many respects to the
Trout in its habits and wants, there are numbers of rivers
abounding with Trout that do not produce Grayling. In
the southern counties of Hampshire and Wiltshire, the
Grayling is found in the Test and both the Avons. In
Herefordshire, in the Dove, the Lug, the Wye, and the
Irvon. In Shropshire, in the Teme and the Clun. In
Staffordshire, in the Hodder, the Trent, the Dove, and the

Wye. In Derbyshire, in the Dove. In Merionethshire, in the Dee, between Curwen and Bala. In Lancashire, in the Ribble. In Yorkshire, in the Derwent, the Ure, the Wharfe, and the Wiske, near Northallerton. Dr. Heysham says it is occasionally taken in the Eden and the Esk in Cumberland. It is not found, that I am aware, either in Ireland or Scotland; Mr. Low, however, includes this fish in his *Fauna Orcadensis*, and it is known to be plentiful in Sweden, Norway, and Lapland. The peculiarity of the local distribution in this country gave rise to the supposition that the Grayling had been originally introduced by the monks, as a fish worth cultivating; many of the rivers containing the Grayling being near the remains of great monasteries. But two circumstances affect this solution : it would be very difficult to bring this fish alive from the Continent to this country; and it is not found in the rivers of Kent, Dorsetshire, Devonshire, or Cornwall, where monastic establishments were formerly numerous.

The Grayling thrives best in rivers with rocky or gravelly bottoms, and seems to require an alternation of stream and pool. According to Sir Humphrey Davy, who has given a good history of the Grayling in his " Salmonia," this fish was introduced into the Test, in Hampshire, from the Avon ; and the former river, in particular parts, appears to suit it the better of the two. Large Grayling are, however, occasionally taken in both these waters, which are particularly resorted to by the southern anglers. Three Graylings, weighing together twelve pounds, were caught by Thomas Lister Parker, Esq. in the Avon, near Ringwood. A Grayling of four and a half pounds' weight has been killed in the Test, and one of five pounds is recorded to have been caught near Shrewsbury.

However fastidious in the quality of the water or the choice of situation in the stream the Grayling is known to be, experiment has proved that this fish will live in ponds that have been newly made in hard soil, or in such as have been very recently and carefully cleaned out; but in these situations the Grayling does not breed, and they will not continue to live in old muddy ponds. The ova of this fish are numerous, large, and of a deep orange colour: the spawning season is in April, or the beginning of May; in this respect differing from the other *Salmonidæ*, most, if not all, of which spawn towards the end of the year, and generally in cold weather. The Grayling, however, is in the finest condition in October and November, when Trout are out of season, not having then recovered the effects of their recent spawning, while the young Grayling of that year are about seven inches in length.

The food of the Grayling, as ascertained by examination, besides the various flies—imitations of which are successfully used by anglers,—consists also of the larvæ of *Phryganea Ephemera* and *Libellula;* the remains of the cases of the former, and the tough skins of all of them, being frequently found in their stomachs. I have found also several small shells, examples of the genus *Physa,* and *Neritina fluviatilis.* Dead shells and small pebbles are also found; but whether these last are taken up by the fish to serve any useful purpose, as in the stomachs of gallinaceous birds, or have only formed part of the cases of the *Phryganea,* may be questioned.

Some English authors have considered the Grayling a migratory fish, passing the winter in the sea, and the summer in fresh water. " Early in spring," says Mr. Donovan, " they ascend the rivers, where they remain till autumn,

and then return to their former element." This may apply
to Grayling on some parts of the European continent,* but
is not the case certainly with our fish in this country, in the
rivers of which it is found in the most perfect condition,
and in consequence most eagerly sought after, in October
and November. The finest specimens I ever saw were
taken in November; and Sir H. Davy states in his " Salmonia," he had proved that the Grayling of England would
not bear even a brackish water without dying.

The term *Thymallus* is said to have been bestowed upon
this fish on account of the peculiar odour it emits when fresh
from the water, which is said to resemble that of thyme ;
and from its agreeable colour as well as smell, St. Ambrose
is recorded to have called the Grayling the flower of fishes.
To be eaten in perfection, it cannot be dressed too soon.
The name Grayling is supposed to be a modification of the
words gray-lines, in reference to the dusky longitudinal bars
along the body.

It has been considered that the large dorsal fin of the
Grayling enabled it to rise and sink rapidly in deep pools ;
but this power would rather seem to be afforded by the
large size of the swimming-bladder. The very large dorsal
fin, compared to the small size of all the other fins, renders
the Grayling unable to stem rapid currents : they are much
more prone to go down stream than up, and are never seen
leaping at a fall, like Trout.

In a Grayling of ten inches long, the length of the head
is to the body alone as one to four; the depth of the body
rather more than equal to the length of the head : from the
point of the nose to the commencement of the dorsal fin is
equal to one-third of the length of the whole fish to the end
of the fleshy portion of the tail; the posterior edge of the

---

* Bloch says the Grayling descends to the Baltic in autumn.

dorsal fin half-way between the point of the nose and the
end of the longest caudal rays ; the adipose fin rather nearer
the dorsal fin than the end of the tail : the height of the
dorsal fin equal to half the height of the body, the first
ray short, the next five increasing gradually in length ;
the sixth ray nearly as long as the seventh, and, as well
as the five anterior rays, articulated and simple ; the seventh
ray and all the rays behind it articulated, branched, and
nearly of the same height ; the length of the base of the fin
not equal to twice the length of its longest ray : the pectoral
fin small, narrow, and pointed : the ventral fins commencing
in a vertical line under the middle of the dorsal fin ; the
ventral axillary scale one-fourth of the length of the fin :
the anal fin commences half-way between the origin of the
ventral fin and the end of the fleshy portion of the tail, and
ends on the same plane as the adipose fin above it ; the
longest ray but little longer than the base of the fin : the
tail forked ; the middle rays rather more than half as long as
the longest.   The fin-rays in number are—

D. 20 : P. 15 : V. 10 : A. 13 : C. 20.   Vertebræ 58.

The head is small and pointed, flattened at the top : the
breadth of the eye equal to one-fourth of the length of the
whole head ; irides golden yellow, pupil blue, pear-shaped,
the apex directed forward : the opening of the mouth, when
viewed in front, square ; the teeth small, incurved, numerous ; none on the tongue, and only a few on the most
anterior part of the vomer : behind the head, the nape and
back rise suddenly ; the body deepest at the commencement
of the dorsal fin, then tapering off to the tail ; abdominal line
but slightly convex ; the scales rather large ; the lateral line
in the middle of the body not very conspicuous, with seven
rows of scales on an oblique line above it, and seven rows

below it ; the sides marked with about fifteen dusky longitudinal bands.   The general colour of the body light yellow
brown, beautifully varied with golden, copper, green, and
blue reflections when viewed in different lights, with a few
decided dark spots : the head brown ; on the cheeks and
gill-covers a tinge of blue : all the fins somewhat darker than
the colour of the body ; the dorsal fin varied with square
dusky spots on the membrane between the rays, the upper
part of the fin spotted and streaked with reddish brown
The Grayling appears to become darker by age, and the
pectoral fins are reddish about spawning time, with small
black spots.

The vignette represents two states of the Stone-fly of
anglers.

ABDOMINAL
MALACOPTERYGII. SALMONIDÆ.

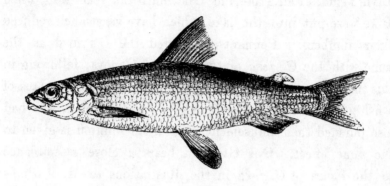

THE GWYNIAD.

SCHELLY. *Cumberland.* — POWAN. *Perthshire.*

Coregonus fera? CUVIER, Règne An. t. ii. p. 307.
    ,,   ,, ? NILSSON, Prod. p. 16, sp. 4.
    ,,   ,, ? JURINE, pl. 7.
Salmo lavaretus, Gwyniad, PENN. Brit. Zool. vol. iii. p. 419, pl. 73.
Coregonus ,,   ,, FLEM. Brit. An. p. 182, sp. 50.

Generic Characters.—Body in appearance herring-like ; with two dorsal fins, the first higher than long, the second adipose ; the scales large ; the mouth small, sometimes with minute teeth on the jaws or tongue, or both.

THE species of the genus *Coregonus* are numerous in Europe, and several of them are so similar to each other, that, without the power of comparing those of this country with foreign specimens, an appropriation of synonymes is at least doubtful. Some authors have even considered the Vendisse of Lochmaben as the same with the Powan of Perthshire, the Schelly of Cumberland, the Gwyniad of Wales, and the Pollan of Ireland : but it will be found that this is not the case ; and, from recent observation, there is now reason to believe that the Pollan of Ireland is distinct from the two species of *Coregonus* found in Great Britain.

The Gwyniad of Wales was formerly very numerous in Llyn Tegid, (Fair Lake,) at Bala, until the year 1803, when Pike were put into the lake, which have very much reduced their numbers. Pennant considered the Gwyniad as the same with the *C. fera* of the Lake of Geneva, following in this the opinion of Willughby ; and in the manuscript notes of a fishing tour in Wales, by two excellent fishermen, who had also pursued their amusement abroad, an opinion is given to the same effect. Our Gwyniad bears a close resemblance to the figure of *C. fera* in the illustrations to M. Jurine's Memoir on the Fishes of Lake Leman : his description I have not seen. The British fish accords also with the short description of the *C. fera* in Professor Nilsson's Prodromus of the Fishes of Scandinavia. It also resembles the *S. Wartmanni* of Bloch, pt. iii. pl. 105 ; but is decidedly distinct from his *S. lavaretus*, pt. i. pl. 25, which is the *C. oxyrhinchus* of Cuvier and Nilsson.

The Gwyniad is very numerous in Ulswater and other large lakes of Cumberland, where, on account of its large scales, it is called the Schelly. Dr. Heysham, the natural historian of Cumberland, and Pennant also, in his British Zoology, have recorded that many hundreds are sometimes taken at a single draught of the net. They are gregarious, and approach the shore in vast shoals in spring and summer. Pennant says, they die very soon after they are taken out of the water, are insipid in taste, and must be eaten soon, for they will not keep long. The poorer classes, who consider, and even call them the Fresh-water Herring, preserve them with salt. The fish is not unlike a Herring in appearance, and the Welsh term Gwyniad has reference to their silvery white colour. They spawn towards the end of the year, and the most usual length of the adult fish is from ten to twelve inches.

The length of the head is about one-fifth of the whole length of the fish; the depth of the body rather exceeding the length of the head : the dorsal fin commences about half-way between the point of the nose and the end of the fleshy portion of the tail ; its longest ray one-third longer than the base of the fin, and equal to three-fourths of the depth of the body : the adipose fin rather nearer the end of the tail than the posterior edge of the dorsal fin ; the pectoral fins narrow, pointed, and a little shorter than the head, inserted low down on the body : the ventral fins arising in a line under the middle of the dorsal fin ; the ventral axillary scale one-third the length of the fin : the anal fin commences half-way between the origin of the ventral fin and the end of the short middle rays of the tail, and ends on the same plane with the adipose fin; the longest anterior ray about equal to the length of the base of the fin ; the other rays diminishing gradually : the tail forked.   The fin-rays in number are—

D. 13 : P. 17 : V. 11 : A. 16 : C. 19.

The head is triangular ; the snout rather truncated ; the jaws nearly equal, the lower just shutting within the upper ; a very few minute teeth on the tongue only ; the eyes large, the breadth more than one-fourth of the length of the head ; the form of the body very like that of a Herring ; the dorsal and abdominal lines but moderately convex ; the scales large ; the lateral line very near the middle of the side.   The irides silvery, the pupils dark blue ; the upper part of the head and back dusky blue, becoming lighter down the sides, with a tinge of yellow ; cheeks, gill-covers, lower part of the sides and belly silvery white ; all the fins more or less tinged with dusky blue, particularly towards the edges.

According to Mr. Thompson of Belfast,* the Pollan, or Lough Neagh *Coregonus*, differs from the Gwyniad of Bala in the following particulars : in the snout not being produced ; in the dorsal fin being nearer the head ; in having fewer rays in the anal fin, and in its position being rather more distant from the tail ; in the dorsal, anal, and caudal fins being of less dimensions ; in the third ray of the pectoral fin being the longest, the first being of the greatest length in the Gwyniad; and in the ventral axillary scale being longer. The numbers of the fin-rays in the Pollan are—

D. 14 : P. 16 : V. 12 : A. 13 : C. 19.   Vertebræ 59.

In the stomach of a Pollan I found one example of a species of *Gammarus*.

The vignette is a view of Whitewell, in the Forest of Bowland, Yorkshire.

* Reports of Proceedings of the Zoological Society of London for 1835, p. 77.

*ABDOMINAL
MALACOPTERYGII.*   *SALMONIDÆ.*

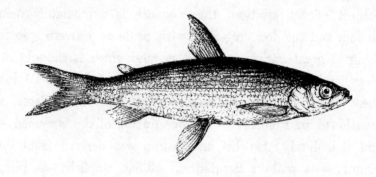

THE VENDACE, OR VENDIS.

Coregonus Willughbii,   *Vendace,* Jardine.
                Vangis and Juvangis, Penn. Brit. Zool. vol. iii. p. 420.
                Vendace, Knox, Trans. R. S. E. vol. xii. p. 503.

But little is known of this delicate fish beyond what has been published by Sir William Jardine, Bart. in the third volume of the Edinburgh Journal of Natural and Geographical Science, and by Dr. Knox, in the Transactions of the Royal Society of Edinburgh. Sir William Jardine, in his original communication, considered this species very closely allied to the *Salmo albula* of Linnæus; but the difficulty of fixing synonymes satisfactorily from the short descriptions of the older authors has since led to a request from him that the name of our distinguished British naturalist should be attached to it, and I with pleasure adopt the suggestion.

The localities inhabited by this species of *Coregonus* are as limited as the range of the species last described was shown to be extensive. The Vendace is only known in the lochs in the neighbourhood of Lochmaben, in Dumfries-shire; and in this district some traditions and curious opinions exist regarding it.

VOL. II.   H

" The Vendace is well known," says Sir William Jardine, " to almost every person in the neighbourhood ; and if, among the lower classes, fish should at any time form the subject of conversation, the Vendace is immediately mentioned, and the loch regarded with pride as possessing something of great curiosity to visiters, and which is thought not elsewhere to exist. The story that it was introduced into these lochs by the unfortunate Mary Queen of Scots, as mentioned in Pennant in his description of the Gwyniad,— and it is likely that his information was derived from this vicinity,—is still in circulation. That the fish was introduced from some Continental lake, I have little doubt ; but would rather attribute the circumstance to some of the religious establishments which at one time prevailed in the neighbourhood, and which were well known to pay considerable attention both to the table and the cellar. Mary would scarcely prefer a lake so far from even her temporary residence for the preservation of a luxury of troublesome introduction, and leave her other fish-ponds destitute of such a delicacy."

" An idea prevails that this fish, if once taken from the water, will die, and that an immediate return will be of no avail ; and it is also believed that it will not exist in any other water except that of the castle loch. These are of course opinions which have gradually, from different circumstances, gained weight, and have at last been received as facts. The fish is of extreme delicacy ; a circumstance which may have given rise to the first notion ; and the introduction of it must have taken place by means of the spawn : the fish themselves, I am confident, could not be transported alive even a few miles. As to the second opinion, they are not confined to the castle loch, but are found in several others, some of which have no communication with that where they are thought to be peculiar."

" In general habits the Vendace nearly resemble the Gwyniad, and indeed most of the allied species of the genus. They swim in large shoals; and during warm and clear weather retire to the depth of the lakes, apparently sensible of the increased temperature. They are only taken with nets, a proper bait not being yet discovered ; and the fact that little excrement is found in their intestines has given rise to another tradition, that they are able to subsist without food. They are most successfully taken during a dull day and sharp breeze, approaching near to the edges of the loch, and swimming in a direction contrary to the wind. They spawn about the commencement of November, and at this time congregate in large shoals, frequently rising to the surface of the water, in the manner of the common Herring, and making a similar noise by their rise and fall to and from the surface. The sound may be distinctly heard, and the direction of the shoal perceived, during a calm and clear evening. They are very productive. The lochs abound with Pike, of which they are a favourite food ; but their quantity seems in no degree to be diminished, notwithstanding that immense numbers must be destroyed. They are considered a great delicacy, resembling the Smelt a good deal in flavour ; and though certainly very palatable, the relish may be somewhat heightened by the difficulty of always procuring a supply. During the summer, fishing-parties are frequent, introducing some stranger friend to this Lochmaben whitebait ; and a club, consisting of between twenty and thirty of the neighbouring gentry, possessing a private net, &c. meet annually in July, to enjoy the sport of fishing, and feasting upon this luxury."

The circumstance that this fish is never caught by anglers made a knowledge of its food a matter of interest in several points of view. Dr. Knox ascertained that this consists

H 2

principally of very minute entomostracous animals, not exceeding seven-twelfths of a line in size. I have been favoured with specimens of the Vendace by Sir William
Jardine and T. S. Bushnan, Esq. which have afforded me
several opportunities of examining the contents of the
stomach and intestines. The contained mass, which is frequently in considerable quantity, has a brownish yellow colour, appearing slightly granulated to the unassisted eye.
A very small portion being placed on a slip of glass, and
agitated gently in conjunction with a drop of water, which
separates the particles, on placing the slip of glass under a
good microscope, two species in various states of perfection
are almost constantly found. The vignette at the end of the
description of this fish represents these two forms. The
first and second figure on the left hand are a back and side
view of a species of the genus *Lynceus* of Muller and others;
the third and fourth figures are a back and side view of a
species of *Cyclops* of Muller. On one occasion, I found a
very small coleopterous insect, the tough skin of a red worm
not much thicker than fine thread, and what appeared to be
a portion of the wing of a dipterous insect.

Dr. Knox found that the females of the Vendace were
more numerous as well as larger than the males, frequently
exceeding eight inches in length; the males not measuring
more than seven inches, which was the length of the specimen here described.

The length of the head compared to that of the body only
was as two to seven; the depth of the body at the commencement of the dorsal fin not quite equal to one-fourth of
the length of the body without the caudal rays: the body elegantly shaped; the convexity of the dorsal and abdominal lines
about equal; the lateral line passes straight along the middle
of the side, with six rows of scales in an oblique line between

the dorsal fin and the lateral line, and the same number between the line and the ventral axillary scale : the dorsal fin commences half-way between the nose and the origin of the upper caudal rays ; the longest ray double the length of the base of the fin : the adipose fin very near the tail ; pectoral fin not quite equal to the length of the head ; the ventral fin commences in a line under the first ray of the dorsal fin ; the ventral axillary scale one-third the length of the fin ; the anal fin commences half-way between the origin of the ventral fin and the end of the fleshy portion of the tail ; the longest ray about equal to the base of the fin : the tail deeply forked ; all the fins large. The fin-rays in number are—

D. 11 : P. 16 : V. 11 : A. 15 : C. 19.   Vertebræ 52.

In form the under jaw is the longest ; the mouth small, the opening square ; a few very minute teeth on the tongue only : the breadth of the eye one-third of the whole head, the posterior part of the iris the broadest ; the colour silvery tinged with yellow, the pupil blue : the upper parts of the body of a delicate greenish brown, shading gradually towards the belly into a clear silver ; the dorsal fin greenish brown ; the lower fins are all bluish white.

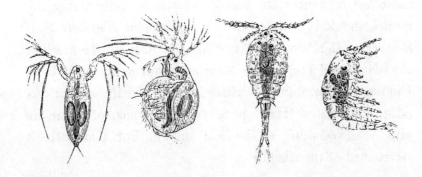

ABDOMINAL
MALACOPTERYGII.   *SALMONIDÆ.*

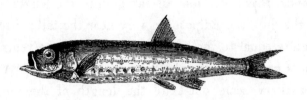

THE ARGENTINE.

Scopelus Humboldtii,	Cuvier, Règne An. t. ii. p. 315.
,, *borealis,*	Nilsson, Prod. p. 20.
Serpes Humboldtii,	Risso, Ich. p. 358, tab. X. f. 38.
Scopelus ,,	,, Hist. t. iii. p. 467.
Argentina sphyræna,	*Argentine,* Penn. Brit. Zool. vol. iii. p. 432, pl. 76.
,, ,,	,, Flem. Brit. An. p. 182.

Generic Characters.—Body long, slender ; the principal dorsal fin over the interval between the ventral and anal fins ; a second dorsal fin, so small as to be scarcely perceptible : the head short ; the mouth and gill-aperture large ; small teeth on both jaws ; palate and tongue smooth.

Pennant, and the Rev. Mr. Low of Orkney, appear to be the only British observers who have met with, on our coast, examples of this brilliant little fish, which Cuvier considers to belong to the genus *Scopelus*, as here stated ; and other references are added, to assist in determining the species should it again occur. The *Scopelus Humboldtii*, if identical with Pennant's Argentine, is taken to the north of our island, and also in the Mediterranean, as the remarks of Professor Nilsson and M. Risso imply ; and the latter natu-ralist enumerates three species of the genus, of which he says *S. Humboldtii* is the best known, but that little is ascertained of their habits.

Pennant's specimen was taken in the sea near Downing in Flintshire: Mr. Low's fish was brought to him by a boy, who said he found it at the edge of the water among sea-weed. The receipt of an additional portion of MS. recently confided to me by William Walcott, Esq. furnishes a notice, written by his late father, of a third instance of the occurrence of the Argentine, which was found stranded on the shore near Exmouth: length two inches and a half. Pennant's description is, " Length two inches and a quarter ; the eyes large, the irides silvery ; the lower jaw sloped much ; the teeth small ; the body compressed, and of an equal depth almost to the anal fin ; the tail forked : the back was of a dusky green ; the sides and covers of the gills as if plated with silver ; the lateral line was in the middle, and quite straight : on each side of the belly was a row of circular punctures ; above them another, which ceased near the vent." The formula of the fin-rays appears to be—

D. 9 : P. 17. : V. 8 : A. 15 : C. 19.

The figure of this fish referred to in M. Risso's work represents the anal fin as containing many more rays than are apparent in the figure by Pennant, from which the representation at the head of this article is copied.

ABDOMINAL
MALACOPTERYGII.   *CLUPEIDÆ.**

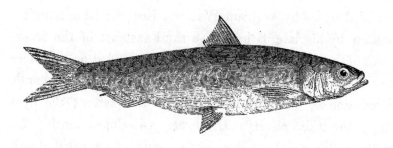

THE PILCHARD.

GIPSEY HERRING.   *Scotland.*

Clupea pilchardus, BLOCH, pt. xii. pl. 406.
  ,, ,, WILLUGHBY, p. 223, tab. P. 1. fig. 1.
  ,, ,, CUVIER, Règne An. t. ii. p. 319.
  ,, ,, *Pilchard,* PENN. Brit. Zool. vol. iii. p. 453, pl. 79.
  ,, ,, ,, DON. Brit. Fish. pl. 69.
  ,, *pilcardus,* ,, FLEM. Brit. An. p. 183, sp. 52.

Generic Characters.—Body compressed ; scales large, thin, and deciduous ;
head compressed ; teeth minute, or wanting ; a single dorsal fin ; abdominal
line forming a sharp keel-like edge, which in some species is serrated ;
branchiostegous rays 8.

THE following account of the Pilchard is derived from the
MS. of Mr. Couch, from whose various scientific acquirements, habits of observation and locality, it may be fairly
inferred that no better authority could be quoted.

 The older naturalists considered the Pilchard, like the
Herring, as a visiter from a distant region ; and they assigned to it also the same place of resort as that fish, with

* The family of the Herrings.

which indeed the Pilchard has been sometimes confounded. To this it will be a sufficient reply, that the Pilchard is never seen in the Northern Ocean, and the few that sometimes wander through the Straits of Dover, or the British Channel, have evidently suffered from passing so far out of their accustomed limits. They frequent the French coasts, and are seen on those of Spain; but on neither in considerable numbers, or with much regularity; so that few fishes confine themselves within such narrow bounds. On the coast of Cornwall they are found through all the seasons of the year, and even there their habits vary in the different months. In January, they keep near the bottom, and are chiefly seen in the stomachs of ravenous fishes; in March, they sometimes assemble in schulls, and thousands of hogsheads have in some years been taken in seans: but this union is only partial, and not permanent; and it is not until July that they regularly and permanently congregate so as to be sought after by the fishermen.

The sean-fishery commences in August, and continues until the shortened days and stormy weather of the equinox render its further prosecution impracticable; but the fish continue to appear, sometimes in great numbers, until the conclusion of the year. The season and situation for spawning, and the choice of food, are the chief causes which influence the motions of the great bodies of these fish; and it is probable that a thorough knowledge of these would explain all the variations which have been noticed in the actions of the Pilchard, in the numerous unsuccessful seasons of the fishery. In some years, at least, a considerable body of Pilchards shed spawn in the month of May—perhaps in the middle of the Channel, where I have known them taken, heavy with roe, in drift-nets shot for Mackerel; yet it seems certain that they do not breed twice in the year, and that the larger body do not perform this function until October, and then at no great distance from the shore. I have known an equally great variation to occur in other fishes, which have in consequence visited us, and been in season, at a time not expected by the fishermen.

They feed with voracity on small crustaceous animals ; and I have found their stomachs crammed each with thousands of a minute species of shrimp, not larger than a flea. It is probably when they are in search of something like this, that fishermen report they have seen them lying in myriads quietly at the bottom, examining with their mouths the sand or small stones in shallow water. The abundance of this food must be enormous, if, as there can be no doubt was the case, all the schulls on the coast were as well fed as the individuals I examined. The Pilchard has been known to swallow a hook baited with a worm ; and it is probable that they devour the roe of fish ; for a gentleman who resided on the shores of the Bay of Biscay informed me that it is the custom of the French fishermen to throw large quantities of the salted pea-roe of fish about their nets, to attract Pilchards, and that he had seen much of this spawn in the stomachs of Pilchards so taken. Large quantities of the roe of fish are imported into France for this purpose from northern nations.

When near the coast, the assemblage of Pilchards assumes the arrangement of a mighty army, with its wings stretching parallel to the land ; and the whole is composed of numberless smaller bodies, which are perpetually joining together, shifting their position, and separating again. There are three stations assumed by this great body, that have their separate influence on the success of the fishery. One is to the eastward of the Lizard, the most eastern extremity reaching to the Bay of Bigbury in Devonshire, beyond which no fishery is carried on, except that rarely it extends to Dartmouth; a second station is included between the Lizard and Land's End; and the third is on the north coast of the county, the chief station being about St. Ives. It is common for one of these districts to be full of fish, while in neither of the others is a schull to be seen; but towards the end of the season they often move from one station to another, or perhaps traverse in succession all the shores of the county. The subordinate motions of the schulls are much regulated by the tide, against the current of which they are rarely known to go; and the whole will sometimes remain parallel to the coast for several weeks, at the distance of a few leagues, and then, as if by general consent, will advance close to the shore, sometimes without being discovered till they have reached it. This usually happens when the tides are strongest, and is the period when the principal opportunity is afforded for the prosecution of the sean-fishery.

The fishery for Pilchards is carried on by drift or driving nets, and with seans. The outfit of the former, which somewhat resembles that already described for Mackerel, consists of a number of nets, great in proportion to the wealth of the proprietor and the size of the boat, but commonly about twenty, each from eighteen to twenty fathoms long, and seven fathoms deep; so that a string of driving nets will sometimes reach three-quarters of a mile. These nets are fastened to each other in length, and to a head-line, appropriated to each, along which runs a row of corks; another line runs loosely along the middle of the nets to afford additional strength, but no lead is used at the bottom. The nets are carried in common fishing-boats, some of which, as at Mount's Bay, are luggers, and most of the others with spritsails: the crews consist each of four men and a boy.

The fishery begins a little before sunset, and the nets are drawn in about two hours, to be again shot as morning approaches; for Pilchards enter the nets better at these seasons. A rope from one end of the string is fastened over the quarter of the boat, and the nets are left to float with the tide, no sails being set, except rarely in very calm weather, to prevent the nets being folded together. Within a few years an improvement has been made, derived, it would appear, from the practice of the herring-fishers, by which more fish have been taken, and much of the hazard obviated to which the nets were exposed by ships passing over them. It consists in diminishing the number and size of the corks along the head-line, and in fixing cords at proper distances, each of which has attached to it a stout buoy. These cords are from two to two and a half fathoms long, and consequently allow the upper edge of the nets to sink to that depth below the surface; but even now it is found that the fish are principally caught in the lower part of the net.

The number of fish taken by a drift-boat in a night's fishing varies exceedingly: from five to ten thousand is considered moderate; it often amounts to twenty thousand. For the season's fishing, about one hundred and fifty thousand fish would be deemed favourable.

For the sean-fishing, three boats are provided, of which two are about forty feet long, and ten wide at the beam, with flat timbers and a sharp bow. The first is termed the sean-boat, and is furnished with a sean two hundred and twenty fathoms in length, and twelve fathoms deep, which is buoyed along the head-rope with corks, and weighed down with leads. The second boat is called the volyer, a term supposed to be a corruption of the word, follower. This boat has a sean from one hundred to one hundred and twenty fathoms in length, and eighteen fathoms deep at its deepest part, and is termed the tuck-sean : it differs from the former, called the stop-sean, as well in shape as in dimensions, the middle being formed into a hollow or bunt. A third boat, called a lurker, is less than the others, and has no sean. The crew attending a sean consists of eighteen men and one or two boys. Seven of these are assigned to each of the larger boats, and the remaining four, including the master seaner, to the lurker. This fishery commences in August, three weeks or a month after the drivers, whose success, or the want of it, has much influence. The three boats proceed in the afternoon to some sandy bay, and cast anchor, keeping a good look-out for the appearance of fish, which are discovered either by the rippling of the water, by the stoiting or leaping of the fish, or by the colour they impart to the sea. In these respects, as marks of the difference between the habits of the Herring and the Pilchard, fishermen observe that the former rarely springs from the water, or stoits, as it is called, except when alarmed or driven; but the Pilchard does this often, and apparently from wantonness. When alarmed, both these fish will rush along the distance of five or six feet, as marked by the briming ;* but the Pilchard does this with more celerity than the Herring.

When the presence of fish is discovered, the lurker proceeds to the place to ascertain the magnitude of the schull, and the direction in which it is moving. The depth of water, clearness of ground from rocks and other obstructions, and the force and direction of the tide, enter also into the calculation of the master before he makes the signal for preparation. All the proceedings are directed by signs, for the fish are alarmed at noise, and when everything is favour-

* The flash of light seen in the sea when disturbed in the night, and supposed to proceed from minute molluscous or crustaceous animals.

able, a warp from the end of the sean is handed to the volyer, whose place it is to keep all taut ; the lurker continuing on the fish to watch their motions, and to point to the sean-boat what is to be enclosed. The sean-boat is rowed by four men, the other three being employed in throwing the net ; and such is the vigour exerted on this occasion, that this great body of net, rope, corks, and lead is thrown into the sea in less than five minutes. The sean at first forms a curved line across the course of the fish ; and while the two larger boats are employed in warping the ends together, the lurker's station is in the opening, where, by dashing the water, the fish are kept away from the only place of escape. When the sean is closed and the ends are laced together, if the body of the fish be great and the sea or tide strong, the net is secured by heavy grapnels, which are attached to the head-ropes by hawsers. It will appear from this account that it is not more difficult to take a thousand hogsheads of fish than a single hogshead ; the only difference being, that with the greater quantity the sean is regularly moored, which with the smaller is unnecessary : it may even be said that the capture of the larger body is most easily effected ; for as its motion is slow, its course is not so speedily altered.

When the evening has closed in, and the tide is low, they proceed to take up the fish. For this purpose, leaving the stop-sean as before, the volyer passes within it, and lays the tuck-sean round it on the inner side : it is then drawn together so as gradually to contract the limits of the fish, and raise them from the bottom. When disturbed, they become exceedingly agitated ; and so great is the force derived from their numbers and fear, that the utmost caution is used lest the net should either sink or be burst. When the tuck-sean is thus gradually contracting and the boats surround it, stones suspended from ropes, called minnies, are repeatedly plunged into the water at that part where escape alone is practicable, until the fish then to be taken up are supported in the hollow or bunt of the sean.

When brought to the surface, the voices of the men are lost in the noise made by the fish as they beat the water. The seaners fix themselves in pairs on the gunwales of the boats, with flaskets to lade the fish on board. When the quantity enclosed in the stop-sean is large, the tuck-sean is made to enclose no more than the boats can carry, of which a master seaner commonly forms a correct judgment by the extent of the briming in his sean, as the fish move in it; and many advantages result from taking up only a portion at one time, for the whole can thus be salted in proper condition, without fatigue or extraordinary expense : thus a week may possibly elapse before the whole of the capture is secured, part being taken up every night.

The description here given of the manner in which the Pilchard fishery is conducted applies to the greater part of the coast, but some variation occurs in particular districts. In Mount's Bay the men and boats employed to take the fish are not the same that convey it to land ; a mode of proceeding rendered necessary by the distance from shore at which it is taken. The fishery at St. Ives is regulated by a particular act of parliament, and there alone huers* continue to be employed, which were formerly common everywhere. The sean-fishery, as practised formerly, resembled that carried on at St. Ives ; and in one of Norden's maps is a representation of the taking of Pilchards by means of a sweep-net, of which one end continues near the shore, as then employed in St. Austle's Bay. The capture was drawn on shore in the mode now used with ground-seans for other

---

* Huers are men posted on elevated situations near the sea, who by various concerted signals, made with a bunch of furze in each hand, direct the fishermen how best to surround a schull of fish.

fish, and consequently none could be taken unless they approached near to an open beach ; and one end of a sean is now termed the pole end, from the pole shod with lead then used to elevate and spread the part to which the warp was attached.

Old and experienced fishermen have stated as the result of long observation, that, besides the well-known fact of the fish being most abundantly taken within a few days after the spring-tides, the direction of the tide has great effect on the motions of the schull. Its progress is always towards the same point, and in drift-nets all the heads of the fish point in one way, unless the tide has turned while the nets were afloat. In a bay where the tide comes round a headland and circles the bay, the fish take the same route, and a man aware of this may know in what direction to watch, and whither the schull is proceeding ; and as, especially when the tide is rapid, he must be careful that the sean is not carried on the back of the schull, the net must be so shot as to have the benefit of the tide, and yet be laid across the front of the fish. A schull will not turn back directly contrary to its former course, although, when alarmed, its direction may be considerably changed. In the open sea, drift-nets are commonly cast in the direction of the tide, because the nets are most easily kept in that course ; but when near land, or the entrance of a bay, a favourite position is parallel to it, by which the fish are intercepted in their advance or retreat. I have seen drift-boats shoot their nets in the midst of a multitude of fish, one in the direction in which they were going, and another across their course, and in less than two hours the second had taken nine thousand, the other not a fish ; and yet the boats frequently prefer the first plan. The most successful time for the drift-net fishery is during hazy nights, with some motion of the wave, for the fish then enter the nets freely, whereas in clear moonlight they are shy; and in very dark nights such is the brightness of the briming, that the nets look like a wall of fire, and deter the fish.

As an object of adventure, the Pilchard fishery is popular in Cornwall, and beyond a doubt the community is greatly benefited by it; yet it frequently happens that the success is partial, and the price low; and it may be questioned whether in any year the greater part of the seans obtain more than their expenses: but when there is a profit it is commonly considerable, and in this lottery every one is led by the hope of being among the fortunate.

The following is a statement, perhaps nearly approaching to the truth where absolute certainty is unattainable, of the amount of property engaged in the Pilchard fishery in the year 1827, when the bounty began to be withdrawn:—Number of seans employed, 186; not employed, 130; total number of seans, 316: number of drift-boats, 368: men employed on board drift-boats, 1600; number of men employed on seans at sea, 2672; number of persons on shore to whom the fishery affords direct employment, 6350; total number of persons employed in the fishery, 10,521: cost of seans, boats, &c. used in the fishery, 209,840*l.*; cost of drift-boats and nets, 61,400*l.*; cost of cellars for curing, and other establishments on shore for carrying on the fishery, 169,175*l.*; total capital invested directly in the Pilchard fishery, 441,215*l.* The outfit of a sean amounts to about 800*l.*; a string of drift-nets will cost about 6*l.*; the net and the boat from 100*l.* to 150*l.*; but this is used throughout the year for the other purposes of fishing. The nets are supposed to last about six years, and ought, of course, to produce their own value within that time, together with an adequate profit; but it is the complaint of the fishermen

VOL. II.   I that this is not the case.   The profit of the men depends on the share of the fish, which is divided into eight parts, of which the boat has one-eighth part, the nets three, and the men four: a boy that accompanies them is rewarded with the fish that may fall into the sea as the nets are drawn, to secure which he is furnished with a bag-net at the end of a rod, termed a keep-net.

The quantity of Pilchards taken is sometimes incredibly large.   A fisherman now alive was present once at the taking of two thousand two hundred hogsheads of Pilchards in one sean; but the greatest number heard of as taken at one time is stated by Borlase at three thousand hogsheads; in reference to which Pennant has made an astounding error, in reckoning by mistake thirty-five thousand fish to a hogshead, instead of three thousand five hundred.   The number since allowed has been three thousand, and is now two thousand five hundred fine fish; but it is scarcely necessary to say that they are not counted.   An instance has been known where ten thousand hogsheads have been taken on shore in one port in a single day, thus providing the enormous multitude of twenty-five millions of living creatures drawn at once from the ocean for human sustenance.

The different modes of curing the fresh fish are detailed elsewhere.   The various ports on the northern shore of the Mediterranean are the principal places to which the preserved fish are exported.

Our term Pilchard is said to be derived from *Peltzer*, a name by which this fish was known to some early Northern Continental authors.   A few Pilchards make their appearance occasionally in the Forth about October, generally preceding the Herrings; but the great shoals appear to belong almost exclusively to our south-western shores.   They are seldom seen east of Bigbury Bay; but in August 1834 a shoal of Pilchards were observed in Poole Harbour, and so many fish were taken that they were sold in the market at a penny a dozen.

Smith's History of the County of Cork contains a full and interesting account of the Pilchard fishery in Bantry Bay. They have been noticed also on the coast of the county of Cork, and taken at Dublin and Belfast. On our eastern coast, a few are taken every year at Yarmouth with the Herrings. They were more than usually abundant there in the years 1780, 1790, and 1799.

Specimens of the Pilchard sometimes measure eleven inches in length; the fish described measured nine inches. It much resembles the Herring, but is smaller and thicker. The length of the head is to the whole length as one to five; the depth of the body equal to the length of the head; the transverse thickness of the body equal to half its depth : the form of the head triangular, the upper surface flat; the dorsal and abdominal lines slightly and equally convex; no perceptible lateral line; the body across the back obtusely rounded; the line of the abdomen smooth; the edges of the scales of the two sides leaving a longitudinal groove from the branchiostegous rays to the vent, along which groove extends a row of scales of a peculiar shape, of which the woodcut here placed is a representation; the two long narrow lateral arms extending up each side under the scales, the shortest projection pointing backward: the scales of the body are very large, deciduous, and ciliated at the free edge.

The distance from the point of

I 2

the nose to the base of the last ray of the dorsal fin, and from thence half-way along the caudal rays, nearly equal: the commencement of the dorsal fin is therefore anterior to the middle of the fish by the whole length of the base of the fin ; the first and second rays shorter than the third, which is equal to the length of the base of the fin ; these first three rays articulated, but simple ; all the other rays branched : pectoral and ventral fins small, the latter commencing in a line under the middle of the dorsal fin ; the axillary scales very long : the anal fin commencing half-way between the origin of the ventral fins and the end of the fleshy portion of the tail ; the first ray short, the second and the last two rays the longest : the tail deeply forked ; the scales at the end of the fleshy portion of the body extending far over the bases of the caudal rays, particularly two elongated scales above and below the middle line.   The fin-rays in number are—

D. 18 : P. 16 : V. 8 : A. 18 : C. 19.   Vertebræ 55.

The mouth is small, without teeth, the under jaw the longest : the breadth of the eye one-fourth of the length of the head, and placed at rather more than its own breadth from the point of the nose ; the irides yellowish white : the cheeks and all the parts of the gill-covers tinged with golden yellow, and marked with various radiating striæ : the posterior edge of the operculum nearly vertical and straight : the upper part of the body bluish green ; the sides and belly silvery white ; the dorsal fin and tail dusky.   Mr. Couch says the Pilchard is sometimes found with a row of spots on the side, like the Shad ; which seems the result of disease, these fish being small, soft, and unfit for curing.

As an appropriate conclusion to this account of the Pilchard fishery of Cornwall, derived principally from the MS. of Mr. Couch, the vignette at the bottom of the page is a representation of the harbour of Polperro, near which Mr. Couch has long resided : and I take this opportunity of recording my obligations to that gentleman, not only for his great liberality in allowing me the unlimited use of his voluminous MS. of the Natural History of the Fishes which have been found on the coasts and in the rivers of Cornwall, with an extensive series of characteristic drawings, but also for the warm interest and substantial support afforded to this work during its progress.

While this sheet was going through the press, the London newspapers noticed the appearance of numerous large shoals of Pilchards on the south coast of Ireland, which the poor fishermen were unable to take advantage of from the want of proper nets and salt.

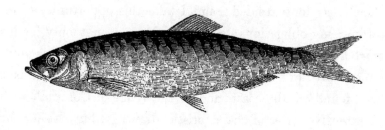

THE HERRING.

Clupea harengus, Linnæus. Bloch, pt. i. pl. 29.
    ,,   ,,   Willughby, p. 219, pl. P. 1, fig. 2.
    ,,   ,,   *Herring,* Ray, Syn. p. 103.
    ,,   ,,   Cuvier, Règne An. t. ii. p. 317.
    ,,   ,,   *Herring,* Penn. Brit. Zool. vol. iii. p. 444, pl. 79.
    ,,   ,,   ,, Flem. Brit. An. p. 182, sp. 51.

Anderson and Pennant were certainly mistaken in supposing that the great winter rendezvous of the Herring is
within the Arctic Circle: " there they continue," says Pennant, " for many months, in order to recruit themselves after
the fatigue of spawning; the sea within that space swarming
with insect food, in a degree far greater than in our warmer
latitudes."

" This mighty army begins to put itself in motion in the
spring. We distinguish this vast body by that name; for the
word Herring is derived from the German *Heer*—an army, to
express their numbers.   They begin to appear off the Shetland Islands in April and May.*   This is the first check this

* In another part of his account, Pennant says the Herrings continue on the
Welsh coast till February. (P. 447.)

army meets with in its march southward. Here it is divided into two parts : one wing of those destined to visit our coasts takes to the east, the other to the western shores of Great Britain, and fill every bay and creek with their numbers ; others proceed towards Yarmouth, the great and ancient mart of Herrings ; they then pass through the British Channel, and after that in a manner disappear. Those which take to the west, after offering themselves to the Hebrides, where the great stationary fishery is, proceed towards the north of Ireland, where they meet with a second interruption, and are obliged to make a second division : the one takes to the western side, and is scarcely perceived, being soon lost in the immensity of the Atlantic; but the other, which passes into the Irish Sea, rejoices and feeds the inhabitants of most of the coasts that border on it. These brigades, as we may call them, which are thus separated from the greater columns, are often capricious in their motions, and do not show an invariable attachment to their haunts."

This is Pennant's account as it regards our own islands. To show that this supposed migration to and from high northern latitudes does not exist, it is only necessary to state, that the Herring has never been noticed, that I am aware, as abounding in the Arctic Ocean : it has not been observed in any number in the proper icy seas ; nor have our whale-fishers or arctic voyagers taken any particular notice of them. There is no fishery for them of any consequence either in Greenland or Iceland. On the southern coast of Greenland the Herring is a rare fish ; and only a small variety of it, according to Crantz, is found on the northern shore. This small variety or species was found by Sir John Franklyn, on the shore of the Polar basin, on his second journey.

" That the Herring is, to a certain degree, a migratory

fish," says Dr. M'Culloch,* " may be true ; but even a much more limited migration is far from demonstrable. It is at any rate perfectly certain that there is no such progress along the east and west coasts from a central point." There can be no doubt that the Herring inhabits the deep water all round our coast, and only approaches the shores for the purpose of depositing its spawn within the immediate influence of the two principal agents in vivification—increased temperature and oxygen ; and as soon as that essential operation is effected, the shoals that haunt our coast disappear: but individuals are to be found and many are caught throughout the year. So far are they from being migratory to us from the North only, that Herrings visit the west coast of the county of Cork in August, which is earlier than those which come down the Irish Channel arrive, and long before they make their appearance at other places much further north. " In former times, the fishery of the east coast did not commence till that on the west had terminated. It is remarkable also that the eastern fishery has become so abundant as quite to have obscured the western." And Dr. M'Culloch, from other examples, confirms a statement previously made, that the fishery has commenced soonest on the southern part of the shore ; and, what is also remarkable, that for some years past it has become later every year.

The Herring is in truth a most capricious fish, seldom remaining long in one place ; and there is scarcely a fishing station round the British Islands that has not experienced in the visits of this fish the greatest variations both as to time and quantity, without any accountable reason.

" Ordinary philosophy is never satisfied," adds Dr. M'Culloch, " unless it can find a solution for everything ;

* See an excellent paper on the Herring in the 32nd number of the Journal of the Royal Institution, for January 1824.

and is satisfied, for this reason, with imaginary ones. Thus, in Long Island, one of the Hebrides, it was asserted that the fish had been driven away by the manufactory of kelp; some imaginary coincidence having been found between their disappearance and the establishment of that business. But the kelp fires did not drive them away from other shores, which they frequent and abandon indifferently without regard to this work. It has been a still more favourite and popular fancy, that they were driven away by the firing of guns; and hence this is not allowed during the fishing season. A gun has scarcely been fired in the Western Islands, or on the west coast, since the days of Cromwell; yet they have changed their places many times in that interval. In a similar manner, and with equal truth, it was said that they had been driven from the Baltic by the battle of Copenhagen. It is amusing to see how old theories are revived. This is a very ancient Highland hypothesis, with the necessary modification. Before the days of guns and gunpowder, the Highlanders held that they quitted coasts where blood had been shed: and thus ancient philosophy is renovated. Steam-boats are now supposed to be the culprits, since a reason must be found: to prove their effect, Loch Fyne, visited by a steam-boat daily, is now their favourite haunt, and they have deserted other lochs where steam-boats have never yet smoked." A Member of the House of Commons, during the sessions of 1835, in a debate on a tithe bill, stated, that a clergyman having obtained a living on the coast of Ireland, signified his intention of taking the tithe of fish; which was, however, considered to be so utterly repugnant to their privileges and feelings, that not a single Herring had ever since visited that part of the shore!

Our common Herring spawns towards the end of October or the beginning of November; and it is for two or three months previous to this, when they assemble in immense numbers, that the fishing is carried on, which is of such great and national importance. " And here," Mr. Couch observes, " we cannot but admire the economy of Divine Providence, by which this and several other species of fish are brought to the shores, within reach of man, at the time when they are in their highest perfection, and best fitted to be his food."

The mode of fishing for Herrings is by drift-nets, very similar to those employed for taking Mackerel and Pilchard, with a slight difference in the size of the mesh. The net is suspended by its upper edge from the drift-rope by various shorter and smaller ropes, called buoy-ropes; and considerable practical skill is required in the arrangement, that the net may hang with the meshes square, smooth, and even, in the water, and at the proper depth; for, according to the wind, tide, situation of their food, and other causes, the Herrings swim at various distances below the surface.

The size of the boat used depends on the distance from shore at which the fishery is carried on; but, whether in deep or in shallow water, the nets are only in actual use during the night. It is found that the fish strike the nets in much greater numbers when it is dark than while it is light: the darkest nights, therefore, and those in which the surface of the water is ruffled by a breeze, are considered the most favourable. It is supposed that nets stretched in the day-time alarm the fish, and cause them to quit the places where that practice is followed; it is therefore strictly forbidden.

The Herring having spawned, retires to deep water, and the fishing ends for that season. While inhabiting the depths of the ocean, its food is said by Dr. Knox to consist principally of minute entomostracous animals; but it is certainly less choice in its selection when near the shore. Dr. Neill found five young Herrings in the stomach of a large female Herring ; he has also known them to be taken by the fishermen on their lines, the hooks of which were baited with limpets ; and they have been repeatedly caught by anglers with an artificial fly. The young abound in the shallow water all round our shores during the summer months. I have seen them taken off Brighton in the small-meshed nets which are there used to draw for Atherine ; and they are caught by boys while angling from piers and rocks at various places along the southern coast. They are very abundant on the Yorkshire coast, where they are called Herring-sile ; and they swarm among the Orkney and Shetland Islands during the whole of the summer. They remain at the mouth of the Thames during their first autumn and winter : many are caught on the coasts of Essex and Kent in the nets used for taking sprats. From repeated examinations, I am induced to believe these young fish do not mature any roe during their first year.

The length of the head compared to the length of the body alone, without the head or caudal rays, is as one to four ; the depth of the body compared to the whole length of the fish, as one to five : the commencement of the dorsal fin half-way between the point of the upper jaw and the end of the fleshy portion of the tail ; the longest ray nearly as long as the base of the fin : the pectoral fin rather large compared to the size of the other fins. The ventral fin arises considerably behind the line of the commencement of the dorsal fin : this fin is small, with elongated axillary scales ; its origin half-way between the point of the lower jaw and the end of the short central caudal rays. The anal fin begins half-way between the origin of the ventral and the end of the fleshy portion of the tail, and extends over half the distance between its origin and the end of the fleshy portion, thus occupying the third quarter division of the distance between the origin of the ventral fin and the end of the fleshy portion of the tail; the rays very short. The tail considerably forked; the outer rays as long again as those of the middle. The fin-rays in number are—

D. 17 : P. 15 : V. 9 : A. 14 : C. 20 : Vertebræ 56 ;

varying in some specimens to

D. 19 : P. 17 : V. 9 : A. 16 : C. 18.

The lower jaw is by much the longer, with five or six small teeth extending in a line backwards on each side from the anterior point; four rows of small teeth on the central upper surface of the tongue ; a few small teeth on the central portion of the upper jaw, and the inferior edges below the gape finely serrated : the eye large ; its diameter compared to the length of the head as two to seven, and placed at the distance of its own breadth from the end of the nose : the dorsal and abdominal lines of the body slightly convex; the belly carinated, but not serrated ; the scales moderate in size, oval, and thin.   The upper part of the fish a fine blue, with green and other reflections when viewed in different lights ; the lower part of the side and belly silvery white; cheeks and gill-covers silvery, exhibiting the appearance of extravasation when the fish has been dead twenty-four hours.   Dorsal and caudal fins dusky ; the fins on the lower parts of the body almost white.

*ABDOMINAL
MALACOPTERYGII.*   *CLUPEIDÆ.*

LEACH'S HERRING.

Clupea Leachii, YARRELL, Zoological Journal, vol. v. p. 277, pl. 12.

THE examination of considerable quantities of the various sorts of fish caught at the mouth of the Thames during winter by fishermen engaged in taking Sprats, has enabled me to select what I believe to be a second species of British Herring.

The common Herring, when it visits our coast in autumn, is taken heavy with roe, which it deposits towards the end of October. It is certain that the fishing for them is abandoned about that time, as no purchasers could be found for the " shotten Herring ;" and it is also well known that the Herrings, having cast their roe, retire from the shore to deep water. Numbers of the young of the common Herring are taken with the Sprats. These are called Yawlings by some fishermen,—a term probably derived from yearling. But these young Herrings differ materially from the Herring which I believe to be new. The yearling fish have the elongated form of the adult common Herring : if seven inches long, which is about their average length, they are only one inch and three-eighths in depth, and are without roe. Having examined them repeatedly during the winter months, I am induced to believe they do not mature any roe during their first year; and the fact of their remaining in large shoals at the mouth of the Thames after the Herrings that have recently spawned have left the shore, may be taken in corroboration; for had they matured and deposited any roe, they would, like the more adult fish of their own species, have experienced the same necessity for retiring to deep water.

The Herring, however, which I now refer to, is found heavy with roe at the end of January, which it does not deposit till the middle of February. Its length is not more than seven inches and a half, and its depth near two inches. It is known that Dr. Leach had often stated that our coast produced a second species of Herring; but I am not aware that any notice of it has ever appeared in print. In order, however, to identify the name of that distinguished naturalist with a fish of which he was probably the first observer, I proposed for it the name of *Clupea Leachii.*

Dr. Leach's observations on the Herring were made during his visit to the extended line of our southern coast in the year 1808; and Mr. Jesse, in his " Gleanings in Natural History," has noticed the superiority and consequent partiality that is said to exist in favour of the Herrings of Cardigan Bay over those that are taken at Swansea.

Of the existence of a second species of Herring on our shores, further proof may be adduced in the following extracts.

" In former times," says Dr. M'Culloch, " the fishery of the east coast of Scotland did not commence till that on the west had terminated. It was then supposed, and not very unreasonably, that the fish had changed their ground, and that these were the western Herrings. Yet it ought to have been plain that this was not the case, as the eastern fish were entirely different in quality from the western, and very inferior. At the same time, they were in that condition as to spawning which proved that they could not have been the same fish. The fact of their being entirely different fish is now at least fully proved, because on both shores the period of the fishery has been the same."—*Journal of the Royal Institution, No. 32, for January* 1824, *p.* 217.

" A smaller but superior species of Herring is found occasionally in Loch Eriboll; but it is chiefly used for home consumption."—*Scotch Statistics, Durness.*

There are three species of Herrings said to visit the Baltic, and three seasons of roe and spawning. The Stromling, or small Spring Herring, spawns when the ice begins to melt; then a larger Summer Herring; and lastly, towards the middle of September, the Autumn Herring makes its appearance, and deposits its spawn.

The length of the head compared to that of the body alone, without the head or caudal rays, is as one to three; the depth of the body greater than the length of the head, and compared to the length of the head and body together is as one to three and a half; it is therefore much deeper in proportion to its length than our common Herring, and has both the dorsal and abdominal lines much more convex: the under jaw longer than the upper, and provided with three or four prominent teeth just within the angle formed by the symphysis; the superior maxillary bones have their edges slightly crenated: the eye is large, in breadth full one-fourth of the length of the whole head; irides pale yellow: the dorsal fin is placed behind the centre of gravity, but not so much so as in the common Herring ; the scales are smaller ; the sides without any distinct lateral line : the edge of the belly carinated, but not serrated ; the fins small. . The fin-rays in number are—

D. 18 : P. 17 : V. 9 : A. 16 : C. 20.   Vertebræ 54.

The back and upper part of the sides are deep blue, with green reflections, passing into silvery white beneath. The flesh of this species differs from that of the common Herring in flavour, and is much more mild.

*ABDOMINAL
MALACOPTERYGII.*

CLUPEIDÆ.

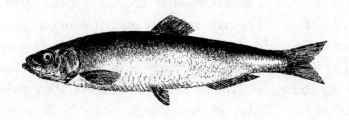

THE SPRAT.

GARVIE HERRING AND GARVIE. *Scotland.*

Clupea sprattus, Linnæus. Bloch, pt. i. pl. 29, fig. 2.
   ,,   ,,   Cuvier, Règne An. t. ii. p. 318.
   ,,   ,,   *Sprat,* Penn. Brit. Zool. vol. iii. p. 457.

Willughby and Ray, deceived apparently by the mis-application of the name by the fishermen of Cornwall, with which the latter became acquainted during his journey in that county, considered that the word Sprat was only a name for the young of the Herring and of the Pilchard, and others have been misled by their authority : but so well is this fish distinguished from both by the strongly serrated edge of the abdomen, that there is not a fisherman round those parts of our coast where the Sprat is taken that cannot immediately distinguish it from either, even in the midst of the darkest night. Its characters being now sufficiently appreciated, it is by some, and ought to be by all, admitted as a good and distinct spécies.

Though a much less valuable fish than the Herring, it is still a very useful one. Coming into the market in im-

VOL. II.   K

mense quantities and at a very moderate price immediately after the Herring season is over, it supplies during all the winter months of the year a cheap and agreeable food. Immense quantities are eaten; and from their rich quality and flavour, the consumption is not solely confined to the lower classes. They are generally eaten fresh, but are also preserved in various ways.

The Sprat is included by Linnæus in his *Fauna Suecica*, and by Professors Nilsson and Reinhardt in their publications on the Fishes of Scandinavia. Dr. Neill says the Sprat is sold in Edinburgh market by the dozen; and I have received specimens that were taken near the Forth, where they are called Garvie Herrings and Garvies. Farther south, they are most plentiful on the Norfolk, Suffolk, Essex, and Kentish coasts. I have taken them on the Dorsetshire coast in June, and they were then in roe. They inhabit the deep water round our southern coast during the summer months, and may be found in the stomachs of many of our voracious fishes every month in the year. I have taken three Sprats from the stomach of a Whiting, and have caught young Sprats off Ramsgate, Hastings, and Weymouth, in the months of August and September. Like the other species of the genus *Clupea*, they are wanderers: the shoals are capricious in their movements, and exceedingly variable in their numbers. " Upwards of a ton weight of Sprats was sold in our market last Saturday." (*Taunton Courier*, January 1832.) " It is nearly fifty years since this useful fish visited the neighbouring coast, and they now appear in exhaustless shoals close in shore on the south coast of Devon."

The Sprat is occasionally taken in Cornwall; and in Ireland, on the coasts of Cork, Dublin, and Belfast.

In Cornwall the true Sprat is, however, very rare; and

the name is appropriated, as it was by the old fishermen whom Ray consulted one hundred and fifty-six years ago, to the fry of the Herring and of the Pilchard. An analogous misapplication of a name exists on the eastern coast, where the true Pilchard rarely occurs, and where the name of Pilchard is given to the fry of the Shad and the half-grown Herring.

The fishing season begins early in November, continuing through the winter months; and the largest quantities are taken when the nights are dark and foggy. A few, and those of the best description, are taken in the same manner as the Mackerel, the Pilchard, and the Herring, by drift-nets of fine twine and suitable small mesh; a mode of fishing peculiarly adapted for the capture of those species which rove in shoals through the water. But the most destructive plan pursued against Sprats is by a mode called stow-boat fishing. The stow-boat net goes with two horizontal beams: the lower one, twenty-two feet long, is suspended a fathom above the ground; the upper one, a foot shorter in length, is suspended about six fathoms above the lower one. To these two beams, or balks, as they are called, a large bag-net is fixed, towards the end of which, called the hose, the mesh is fine enough to stop very small fry. The mouth of the net, twenty-two feet wide and thirty-six feet high, is kept square by hanging it to a cable and heavy anchor at the four ends of the beams. The net is set under the boat's bottom; and a rope from each end of the upper beam brought up over each bow of the boat, raises and sustains the beam, and keeps the mouth of the net always open, and so moored that the tide carries everything into it. A strong rope, which runs through an iron ring at the middle of the upper beam, and is made fast to the middle of the lower beam, brings both beams together parallel, thus closing the mouth

K 2

of the net when it is required to be raised. In this way an enormous quantity of Sprats, with the fry of many other species, are taken, which are principally sold by measure to manure land near the coast.

From four to five hundred boats are thus employed during the winter. Many thousand tons in some seasons are taken and sold at sixpence and eight-pence the bushel, depending on the supply and demand, to farmers, who distribute about forty bushels of Sprats over an acre of land, and sometimes manure twenty acres at the cost of twenty shillings an acre. In the winter of 1829-30, Sprats were particularly abundant: barge-loads, containing from one thousand to fifteen hundred bushels, bought at sixpence a bushel, were sent up the Medway as far as Maidstone to manure the hop-grounds. The coasts of Kent, Essex, and Suffolk are the most productive. So great is the supply thence obtained, that notwithstanding the immense quantity consumed by the million and a half inhabitants of London and its neighbourhood, there is yet occasionally a surplus to be disposed of at so low a price as to induce the farmers even so near the metropolis as Dartford to use them for manure.

A full-sized Sprat measures six inches in length, and rather more than one inch and one-eighth in depth. The length of the head compared to that of the body alone is as one to four ; compared to the whole length of the fish, as one to six : the depth of the body is to the whole length as one to five. The dorsal fin commences exactly half-way between the point of the lower jaw and the end of the caudal rays : the ventral fins arise in a vertical line under the first dorsal fin-ray, and have no axillary scales ; the ventral fins in the Pilchard and Herrings begin under the middle of the dorsal fin, and both have axillary scales,—these are two other external distinctions : the under jaw is the longest ; the diameter of the eye less than one-fourth of the whole head : considerable convexity of the dorsal and abdominal lines ; the latter serrated before the ventral fins, and still more strongly so behind them : the tail deeply forked ; the scales large, round, and deciduous ; the upper part of the head and back dark blue, with green reflections passing into silvery white on the gill-covers, sides, and belly ; the dorsal and caudal fins dusky ; pectoral, ventral, and anal fins white. The fin-rays in number are—

D. 17 : P. 15 : V. 7 : A. 18. : C. 19. Vertebræ 48.

The vignette below represents the mode of fishing for Whitebait in the Thames : the account of the fish commences on the next page.

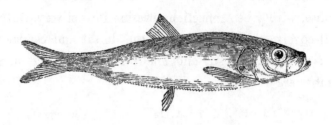

THE WHITEBAIT.

Clupea alba,   Yarrell, Zool. Journ. vol. iv. p. 137 and 465, pl. 10.
   ,,   ,,   *Whitebait*, Penn. Brit. Zool. vol. iii. p. 465, pl. 80.
Clupea alosa,   *Young Shad*, Don. Brit. Fish. pl. 98.

In the papers on the subject of the Whitebait published in the fourth volume of the Zoological Journal, I endeavoured to prove, historically and anatomically, that this little fish was not, as had been supposed, the young of the Shad, but a distinct species. In its habits it differs materially from all the other British species of *Clupea* that visit our shores or our rivers. From the beginning of April to the end of September this fish may be caught in the Thames as high up as Woolwich or Blackwall, every flood-tide, in considerable quantity, by a particular mode of fishing to be hereafter described. During the first three months of this period, neither species of the genus *Clupea*, of any age or size, except occasionally a young Sprat, can be found and taken in the same situation by the same means. The young Shad of the year are not two inches and a half long till

November, when the Whitebait season is over; and these young Shad are never without a portion of that spotted appearance behind the edge of the upper part of the operculum, which in one species particularly is so marked a peculiarity in the adult fish. The Whitebait, on the contrary, never exhibits a spot on the side at any age; but from two inches long up to six inches, which is the length of the largest I have seen, the colour of the sides is uniformly white.

About the end of March or early in April, Whitebait begin to make their appearance in the Thames, and are then small, apparently but just changed from the albuminous state of very young fry.* During the fine weather of June, July, and August, immense quantities are consumed by visiters to the different taverns at Greenwich and Blackwall. Pennant says, " They are esteemed very delicious when fried with fine flour, and occasion during the season a vast resort of the lower order of epicures to the taverns contiguous to the places where they are taken." What might have been the particular grade of persons who were in the habit of visiting Greenwich to eat Whitebait in the days when Pennant wrote, I am unable to state; but at present, the fashion of enjoying the excellent course of fish as served up either at Greenwich or Blackwall is sanctioned by the highest authorities, from the court at St. James's Palace in the West, to the Lord Mayor and his court in the East, including the Cabinet Ministers† and the philosophers of the Royal Society. As might be expected, examples so numerous and

* The Shad do not deposit their spawn till the end of June or the beginning of July.

† In the Morning Post of the day on which this account of the Whitebait was written, September 10th, 1835, the following paragraph appeared :—

" Yesterday the Cabinet Ministers went down the river in the Ordnance barges to Lovegrove's West India Dock Tavern, Blackwall, to partake of their annual fish dinner. Covers were laid for thirty-five gentlemen."

influential have corresponding weight ; and accordingly there
are few entertainments more popular or more agreeable than a
Whitebait dinner.

The fishery is continued frequently as late as September ;
and specimens of young fish of the year, four and five inches
long, are then not uncommon, but mixed, even at this late
period of the season, with others of very small size, as though
the roe had continued to be deposited throughout the summer ; yet the parent fish are not caught, and are believed by
the fishermen not to come higher up than the estuary, where,
at this season of the year, nets sufficiently small in the mesh
to stop them are not in use.

The particular mode of fishing for Whitebait, by which
a constant supply during the season is obtained, was formerly
considered destructive to the fry of fishes generally, and great
pains were taken to prevent it by those to whom the conservancy of the fishery of the Thames was entrusted ; but since
the history and habits of this species have been better understood, and it has been ascertained that no other fry of any
value swim with them,—which I can aver,—the men have been
allowed to continue this part of their occupation with little
or no disturbance, though still using an unlawful net.

When investigating the subject of the Whitebait, I was
occasionally engaged in witnessing the mode by which such
numbers were taken. The mouth of the net is by no means
large, measuring only about three feet square in extent ; but
the mesh of the hose, or bag-end of the net, is very small.
The boat is moored in the tide-way, where the water is from
twenty to thirty feet deep ; and the net with its wooden
frame-work is fixed to the side of the boat, as shown in
the vignette at page 125. The tail of the hose, swimming
loose, is from time to time handed into the boat, the end
untied, and its contents shaken out. The wooden frame

forming the mouth of the net does not dip more than four feet below the surface of the water; and, except an occasional straggling fish, the only small fry taken with the Whitebait are the various species of Sticklebacks, and the very common Spotted or Freckled Goby, described in vol. i. page 258; neither of which are of sufficient value or importance to require protection.* The farther the fishermen go down towards the mouth of the river, the sooner they begin to catch Whitebait after the flood-tide has commenced. When fishing as high as Woolwich, the tide must have flowed from three to four hours, and the water become sensibly brackish to the taste, before the Whitebait will be found to make their appearance. They return down the river with the first of the ebb-tide; and various attempts to preserve them in well-boats in pure fresh water have uniformly failed.

The Hamble, which runs into the Southampton Water, is the only other river from which I have received Whitebait. But this I believe to be owing rather to the want of a particular mode of fishing by which so small a fish can be taken so near the surface, than to the absence of the fish itself; which, abounding as it does in the Thames, I have very little doubt might be caught in some of the neighbouring rivers on our south and east coasts.

The Thames fishermen who live at and below Gravesend know the Whitebait perfectly, and catch them occasionally of considerable size in the small-meshed nets used in the Upper and Lower Hope for taking shrimps, called trinker-nets, which are like Whitebait nets, only larger; but these

---

* The fifteenth printed rule and order of the Lord Mayor and his Court is, that "no person shall take at any time of the year any sort of fish usually called Whitebait, upon pain to forfeit and pay five pounds for every such offence; it appearing to this court that under pretence of taking Whitebait the small fry of various species of fish are destroyed."—Page 11.

nets, working near the bottom, principally arrest the fry of the ground-swimming fishes.

The Sprat-fishers take Whitebait frequently on the Kentish and Essex coasts throughout the winter.

The length of the head compared with that of the body alone is as two to five ; the depth of the body compared to the whole length of the fish, as one to five : the dorsal fin commences half-way between the point of the closed jaws and the ends of the short middle caudal rays ; the longest ray of the dorsal fin as long as the base of the fin : the ventral fin arises behind the line of the commencement of the dorsal, and half-way between the point of the closed jaws and the end of the longest caudal rays ; the tail long and deeply forked.   The fin-rays in number are—

D. 17 : P. 15 : V. 9 : A. 15 : C. 20.   Vertebræ 56.

The head is elongated ; the dorsal line less convex than that of the abdomen ; the scales deciduous ; the abdominal line strongly serrated from the pectoral fin to the anal aperture.

The lower jaw the longest, and smooth ; the upper slightly crenated : the tongue with an elevated central ridge without teeth : the eye large ; the irides silvery : the upper part of the back pale greenish ash ; all the lower part, the cheeks, gill-covers, sides, and belly, silvery white : dorsal and caudal fins coloured like the back ; the latter tipped with dusky : pectoral, ventral, and anal fins, white.   The only food I could find in the stomach were the remains of minute crustacea.

For a representation of the mode of fishing for Whitebait, see the vignette at page 125, to which the block was transferred for want of sufficient space here.

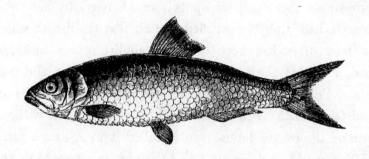

THE TWAITE SHAD.

Alosa finta,   Cuvier, Règne An. t. ii. p. 320.
   ,,   ,,   Willughby, pl. P. 3, fig. 1.
   ,,   ,,   *La Feinte,* Duhamel, sect. iii. pl. 1, fig. 5.
Clupea alosa, Linnæus. Bloch, pt. i. pl. 30.
   ,,   ,,   *Shad,* Penn. Brit. Zool. vol. iii. p. 460, pl. 80.
   ,,   ,,   ,,   Don. Brit. Fish. pl. 57.
   ,,   ,,   ,,   Flem. Brit. An. p. 183, sp. 53.

Generic Characters.—Upper jaw with a deep notch in the centre; in other respects like *Clupea.*

Baron Cuvier, in the last edition of the *Règne Animal,* has advanced the Shads, of which we have two spécies, to the rank of a genus, on account of the deep central notch in the upper lip; and I have followed this example for the additional reason that it will the more easily and effectually afford the means of obtaining a desirable alteration in our nomenclature.

According to Cuvier, most modern authors have misapplied the systematic trivial names of these two species, calling the Shad with teeth, and several spots along each side, *C. alosa;* and the larger Shad without teeth, and with a single spot only behind each gill-cover, or none at all, *C. finta.*

The *Alosa* of Rondeletius is not described or figured as possessing either teeth or spots; and Cuvier, by his usual research, had probably satisfied himself that the fish to which the term *alosa* had been originally applied was a toothless Shad, and that the toothed and spotted Shad was the true *finta.* Pennant, in noticing the second British species of Shad taken in the Thames and the Severn, which is without teeth or the row of lateral spots, called it an Allis; a name which it would be desirable still to retain, in reference to the generic term *Alosa.* The old name for the Shads was *Lachia ;* and hence are derived *Hallachia, Alachia, Alosa, Alose,* and Allis or Allice.

The differences noticed by Pennant and others in the smaller species of Shad, taken also in the Severn, near Gloucester, called the Twaite, induces the belief that it is our common Thames Shad; and the note by the editor of the last edition of the British Zoology, at the foot of page 463, (vol. iii.) is particularly deserving of notice. " I suspect," says the note, " that the Shad and Twaite are distinct species, and correspond with the *Alose* and *Feinte* of Duhamel." This appears to be precisely the case, as a comparison of our two Shads with the representations in Duhamel's work will prove : and Professor Nilsson, in his Prodromus of the Fishes of Scandinavia, which has been frequently referred to, has correctly designated and described our more common Shad of the Thames as the *finta** of Cuvier.

I venture to propose the names of Twaite Shad and

---

* Page 22.—*C. finta* Cuv. *C. maxilla superiore antice profunde incisa ; inferiore vix longiore ; maculis 5—6 lateralibus in serie positis ; dentibus utriusque maxillæ distinctis. Longit. circa 15 poll.*

Allice Shad for our two species, the better in future to distinguish them; thus combining the generic name Shad with a trivial name by which these two fishes have been hitherto, to some extent at least, locally known.

The Twaite Shad then, if I may so call it, is a sea-fish which enters our rivers about May, and in consequence of the time of its annual visit to some of the rivers of the European Continent is called the May-fish. The object of its visit to the fresh water is to deposit its spawn; and, that accomplished, it returns to sea by the end of July. Twaite Shads appear during these three months in abundance in the Thames, from the first point of land below Greenwich, opposite the Isle of Dogs, to the distance of a mile below; and great numbers are taken every season. These fish produce, however, but a small price to the fishermen, being in little repute as food, their muscles being exceedingly full of bones and dry. Formerly great quantities of the Twaite Shad were caught with nets in that part of the Thames opposite the present Penitentiary at Millbank, Westminster. Above Putney Bridge was another favourite spot for them; but the state of the water, it is believed, prevents the fish ascending the river in the same manner as in former years, and but few comparatively are taken. The ordinary size of the adult fish of this species is from twelve to sixteen inches.

Shad are not allowed to be caught in the Thames after the 30th of June, that the remaining fish may cast their spawn without interruption from nets.*

The principal spawning-time of the Twaite Shad in the Thames is about the second week in July, when numbers may be seen and heard frisking at or near the surface. In the language of fishermen, the Shad are said to thrash the

* Whitebait are plentiful throughout May and June.

water with their tails : they appear to disencumber themselves of the matured roe by violent muscular action ; and
on a calm still evening or night the noise they make may be
heard at some distance. I have obtained the young only
two inches and a half long in October ; and suspect they
grow slowly, finding them only four inches long, and the
young of the larger Allice Shad only six inches long, in the
following spring.

The habits and habitat of the two species of Shads have
probably been very frequently confounded. Though both
are common in the Severn during a particular season, Montagu has not noticed the appearance of either on the coast of
Devon. One species has been noticed on the Cornish coast
by Mr. Couch, and has also been taken near Dublin. On
the eastern coast it is common in the Thames ; is occasionally
taken off Yarmouth, on the Norfolk coast, with the Herrings, and also in the Tyne. It appears to have a considerable range to the northward, both Professors Nilsson and
Reinhardt including it among the fishes of Scandinavia.
The food of the Shads is small fish and the softer-skinned
crustacea.

The length of the head compared to the whole length
of the fish is as one to five ; the depth of the body rather
greater than the length of the head ; the distance from the
point of the nose to the commencement of the dorsal fin, measured again from thence backwards, falls far short of the end
of the fleshy portion of the tail ; the base of the last dorsal fin-
ray is half-way between the point of the nose and the end of
the caudal rays ; the longest ray of the dorsal fin is as long as
the base of the fin ; the ventral fins, without axillary scales,
are placed a little behind the line of the commencement of
the dorsal fin ; the base of the anal fin, occupying about
two-fifths of the space between the ventral fin and the end

of the fleshy portion of the tail, is shorter than the anal fin in the Allice Shad, and has five rays less, beginning also more forward: the tail deeply forked; the caudal rays with two thin membranous appendages on each side, parallel to the seventh and thirteenth caudal rays, about an inch in length by three-eighths deep; all four membranes opening from the centre, being attached by the outer edge only. The scales of the body rather large in proportion than those of the Allice; the lateral line, as in most of the *Clupeidæ*, scarcely perceptible. The abdomen strongly serrated. The lower jaw the longest, with a few teeth anteriorly; the upper jaw with a deep central notch, and a row of small teeth on the edge down each side. The breadth of the eye equal to one-fourth of the length of the head; the mucous vessels on the surface of the gill-covers beautifully arborescent; the top of the head and back dusky blue, with brown and green reflections in particular points of view; from the upper edge of the operculum a row of five or six dark spots extend in a line backwards, the last generally the most indistinct, the number sometimes more than six; the irides, sides of the head and body, silvery white, with a tinge of copper colour; dorsal and caudal fins dusky; pectoral, ventral, and anal fins white. This species is immediately distinguished from the Allice Shad by possessing teeth, the lateral spots, and the smaller anal fin. The fin-rays in number are—

D. 18 : P. 15 : V. 9 : A. 21 : C. 19. Vertebræ 55.

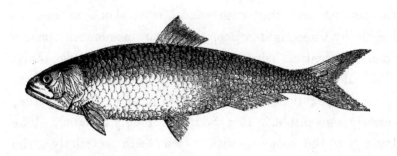

THE ALLICE SHAD.

Alosa communis,   Cuvier, Règne An. t. ii. p. 319.
   ,,   ,,   Willughby, pl. P. 3, fig. 2.
   ,,   ,,   *Allice,* Penn. Brit. Zool. vol. iii. p. 463.
   ,,   ,,   *Alose,* Duhamel, sect. iii. pl. 1, fig. 1.

THE ALLICE SHAD, by far the larger of the two in size, appears to be much more limited in its localities as a British species. It is represented by Pennant and others as abundant in the Severn, but is much less known elsewhere.

Dr. Hastings, in his Illustrations of the Natural History of Worcestershire, at page 77 says, " This is another fish which the Severn affords in great perfection. These fish generally appear in May, though sometimes in April. This, however, depends a good deal upon the quality of the water : if it is clear, they ascend early in the spring ; but if there happens to be a flood, they wait till the waters are restored to their former purity ; and if they meet with a flood in their progress upward, they immediately return, and keep below Gloucester. The weight of this Shad (the Allice of Pennant) is seldom less than four pounds ; they continue in the river about two months, and are succeeded by a variety called the Twaite, which is less than the Shad, never weighing more than two pounds, and is but little esteemed. Dr. Fleming says, that the celebrated Whitebait of the Thames, which appears near Blackwall and Greenwich during the month of July, is the fry of this fish ; but as, although the Shad are plentiful in the Severn, we hear nothing of the Whitebait,* further investigation seems to be required on this point."

In the Thames, the Allice Shad is of rare occurrence. A specimen was brought to me in 1831, that had been caught above Putney Bridge ; and another was taken in 1833, which is noticed by Mr. Jesse in the third series of his Gleanings in Natural History, page 147. " This fish was taken June 25th, opposite Hampton Court Palace ; and its appearance so high up the river is very unusual. On taking it out of the well of the boat, it was full of spawn, and died immediately." I have had opportunities of examining very fine specimens from the Severn, sent to me by T. B. L. Baker, Esq. of Hardwick Court.

The flesh of this species is said to be of good flavour, and the quality is considered to improve the higher the fish ascends the river. Ælian says the Shads appear to take pleasure in the sounds of musical instruments ; but if it happens to thunder when they are ascending rivers, they return rapidly to the sea.

Both species of Shads have great resemblance, except in size, to Herrings, and have been frequently called the mother of Herrings. The large Herrings of two feet in length, so called by Anderson and others, and said to occur in the

* This, it may be remembered, was adduced as one of the proofs that the Whitebait were not the young of the Shad.

VOL. II.   L

Northern Seas, and among our Northern Islands, are no doubt to be considered as referring to our Shads.

The specimen described measured two feet in length; the body deep and compressed; the thickness rather less than one-third of the depth. The length of the head compared to that of the whole fish is as one to six; the depth compared to the whole length, as one to four and a half. The length of the base of the dorsal fin three inches; the fourth ray, which is the longest, is one-third shorter than the whole length of the base of the fin; the first and second rays shorter than the third; these three rays simple, all the others branched: the first ray half-way between the point of the nose and the last ray of the anal fin; the last ray exactly half-way between the point of the nose and the end of the tail. Pectoral fin small; the upper ray the longest, strong, and simple; the others branched: ventral fin also small; the first ray arising in a vertical line under the first ray of the dorsal fin; axillary scales long, narrow, and pointed: anal fin commencing half-way between the ventral fin and the origin of the lower caudal rays, nearly one-fourth longer in the base than the dorsal fin; the first three rays shorter than the fourth, which is the longest, and only one-third the length of the base of the fin: the tail long and slender, deeply forked; the rays of the middle only one-fourth of the length of the longest external rays; the seventh and thirteenth caudal rays furnished with membranous appendages on each side similar to those observed in the Twaite Shad. The fin-rays in number are—

D. 19 : P. 15 : V. 9 : A. 26 : C. 20.

The lower jaw the longest and smooth; the upper jaw with a central notch; the lateral edges crenated: the breadth of the eye rather less than one-fifth of the length of the

head, and placed one diameter and a half from the end of the
nose ; mucous vessels of the gill-covers beautifully distributed ; the nape and shoulders rise suddenly ; the greatest
depth of the body just before the ventral fin ; scales of the
body rather large, nearly circular, and thin ; no distinct lateral line ; abdominal edge strongly serrated, particularly
behind the ventral fins. The colours very similar to those
of the Twaite Shad, with a single dusky patch behind the
operculum, sometimes scarcely visible.

Figure 1 of plate III. in Dr. Fleming's Philosophy of
Zoology is a representation of the Allice Shad.

Intending to make the fishing-boats of several countries
the subjects of some of the vignettes, that at page 120 represents a Dutch boat : the vignette below is a representation of
a French fishing-boat.

L 2

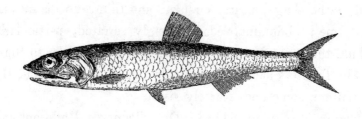

THE ANCHOVY.

Engraulis encrasicolus,	*Anchovy,*	Flem. Brit. An. p. 183, sp. 54.
,, *vulgaris,*		Cuvier, Règne An. t. ii. p. 322.
,, ,,		Willughby, p. 225, P. 2, fig. 2, App. 27.
Clupea encrasicolus,		Linnæus. Bloch, pt. i. pl. 30, fig. 2.
,, ,,	*Anchovy,*	Penn. Brit. Zool. vol. iii. p. 459, pl. 78.
,, ,,	,,	Don. Brit. Fish. pl. 50.

Generic Characters.—Distinguished from the Herrings in having the head pointed; the upper jaw the longest; the mouth deeply divided; the opening extending backwards behind the line of the eyes; the gape and branchial apertures very large; the ventral fins in advance of the line of the commencement of the dorsal; the abdomen smooth; branchiostegous rays 12.

I have followed Dr. Fleming in preserving to the Anchovy the old name by which it was formerly known. It was called *Lycostomus* from the form of its mouth; and *Encrasicholus Engraulis,* because from its bitterness it was supposed to carry its gall in its head. For this reason the head as well as the entrails are removed when the fish is pickled.

The Anchovy is a common fish in the Mediterranean from Greece to Gibraltar; and was well known to the Greeks and Romans, by whom the liquor prepared from it, called

Garum, was in great estimation. Its eastern range is extended into the Black Sea.

The fishing for them is carried on during the night, and lights are used with the nets.

The Anchovy is common on the coasts of Portugal, Spain, and France; it occurs, I have no doubt, at the Channel Islands, and has been taken on the Hampshire coast, and in the Bristol Channel. In the Appendix to Willughby's work, it is mentioned as having been taken on the coast of Wales; Pennant obtained it near his own residence at Downing in Flintshire; and Mr. Bicheno has very recently obtained several on the coast of Glamorganshire. It is said to be sold frequently in Liverpool market, and is reported to be at this time an inhabitant of the piece of water below Blackwall called Dagenham Breach.

Its range to the north is extensive, as it is occasionally taken in the Baltic and on the coast of Norway; but is not included by Linnæus in his *Fauna Suecica.*

The Anchovy appears to attain a much larger size than has usually been accorded to it: from four to five inches in length is the more ordinary size; but Mr. Couch says, " I have seen it in the Cornish seas of the length of seven inches and a half; and I have met with specimens from autumn, through the winter, to the middle of March : it is therefore probable that a fishery might be established with good prospect of success; for though the nets employed for other fish can take but few of them, the numbers found in the stomachs of the Whiting and other ravenous fishes show that they are in considerable abundance."

The Anchovy is immediately recognised among the species of the family to which it belongs, by its sharp-pointed head, with the upper jaw considerably the longest. The length of the head compared with the length of the body alone is as one to three; the depth of the body but two-thirds of the length of the head, and compared to the length of the whole fish is as one to seven: the first ray of the dorsal fin arises half-way between the point of the nose and the end of the fleshy portion of the tail; the third ray of the dorsal fin, which is the longest, is of the same length as the base of the fin: the pectoral fin small; the ventral fins arise, in a vertical line, in advance of the commencement of the dorsal fin, which is over the space between the ventral and anal fins: the base of the anal fin is as long as the distance from its commencement to the origin of the ventral fins; the rays short: the tail deeply forked. The fin-rays in number are—

D. 14 : P. 15 : V. 7 : A. 18 : C. 19.

The breadth of the eye is one-fifth of the length of the whole head; the peculiarity in the comparative length of the jaws has been previously noticed; the gill-covers are elongated; the scales of the body large and deciduous: the colour of the top of the head and back blue, with a tinge of green; irides, gill-covers, sides, and belly, silvery white; the fins delicate in structure, and greenish white; the membranes connecting the rays almost transparent.

In a series of notes on the occurrence of rare fish at Yarmouth and its vicinity, with which I have been favoured by Dawson Turner, Esq. there is mention of a specimen of the Anchovy, taken on the beach, which measured six inches and a half in length : an additional proof of the large size acquired by this fish on our shores.

SUBBRACHIAL
MALACOPTERYGII.   *GADIDÆ.**

THE COMMON COD.

THE KEELING.

Morrhua vulgaris, Cuvier, Règne An. t. ii. p. 331.
    ,,   ,,   *Cod,*   Flem. Brit. An. p. 191, sp. 76.
    ,,   ,,   *Codfish, Keeling,* Willughby, p. 165, L. 1, fig. 4.
    ,,   ,,   Linnæus. Bloch, pt. ii. pl. 64.
    ,,   ,,   *Common Codfish,* Penn. Brit. Zool. vol. iii. p. 231.
    ,,   ,,   *Codfish,* Don. Brit. Fish. pl. 106.

Generic Characters.—Body elongated, smooth, compressed towards the tail ; back furnished with three dorsal fins ; ventral fins pointed ; abdominal line with two fins behind the anal aperture ; the lower jaw with one barbule at the chin ; branchiostegous rays 7.

Baron Cuvier's first division of his second order of fishes, those with flexible fin-rays, and with the ventral fins attached to the abdomen, being concluded, the soft-finned fishes of the second division, or those forming his third order, succeed. These are recognised by having the ventral fins placed very near the pectorals ; the bones supporting the

* The family of the Codfish.

former being attached to the bones of the shoulder supporting the latter: and this disposition of the ventral fins has
been conveniently referred to by the single term subbrachial.

This division includes some of the species most valuable
to man as articles of food and commerce; among which may
be particularly noticed some of those belonging to the first
family, which includes the Common Cod, Haddock, Whiting, and many others to be hereafter particularly noticed, all
more or less remarkable for the excellence of their flesh,
which is white, firm, separates readily into flakes, is agreeable
to the taste, wholesome, and cheap.

The old genus *Gadus* of Linnæus included fishes with one,
two, or three dorsal fins, one or two anal fins, with or without
barbules or cirri about the mouth, and of very different forms
of body. These have been separated by Cuvier, whose first
genus includes only those with three dorsal fins, two anal
fins, and one barbule at the chin, as the generic characters
determine.

The Common Cod is not only one of those species most
universally known, but is also one of the greatest intrinsic
value, whether we consider the quality of the fish itself, the
enormous numbers in which it is taken, or the extensive
range over which it exists. In the seas with which Eu-
ropeans are best acquainted, this fish is found universally
from Iceland very nearly as far south as Gibraltar; but it
does not exist in the Mediterranean: and on the eastern side
of the American continent, and among its numerous islands,
from the 40th degree of latitude up to 66°, it is even still
more abundant.

In this country it appears to be taken all round the coast:
among the islands to the north and west of Scotland it is
abundant; most extensive fisheries are carried on: and it

may be traced as occurring also on the shore of almost every county in Ireland. In the United Kingdom alone, this fish, in the catching, the curing, the partial consumption and sale, supplies employment, food, and profit to thousands of the human race.

The Codfish is very voracious; a favourable circumstance for the fishermen, who experience little difficulty in taking them with almost any bait whenever a favourable locality is ascertained. As these fish generally inhabit deep water, from twenty-five to forty and even fifty fathoms, and feed near the ground on various small fish, worms, crustacea* and testacea, their capture is only attempted with lines and hooks. Two sorts of lines, adapted for two very different modes of fishing, are in common use. One mode is by deep sea-lines, called bulters, on the Cornish coast: these are long lines, with hooks fastened at regular distances along their whole length by shorter and smaller cords called snoods; the snoods are six feet long each, and placed on the long line twelve feet from each other, to prevent the hooks becoming entangled. Near the hooks these shorter lines, or snoods, are formed of separate threads loosely fastened together, to guard against the teeth of the fish. Some variations occur at different parts of the coast, as to the number of hooks attached to the line, as well as in the length of the snood; but the distance on the long line between two snoods is always double the length of the snood itself. Buoys, buoy-ropes, and anchors or grapples, are fixed one to each end of the long line; the hooks are baited with sandlaunce, limpet, whelk, &c.: the lines are always laid, or, as it is termed, shot, across the tide; for if the tide runs upon the end of the line, it will force the hooks together, by which the whole tide's fishing is irrecoverably lost: they

* Mr. Couch has taken thirty-five crabs, none less than the size of a half-crown piece, from the stomach of one Cod.

are deposited generally about the time of slack water, between each ebb and flow, and are taken up or hauled for examination after being left about six hours, or one flood or ebb.

An improvement upon this more common plan was some years ago suggested by Mr. Cobb, who was sent to the Shetlands by the Commissioners appointed for the improvement of the fisheries. He fixed a small piece of cork within a certain distance of the hook, about twelve inches, which suspended and floated the bait so as to prevent its falling on the ground; by which method the bait was more freely shown to the fish, by the constant and variable motion produced upon it by the tide. In the old way, the bait was frequently hid from the fish by being covered with seaweed, or was consumed by some of the numerous Star-fish and Crabs that infest the ground.

The fishermen, when not engaged in shooting, hauling, or rebaiting the long lines, fish with hand-lines, armed with two hooks kept apart by a strong piece of wire : each fisherman manages two lines, holding one line in each hand ; a heavy weight is attached to the lower end of the line not far from the hooks, to keep the bait down near the ground, where the fish principally feed. These two modes of line-fishing are practised to a great extent nearly all round the coast ; and enormous quantities of Cod, Haddock, Whiting, Coalfish, Pollack, Hake, Ling, Torsk, and all the various flat-fish, usually called by the general name of whitefish, are taken. Of Codfish alone, the number taken in one day is very considerable ; from four hundred to five hundred and fifty fish have been caught on the banks of Newfoundland in ten or eleven hours by one man ; and a master of fishing-vessels trading for the London market told me that eight men, fishing under his orders off the Dogger Bank, in twenty-five fathoms water, have taken eighty score of Cod in one day. These are brought to Gravesend in stout cutter-rigged vessels of eighty or one hundred tons burthen, called storeboats, built for this traffic, with a large well in which the fish are preserved alive; and of these a portion is sent up to Billingsgate market by each night-tide.

Well-boats, for preserving alive the fish taken at sea, came into use in this country early in the last century. They are said to have been first built at Harwich about 1712. The storeboats remain as low down as Gravesend, because the water there is sufficiently mixed to keep the fish alive : if they were to come higher up, it would kill them.

A change has lately taken place from the Cod having shifted their ground. Formerly the Gravesend and Barking fishermen obtained no Cod nearer than the Orkneys or the Dogger Bank; but for the last two or three years the supply for the London market has been obtained by going no farther than the Lincolnshire and Norfolk coasts, and even between that and London, where previously very few fish could be obtained.

Cod have been kept in salt-water ponds in different parts of Scotland, and found to maintain their condition unimpaired. Of these ponds there are three; one in Galloway, another in Fife, and a third in Orkney. That in Galloway is at Logan, the seat of Colonel M'Dowall: it is a basin of thirty feet in depth, and one hundred and sixty in circumference, hewn out from the solid rock, and communicating with the sea by one of those fissures that are common to bold and precipitous coasts. A fisherman is attached to this preserve, whose duty it is constantly to supply the fish with the necessary quantity of food, which several species soon learn to take eagerly from the hand. In the course of the fishing for this daily supply, such fish as are not too much injured are placed in the reservoir; the others are cut in pieces for food for the prisoners. The whelks, limpets, and other testacea, are boiled to free them from the shells; and no sooner does the keeper or his son appear with the well-known basket of prepared food, than a hundred mouths are simultaneously opened to greet the arrival. The Cod-fish are the most numerous in this preserve; one of which has lived twelve years in confinement, and attained a large size.

In a natural state the Cod spawns about February; and nine millions of ova have been found in the roe of one female. The Cod is in the greatest perfection as food from the end of October to Christmas. It may, in fact, be said of the whole of the family of *Gadidæ,* that they are in the best condition for the table during the cold months of the year. The young of the Cod, about six inches long, abound at the mouth of the Thames and Medway throughout the summer: as autumn advances, they gain size and strength, and are caught from twelve to sixteen inches in length by lines near the various sandbanks in the Channel. When of Whiting size, they are called Codlings and Skinners; and when larger, Tamlin Cod.

On the coast of Durham and Northumberland, and at the Isle of Man, the Cod acquire a dark red or reddish brown colour; and are called Red Cod, Ware Cod, and Red Ware Cod, when of this particular colour. I saw a considerable quantity in this state in Berwick market, and have had others sent to me by Dr. Johnston. Both the varieties of our Common Cod—for there appears to be two well-marked varieties—were equally red. This colour is considered to be the consequence of particular food obtained in certain localities. At a short distance only from the situations named, the Codfish are of the usual ash-green colour.

The largest Codfish I have a record of weighed sixty pounds, was caught in the Bristol Channel, and produced five shillings : it was considered cheap there at one penny the pound. Pennant, however, states that a Codfish of seventy-eight pounds' weight was caught at Scarborough, and sold for one shilling.

There appears to be two well-marked varieties of the Common Cod ; one with a sharp nose, elongated before the eye, and the body of a very dark brown colour, which is usually called the Dogger Bank Cod. This variety prevails also along our southern coast. The other variety has a round blunt nose, short and wide before the eyes, and the body of light yellowish ash-green colour, and is frequently called the Scotch Cod. Both sorts have the lateral line white. I believe the distinction of more southern and northern Cod to be tenable, and that the blunt-headed lighter-coloured fish does not range so far south as the sharper-nosed dark fish. In this view I may quote the authority of Dr. Mitchell, who, in his paper on the Fishes of New York, says of the broad-nosed fish, " We get him, however, only in the cool season, for the summer temperature of our waters kills him : he is, therefore, only found here between November and April." Our fishermen now finding plenty of Codfish much nearer home, the London shops for the last year or two have only now and then exhibited specimens of the short-nosed northern Cod : both varieties are equally good in quality, and both are frequently taken on the same ground.

The length of the specimen described was three feet, and the weight about twelve pounds. The length of the head compared to the length of the body alone, without the caudal rays, is as one to two and a half ; the depth of the body equal to the length of the head : the first dorsal fin com-

mences in a vertical line just behind the origin of the pectorals; the second dorsal commences in a line over the anal
aperture, and ends on the same plane as the first anal fin;
the third dorsal fin and the second anal fin begin and finish
on the same plane: the tail nearly square; all the rays of
the fins covered with an extension of the skin of the body.

The fin-rays in number are—

D. 10. 20. 18 : P. 20 : V. 6 : A. 20. 16 : C. 26.   Vertebræ 50.

The head is large; the belly tumid and soft; the body
tapering gradually throughout the latter half; the cavity of
the abdomen extended internally behind the anal aperture,
the intestine being recurved: the upper part of the head,
cheeks, back, and sides, mottled and spotted with greenish
ash; the belly white; the lateral line white, broadest along
the posterior half; all the fins dusky, the first and second
dorsal being rather lighter in colour than the rest: a broad
band of short teeth on the upper jaw, which is the longest,
and on the anterior part of the vomer; a narrower band on
the lower jaw, with one elongated barbule at the chin: the
irides silvery, the pupil blue; the breadth of the orbit one-
sixth of the length of the head.

The vignette represents a Scheveling fish-cart.

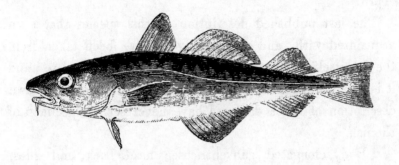

THE DORSE,

OR VARIABLE COD.

Morrhua callarias,	CUVIER, Règne An. t. ii. p. 332.
,, ,,	FLEM. Brit. An. p. 191.
,, ,,	WILLUGHBY, p. 172, L. 1, fig. 1.
Gadus ,,	LINNÆUS. BLOCH, pt. ii. pl. 63.
,, ,,	BERKENHOUT, Syn. edit. 1795, p. 67, sp. 2.
,, ,,	*Variable Cod,* PENN. Brit. Zool. vol. iii. p. 239.

THE authority upon which this species was originally introduced into the catalogue of British fishes seems now to be questionable. Neither Berkenhout nor the naturalists who have followed him, in including the name of it, appear to have seen any British example ; and **Dr.** Fleming, who from his northern locality was the most likely to have seen speci-mens, mentions it only on the authority of others, and does not number it in his series of species.

It appears to be a fish well known in the Baltic, and fre-quently called the Baltic Cod. It is included by Professor Nilsson in his Fishes of Scandinavia, and seems to be fully entitled to one of its names, that of Variable Cod, four northern varieties appearing to be well known, which are each distinguished there by a particular term referring to peculiarities in the colouring. It spawns in March and April.

The last published description of this species that I am acquainted with, and most likely to have been taken from the fish itself, is that by M. Nilsson, before referred to ; and it is here given rather than multiply in print any well-known description of older date. I have never seen a specimen of the fish.

" Body elongated, subventricose ; head, back, and sides, more or less spotted ; lateral line white, bent ; tail square ; upper jaw much the longer ; snout prominent, sharp ; under jaw only half as long as the head, and ending on a line half-way between the nose and the eye."

The fin-rays in number are—

D. 15. 18. 20. : P. 20 : V. 6 : A. 19. 18. : C. 24.

Length from twelve to twenty-four inches.

SUBBRACHIAL
MALACOPTERYGII.   *GADIDÆ.*

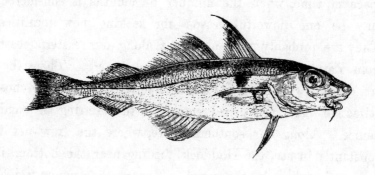

THE HADDOCK.

Morrhua æglefinus, Cuvier, Règne An. t. ii. p. 331.
    ,,   ,,   Haddock, Flem. Brit. An. p. 191, sp. 77.
    ,,   ,,   Hadock, Willughby, p. 170, L. 2.
Gadus   ,,   Linnæus. Bloch, pt. ii. pl. 62.
    ,,   ,,   Hadock, Penn. Brit. Zool. vol. iii. p. 241.
    ,,   ,,   Haddock, Don. Brit. Fish. pl. 59.

The Haddock is almost as well known as the Common Cod ; and from the quantity taken at numerous localities round the coast, and the facility with which the flesh can be preserved, it is a fish of considerable value. Besides frequenting the coast of Great Britain, from the extreme north to the Land's End, the Haddock may be traced nearly all round the shores of Ireland; and the largest examples have been taken in Dublin Bay and off the Nymph Bank. Though ranging over a considerable space both north and south of the geographical situation of this country, the Haddock does not exist either in the Baltic or in the Mediterranean.

Haddocks swim in immense shoals, but are uncertain as to

VOL. II.   M their appearance in places that had been formerly visited, and they are prone to change their ground after having arrived. The enormous consumption of food even in a short space of time, when the number of mouths is considered, may be one powerful reason for seeking new localities. They are probably more abundant along our eastern coast, from Yarmouth to the Tyne, than elsewhere. There they are caught with long-lines and hand-lines, and the most attractive baits are pieces cut from the Herring or Sand-launce. Along our southern shore, where the trawl-net is constantly in use, the Haddock, feeding near the bottom, is frequently taken in the trawl. The most common weight of a Haddock is from two to four pounds. I have seen Haddocks of ten pounds' weight in the London market; the Brixham trawling-ground has produced Haddock of fourteen pounds; but the largest seen for some years past weighed sixteen pounds, and was taken in Dublin Bay.

Haddocks spawn in February and March, and the young are six inches long by the beginning of September. When kept in confinement in the salt water preserve referred to in the account of the Common Cod, the Haddocks were found to be the tamest fishes in the pond, and took limpets one after another from the hand. Their food is small fish, and almost any of the inferior animals of the deep, even the spiny Aphrodita. They are in the best condition for the table during the last three months of the year.

The French fishermen call the Haddock, *Hadot,* whence probably our name was derived.

Pennant says, " Our countryman Turner suggested that the Haddock was the *Onos* or *Asinus* of the ancients. Different reasons have been assigned for giving this name to the species, some imagining it to be from the colour of the fish, others because it used to be carried on the backs of asses to market." A different reason appears to me more likely to have suggested the name: the dark mark on the shoulder of the Haddock very frequently extends over the back and unites with the patch of the shoulder on the other side, forcibly reminding the observer of the dark stripe over the withers of the ass; and the superstition that assigns the mark in the Haddock to the impression St. Peter left with his finger and thumb when he took the tribute-money out of a fish of this species, which has been continued to the whole race of Haddocks ever since the miracle, may possibly have had reference, or even its origin, in the obvious similarity of this mark on the same part of the body of the Haddock and of the humble animal which had borne the Christian Saviour. That the reference to St. Peter is gratuitous, is shown by the fact that the Haddock does not exist in the sea of the country where the miracle was performed.

The length of the specimen described was twenty inches. The length of the head compared to the length of the body, without including the caudal rays, is as one to two and a half; the depth of the body less than the length of the head: the first dorsal fin commences in a line over the origin of the pectorals; the second dorsal fin begins in a line over the anal aperture, and ends nearly on the same plane with the first anal fin; the third dorsal fin, and the second anal fin, commence nearly on the same plane, but the base of the first is longer than that of the second: the caudal rays rather long, and the tail slightly forked. The fin-rays in number are—

D. 15. 21. 19. : P. 18 : V. 6 : A. 24. 8. : C. 25. Vertebræ 54.

The head slopes suddenly from the crown to the point of the nose; the upper jaw much longer than the lower; the nose projecting beyond the opening of the mouth, which is

M 2

small ; a broad band of short teeth on the superior maxillary bones, and a patch of teeth also, of the same character, on the most anterior part of the vomer ; lower jaw furnished with a narrow band of teeth : the barbule at the chin small : the eye large ; the diameter of the orbit more than one-fourth of the whole length of the head ; the irides silvery ; the pupil large, somewhat angular in form, and blue : the head, cheeks, back, and upper part of the sides, dull greyish white ; lower part of the sides and belly almost white, slightly mottled with grey ; the body covered with small scales ; the lateral line strongly marked and black ; under the middle of the first dorsal fin, but below the lateral line, a black patch, which in many specimens extends over the back and unites with the mark on the other side ; the dorsal fins and tail dusky bluish grey ; pectoral, ventral, and anal fins lighter.

SUBBRACHIAL
MALACOPTERYGII.   *GADIDÆ.*

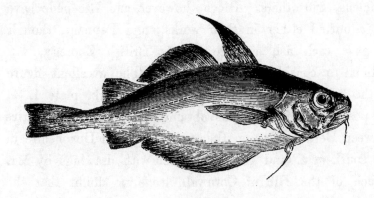

THE BIB, POUT, AND WHITING POUT.

SMELTIE, *Zetland.*—KLEG, *Scarborough.*—BLENS *and*
BLINDS, *Devonshire and Cornwall.*

Morrhua lusca,   *Bib,*   FLEM. Brit. An. p. 191, sp. 78.
   ,,   *barbata,* *Pout,* FLEM. ,,   ,,   ,,   sp. 79.
Asellus luscus,   *Bib & Blinds,* WILLUGHBY, p. 169.
   ,,   *barbatus,* *Whiting Pout,* WILLUGHBY, App. 22, L. 4.
Gadus luscus,   LINNÆUS.
   ,,   *barbatus,* LINNÆUS. BLOCH, pt. v. pl. 166.
   ,,   *luscus*   *Bib,*   PENN. Brit. Zool. vol. iii. p. 247, pl. 34.
   ,,   *barbatus,* *Pout,* PENN. ,,   ,,   ,,   p. 246.
   ,,   *luscus,*   *Bib,*   DON. Brit. Fish. pl. 19.

THE systematic terms which refer to the Bib and Pout
are here brought together in the belief that they are but
different names for the same fish.

Willughby, in his Ichthyology, page 169, first described
his *Asellus luscus,* under its Cornish names of Bib and Blinds.
Ray, in his Appendix to Willughby's work, which he edited,
admitted from Martin Lister, as a species, the Whiting
Pout of the London market, not aware that it was the same

fish as the Bib of Cornwall, which had been already included and described by Willughby himself. Ray continued them as distinct in his own Synopsis, and was followed by Artedi, Linnæus, and others. Bloch, however, and Lacépède have not admitted either in their works; and Pennant, though he gave each fish a place in his British Zoology, was inclined to consider them identical. The excellent figure of the fish given by Ray in Willughby's work, plate L. 4, the possession of specimens obtained from various localities between Berwick Bay on the north-east, and Devonshire in the south-west, and these compared with drawings by Mr. Couch of the Bib of Cornwall, leave no doubt that the *luscus* and *barbatus* of authors are the same fish.

The Bib or Pout, though not abundant, is yet a well-known species, which is found on many parts of our coast, particularly those that are rocky. Northward it appears to range as far as Greenland; and is caught on the coasts of Norway and Sweden. It is taken at Zetland and in the Forth. I have received specimens from Dr. Johnston, taken at Berwick; and it occurs on the coast of Norfolk. It is common about the mouth of the Thames; and on the Dutch as well as the French coast. Along our southern shore as far west as Devonshire, it is very commonly taken in the trawl-nets; but on the rocky coast of Cornwall it is caught by a baited hook. It has been taken on the coast of Carnarvonshire, at Dublin, at Belfast, and Loch Foyle; and I have no doubt may be found all round the coast.

From a dark spot at the origin of the pectoral fin, in which it resembles the Whiting, one of its most common names is Whiting Pout; and from a singular power of inflating a membrane which covers the eyes and other parts about the head, which, when thus distended, have the appearance of bladders, it is called Pout, Bib,* Blens,* and Blinds.* The flesh is excellent; and, like most of the other fishes of this family, it is in the best condition for the table in November and December. Its food is small fish and the various animals allied to the shrimps. It is most frequently caught in spring, because it then approaches the shore for the purpose of spawning. The largest specimen I have seen measured in length sixteen inches.

The length of the head is to the whole length of the fish as one to four; the depth of the body is greater than the length of the head, and compared to the whole length of the fish as one to three and a half: the first dorsal fin commences in a vertical line a little behind the origin of the pectoral fin; the longest ray longer than the base of the fin: the rays of the second dorsal fin are short; the base of it as long again as the base of the first dorsal fin, and ending nearly on the same plane as the first anal fin: the base of the third dorsal fin is nearly as short as that of the first dorsal, commencing and ending on the same plane with the second anal fin, and both are similarly truncated. The ventral fins are considerably in advance of the line of the origin of the pectorals; the first two rays elongated, and divided at the ends: the anal aperture is in a line under the origin of the pectoral fins, but the cavity containing the intestines extends much farther back; the first anal fin commences nearly in a line with the beginning of the first dorsal, and ends on a line with the ending of the second dorsal; the rays forming the middle portion of the fin the longest, the others declining in length towards each extremity: the second anal fin, as before mentioned, in extent of base and form like the third dorsal:

* Probably derived from Bleb and Blain, two old words meaning a blister or a bubble in the water.

the tail long, the end of the rays nearly square. The number of fin-rays:—

D. 11. 20. 16. : P. 18 : V. 6 : A. 33. 19. : C. 21.   Vertebræ 48.

In the tails of the fishes of this family there are many short rays which are not counted, being outside the longest ray above and below.

In form the Whiting Pout is the deepest for its length of the British *Gadidæ* : the upper jaw is the longest; the band of teeth of several rows, those forming the outer row the largest; under jaw with a single row ; the barbule at the chin rather long; various mucous pores about both jaws: the eyes large ; the orbits covered with a loose membrane which the fish has the power of distending ; the diameter of the eye equal to one-third of the length of the head ; the irides orange colour: the dorsal and abdominal lines exhibit considerable convexity ; the body tapers rapidly from the line of the ending of the second dorsal and first anal fins : the colour of the head, back, and upper part of the sides, a yellow reddish brown, becoming lighter on the belly, and tinged in places with bluish grey ; at the base of the pectoral fins a black spot : scales small and deciduous : posterior half of the lateral line straight, then rising in a curve over the pectoral fin ; all the fins, except the ventrals, dusky brown ; the ventrals nearly white ; the first anal fin in large-sized specimens edged with fine blue.

SUBBRACHIAL
MALACOPTERYGII.   *GADIDÆ.*

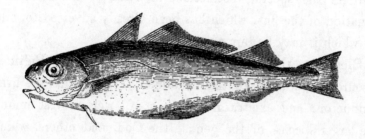

THE POOR, OR POWER COD.

Morrhua minuta,	Poor,	FLEM. Brit. An. p. 191, sp. 80.
Gadus minutus,	LINNÆUS.	BLOCH, pt. ii. pl. 67, fig. 1.
,, ,,	Power,	PENN. Brit. Zool. vol. iii. p. 249, pl. 34.
,, ,,		WILLUGHBY, p. 171.
,, ,,		RAY, Syn. p. 163, fig. 6.

THE POOR, OR POWER COD, though somewhat similar in general appearance to the Pout last described, is yet readily to be distinguished from it by several well-marked characters. It is not so deep when of the same length; the first anal fin does not begin so far forward as in the Pout by nearly the whole length of the base of the first dorsal fin; the longest rays of the third dorsal fin and the second anal fin are shorter than the bases of the respective fins, and do not, therefore, produce the same vertically truncated appearance as in the same relative fins of the Pout; and the barbule at the chin is much shorter.

The Power, or Poor Cod, the smallest of its genus, so called, it is said, on account of its diminutive size, seldom exceeding six or seven inches in length, and therefore comparatively of little value, was first described as an English fish by Dr. Jago, of Cornwall, and was introduced by Ray at the end of his *Synopsis*, with a figure which particularly exhibits the specific distinctions afforded by the form and situation of the fins, which has been already adverted to, and by which it may be immediately recognised.

Bloch says that the appearance of this fish in the Baltic is a source of pleasure to the fishermen. It is called the fish-conductor; and excites great hopes of a rich harvest among the larger species of the genus, the Cod, and others, which follow in the rear, preying relentlessly on their more diminutive generic companions: the fishermen in their turn prey upon them.

Mr. Couch says it frequents the edges of rocks, is caught by the hook, and, though always good for the table, is, on account of its small size, chiefly used for bait. Montagu says it is taken frequently on the Devonshire coast with the hook, and also in the crab-pots. In the nets worked on that coast it is caught along with the Bib, the fishermen selling both as Whiting Pout.

The length of the head compared to the whole length of the fish is as one to five ; the depth of the body rather more than the length of the head : the first dorsal fin begins behind the line of the origin of the pectorals : the longest ray as long as the base of the fin: the second dorsal fin begins and ends on the same planes with the first anal fin ; the base of the second dorsal fin as long again as the base of the first dorsal fin ; the base of the third dorsal fin rather more than half as long as that of the second dorsal ; the third dorsal and the second anal fins begin and end on the same planes, and the peculiarity of their forms has been referred to. The vent, or anal aperture, is in a line under the most posterior portion of the first dorsal fin ; the first anal fin begins immediately behind the vent, and under the commencement of the second dorsal fin ; the second anal fin has been noticed ; the tail is long, with the rays slightly forked. The number of fin-rays :—

D. 12. 19. 17. : P. 14 : V. 6 : A. 25. 17. : C. 18.

The head is short and the nose blunt : the barbule at the chin neither so long nor so slender as in the Pout ; several mucous pores about the mouth : the eye large ; the breadth equal to one-third of the length of the head ; the irides orange : the scales of the body minute and deciduous ; the lateral line but slightly curved, and that only where it rises over the pectoral fin : the upper part of the head and back brownish yellow, becoming lighter on the cheeks and sides ; the belly dirty white ; pectoral, dorsal fins, and tail, yellow brown, darker at the edges ; ventral and anal fins dirty yellowish white.

THE SPECKLED COD.

Morrhua punctata, Speckled Cod, Flem. Brit. An. p. 192, sp. 81.
Gadus punctatus,   ,,   ,,   Turton, Brit. Faun. p. 90, sp. 18.

According to Dr. Turton, the Speckled Cod is frequently taken in the weirs at Swansea. The specific characters
are, " pale brown with golden spots, beneath white, thickly
covered with minute dusky specks : upper jaw longest."
The description is, " Body eighteen inches long, slightly
arched on the back, a little prominent on the belly, covered
above with numerous gold-yellow roundish spots, beneath
with dusky specks, which are stellate under a glass : head
large, gradually sloping : teeth small ; in several rows in the
upper jaw, in the lower a single row : nostrils double ; iris
reddish, pupil black : chin with a single beard ; nape with
a deep longitudinal groove : lateral line nearer the back,
curved as far as the middle of the second dorsal fin, growing
broader and whiter towards the end : upper fins and tail
brown, with obscure yellowish spots, and darker towards the
end ; lower ones tinged with green : vent near the middle of
the body : scales small ; all of them under a glass minutely
speckled with brown : gill-covers of two pieces. The fin-
rays :—

D. 14. 20. 18. : P. 18 : V. 6 : A. 19. 16 : C. 36.

Of the ventral fin the first ray is shorter than the second, and
divided a little way down : the tail even."

" Differs from *G. morrhua* in not having the first anal fin-
ray spinous, and in the lower jaw being considerably longer ;
from *G. luscus* in the first ray of the ventral fin being shorter
than the second ; from *G. barbatus* in wanting the seven dis-

tinct punctures on the lower jaw, in its small scales, and in the first dorsal fin not ending in a long fibre ; and from *G. callarius* in not being spotted with brown, and in having the lateral line white."

No other record of this fish has appeared that I am aware.

Some years since, I obtained from a fisherman at the mouth of the Thames a fresh-caught example of a species of *morrhua,* with the middle dorsal and the first anal fins short ; the body as deep for its length as the *luscus ;* the length of the head compared to the whole length of the fish as one to three. Among the fishermen it was by some considered to be an accidental deformity, with injury of the spine, and their name for it was Lord-fish ; others said it was a fish which they met with occasionally, and believed it distinct from any other. A coloured drawing was made at the time, but the fish was not preserved. The fin-rays were as stated ; and it will be observed, on comparing the numbers, that they do not differ very widely from those of the Common Cod.

D. 14. 19. 18. : P. 14 : V. 6 : A. 17. 11. : C. 24.

The figure below is taken from the drawing referred to, but carefully reduced : upper part of the head, back, and fins, mottled with two shades of brown ; the sides of the body lighter ; the belly white ; the lateral line white, arching high over the pectoral fins : the irides reddish orange

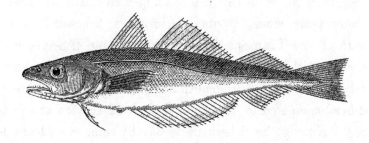

THE WHITING.

Merlangus vulgaris, Cuvier, Règne An. t. ii. p. 332.
,,   ,,   *Whiting,* Flem. Brit. An. p. 195, sp. 91.
,,   ,,   ,,   Willughby, p. 170. L. 5.
Gadus merlangus, Linnæus. Bloch, pt. ii. pl. 65.
,,   ,,   *Whiting,* Penn. Brit. Zool. vol. iii. p. 255.
,,   ,,   ,,   Don. Brit. Fish. pl. 36.

Generic Characters.—The same as those of *Morrhua,* except that they have
no barbule at the chin.

THE WHITING is well known for the excellence of its
flesh, surpassing in delicacy that of any of the other species
of the valuable family of fishes to which it belongs: the
pearly whiteness of its flaky muscles, added to its extreme
lightness as an article of food, recommend it particularly to
invalids who are unable to digest more solid nutriment.

It is caught in great abundance almost all round our
coast, and may be traced from the Orkneys to Cape Clear.
Whitings of several pounds' weight have been caught as far
north as the Dogger Bank; they have been taken also of

nearly equal size on the coast of Cornwall; and on the Nymph Bank, along the extended line of the south coast of Ireland. In that country they have also been found on the eastern coast from Waterford to Antrim, and from thence north and west as far as Lough Foyle.

The fishing for Whiting with lines is pursued nearly all the year through; but the fish is most plentiful in the months of January and February, when it comes in large shoals towards the shore for the purpose of depositing its spawn, and is taken in abundance within half a mile, and seldom exceeding three miles, from land. A much larger quantity than is consumed while fresh being frequently taken, a portion is easily preserved either by salting or drying.

The Whiting is a voracious feeder, and seizes indiscriminately any of the mollusca, worms, small crustacea, and young fishes. I remember to have taken several Sprats from the stomach of a Whiting; and Mr. Couch has known four full-grown Pilchards taken from the inside of a Whiting that weighed four pounds. It appears to prefer sandy banks, but shifts its ground frequently in pursuit of the various fry of other fishes, upon which it principally subsists.

Though occasionally occurring in the London market of three or four pounds' weight, the most usual size is from twelve to sixteen inches in length, and weighing about one pound and a half.

The length of the head compared to that of the body alone is as one to three; the depth of the body not equal to the length of the head, or compared to the whole length as one to six. The first dorsal begins behind the line of the origin of the pectorals and before the line of the vent; the second dorsal and first anal fins end on the same line; the third dorsal and second anal fins begin and end on the same plane ; the ventral fins are placed very forward ; the second ray elongated : the anal aperture is in a line under the middle of the first dorsal fin ; the first anal fin commencing just behind the vent : the tail elongated ; the end nearly square. The fin-rays are—

D. 13. 19. 18. : P. 19 : V. 6 : A. 31. 20. : C. 30.   Vertebræ 55.

The body of the Whiting, like the bodies of those belonging to this division, is longer for its depth than that of the Codfish ; the scales small, oval, and deciduous ; the lateral line dark and straight posteriorly, but rising gradually throughout the anterior half ; the head elongated ; the mouth and gape large ; the tongue white and smooth ; the upper jaw decidedly the longest, with one row of large and sharp-pointed teeth on the outer edge, and several rows of smaller ones within ; the vomer with a few teeth arranged in a semi-circular line on the anterior part ; the lower jaw with various mucous orifices along the under surface, and a single row of sharp teeth along the upper outer edge, which, when the mouth is closed, range within the outer row of teeth on the upper jaw : the eye in breadth less than one-fourth of the head, and placed more than its breadth from the end of the nose ; the irides silvery ; the pupils blue.   The upper part of the head and the back above the lateral line pale reddish ash brown ; sides and belly silvery white ; pectoral, caudal, and dorsal fins, pale brown ; ventral and anal fins almost white ; the pectoral fins each with a decided dark patch at the base.

SUBBRACHIAL
MALACOPTERYGII.   *GADIDÆ.*

THE COALFISH.

Merlangus carbonarius, Cuvier, Règne An. t. ii. p. 332.
    ,,   ,,   *Coalfish,* Flem. Brit. An. p. 195, sp. 93.
    ,,   ,,   *Colefish,* Willughby, p. 168, L. 3.
Gadus   ,,   ,,   Linnæus. Bloch, pt. ii. pl. 66.
    ,,   ,,   *Coalfish,* Penn. Brit. Zool. vol. iii. p. 250.
    ,,   ,,   ,,   Don. Brit. Fish. pl. 13.

The Coalfish is most decidedly a northern fish, but, being a hardy species, is not without considerable range to the southward. It was the only fish found by Lord Mulgrave on the shores of Spitzbergen; and the fry, only four or five inches in length, were caught with the trawl-net on the west coast of Davis's Straits, during the first voyage of Captain Sir Edward Parry. It abounds in all the northern seas and in the Baltic, and may be said to swarm in the Orkneys, where the fry all the months of summer and autumn are the great support of the poor. Dr. Neill, in his tour of the islands of Orkney and Shetland, saw an old man, and perhaps one or two boys, seated upon almost every projecting rock, holding in each hand a wand or fishing-rod, and catching young Coalfish as fast as they could bait their hooks.

VOL. II.   N

As an article of food, it is more prized when small than when of large size. The flesh of specimens weighing from fifteen to thirty pounds is usually preserved, either salted or dried.

This fish has more provincial names than any other species, some of which only refer to it when of a particular size. Among the Scotch islands the Coalfish is called Sillock, Piltock, Cooth or Kuth, Harbin, Cudden, Sethe, Sey, and Grey-Lord. In Edinburgh and about the Forth the young are called Podleys; at Newcastle the fry are called Coalsey; and, when twelve inches long, Poodlers. Many are caught along shore; and frequently, also, from a boat rowed gently, the angler using a rod in each hand, and trailing a fly from each line.

Mr. Couch says, " It is in the highest condition from October to December, at which season it prowls after prey in large companies; so that when met with they prove a valuable capture to the fishermen: for though but coarse food, yet being wholesome, substantial, and cheap, they are eagerly purchased by the poor either fresh or salted. They swim at no great depth, and with great rapidity; but when attracted by bait, will keep near a boat until all are taken; and I have known four men with two boats, two men in each boat, take twenty-four hundred weight with lines in a very few hours. The season for spawning is early in spring; immediately after which this fish becomes so lank as to be worthless, in which state it continues through the summer."

In the Orkneys, according to Mr. Low, the young appear about May; in the Tyne, about June; and on the Cornish coast in July. The adult fish are called Rauning Pollacks by the Cornish fishermen: rauning being the ancient and even the popular modern pronunciation of ravening, used in reference to voracity.

The Coalfish may be traced on the Irish coast from Waterford along the eastern shore to Belfast, under the various names of Black Pollack, Blockin, and Grey-Lord.

When detained and well fed in a salt-water pond, Coalfish acquire large size. " They were bold and familiar; floating about slowly and majestically, till some food was thrown to them ; this they seized voraciously, whether it consisted of shell-fish or ship-biscuit. They would also occasionally approach the margin and take food from the hand."—*Jesse's Gleanings.*

From the point of the lower jaw to the end of the operculum the length is to that of the body and tail as one to three and a half; the depth of the body about equal to the length of the head : the first dorsal fin begins behind the line of the origin of the pectoral fin and before the line of the vent ; the second dorsal and the first anal fins end together nearly on the same plane ; the third dorsal and second anal fins nearly parallel : the fleshy portion of the tail elongated ; the rays forked : the ventral fins small ; and the rays of the pectoral fin only extending as far as the line of the vent. The fin-rays are—

D. 11. 20. 20. : P. 19 : V. 6 : A. 24. 19. : C. 32.

The head and body elegantly shaped ; the scales small and oblong ; the lateral line silvery white and nearly straight ; the upper part of the head and the back above the lateral line almost black ; much lighter in colour below the line, becoming greyish white with golden reflections on the sides and belly ; pectoral, caudal, and dorsal fins, bluish black ; ventral and anal fins greyish white : the upper jaw rather the shortest ; the lips tinged with purple red ; the mouth black ; the teeth very small ; the irides silvery white ; the pupil blue.

N 2

THE POLLACK.

WHITING POLLACK.   LYTHE, *Scotland.*

Merlangus pollachius,	Cuvier, Règne An. t. ii. p. 333.
,,	,,	*Pollack,* Flem. Brit. An. p. 195, sp. 92.
,,	,,	*Whiting Pollack,* Willughby, p. 167.
Gadus	,,	Linnæus. Bloch, pt. ii. pl. 68.
,,	,,	*Pollack,* Penn. Brit. Zool. vol. iii. p. 254.
,,	,,	*Whiting Pollack,* Don. Brit. Fish. pl. 7.

THE POLLACK is much less abundant on some parts of the coast than the Coalfish; but, like that species, is an inhabitant of the seas all round our shores. Mr. Low, in his " Natural History of the Orkneys," says, " They are frequently caught close in with the shore, almost among the sea-ware, and in deep holes among the rocks. They seem to be a very frolicsome fish; and I have been several times fishing for them when they would keep a constant plashing in the water. They bite keenly, scarce allowing the hook to be in the water before one or other jumps at it. They are better eating than the Coalfish; but I do not know whether they are ever dried or preserved otherwise, as the quantity caught is scarce worth curing." Hand-line fishing for Pollacks is called whiffing.

This fish is called Lythe in Scotland, as already quoted; but whether this term is intended to refer to its supple, pliant activity, or is derived from *lithos*, a stone, from its living among rocks, I have not seen stated. Fine specimens of the Pollack are taken about the rocky coast of Scarborough, where they are called Leets.

The Pollack is caught at Hastings and Weymouth. Colonel Montagu says it is frequently taken in Devonshire, where it is bought by the inexperienced as Whiting. When only twelve or fourteen inches long, the flesh possesses a considerable portion of the pearly appearance and delicacy of that fish.

Mr. Couch says, " The Pollack is at all seasons one of our most common fishes, but it is not gregarious except in pursuit of prey; and it rarely wanders far from its usual haunts, which are along the edges of rocks, where, with the head directed towards the coming tide, it is ready for any prey that approaches. The smaller ones, which occupy such a station covered with oreweed, have their colours very bright, and the belly of a saffron yellow; while on clean ground they are less brilliant. In summer evenings, they are often seen eager in pursuit of the Sandlaunce, frequently spring from their element, and are often taken by anglers from the rocks and piers. The Pollack spawns in winter near the land; and the young abound near the edge of the tide in rocky ground at the beginning of summer."

In Ireland, the Pollack may be traced as occurring on the coast of the counties of Cork, Waterford, Dublin, Antrim, Londonderry, and Donegal, under the names of Pollack, Laith, and Lythe.

The length of the head compared to that of the body is as one to three and a half; the depth of the body is to the whole length of the fish as one to four and a half: the first dorsal fin begins, as in the Coalfish, behind the line of the origin of the pectoral fin, and before the line of the situation of the vent ; the second dorsal fin and the first anal fin end on the same line ; the third dorsal fin and the second anal fin begin and end very nearly on the same plane ; the first ray of each of the dorsal fins the longest ; the ventral fin very small ; the anal aperture in a line under the middle of the first dorsal fin ; the fleshy portion of the tail long and slender ; the end of the rays concave.   The fin-rays in number are—

D. 12. 19. 15. : P. 19 : V. 6 : A. 24. 16 : C. 31.

The lower jaw is much the longest ; the mouth and lips red, with various mucous orifices about them ; the irides silvery ; the sclerotic coat cartilaginous ; the upper angle of the operculum produced ; the body elongated ; the upper part of the head and the back above the lateral line olive brown ; the sides dull silvery white mottled with yellow, and in young fish spotted with dull red ; the lateral line dusky, curved over the length of the pectoral fin, then descending and passing in a straight line to the tail ; the dorsal fins and tail brown ; the pectoral and anal fins brown edged and tinged with reddish orange.

THE GREEN COD.

Merlangus virens, Cuvier, Règne An. t. ii. p. 33.
     ,,   ,,   Flem. Brit. An. p. 195, sp. 94.
Gadus   ,,   Linnæus.
     ,,   ,,   *Green Cod,* Penn. Brit. Zool. vol. iii. p. 253.

The Green Cod was first added to the catalogue of
British Fishes by Pennant, on the authority of Sir Robert
Cullum, Bart.; and if a distinct species, which some have
doubted, it is not only abundant, but has an extensive
range.

It is mentioned as an inhabitant of the northern seas by
Linnæus and others, and is included in the recently published works of Professors Nilsson and Reinhardt, who have
devoted particular attention to the fishes of Scandinavia.
Dr. Neill says it is taken in the Frith of Forth during
summer; and Mr. Couch obtains it on the Cornish coast of
eight or ten inches in length.

This fish is by some considered as the young of the Coalfish, and by others as the young of the Pollack. It appears,
however, to be decidedly distinct from the Pollack, in hav-

ing its jaws nearly equal in length: in the Pollack the under jaw is by much the longest; the lateral line in the Green Cod is straight, in the Pollack the lateral line is curved over the whole length of the pectoral fin.   Mr. Couch, in his MS. considers the Green Cod as the young of the Coalfish, with which it certainly agrees in both the particulars in which it differs from the Pollack, but differs also decidedly in colour from the Coalfish.   It seems to combine in itself the colouring of the Pollack with some of the peculiarities of the Coalfish, but appears also to be deeper for its length than either; though if the young of a large species, judging by analogy, that would not be the case.

Following the example of the Northern naturalists, who have opportunities of making constant comparison between this fish and the Coalfish from the abundance of both, and who have hitherto considered them distinct, the Green Cod is here allowed a separate place.   The figure is from a drawing by Mr. Couch, whose opinion is entitled to attention; and the subject invites the investigation of those who are so located as to be able to obtain specimens of both.

Not possessing a specimen, the description here given is derived from the Prodromus of M. Nilsson.   The under jaw scarcely longer than the upper: the tail deeply forked; the lateral line straight, white; the colour of the back dark green, passing by degrees into silvery grey on the sides.

From six to twelve inches is the usual size allowed to the Green Cod; M. Nilsson gives it a length from two to three feet, and adds that it spawns in winter.

The number of fin-rays as given by Linnæus :—

D. 13. 20. 19. : P. 17 : V. 6 : A. 24. 20. : C. 40.

Dr. Fleming adds, " Teeth in the upper jaw, numerous, strong."

SUBBRACHIAL
MALACOPTERYGII.   GADIDÆ.

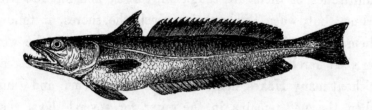

THE HAKE.

Merlucius vulgaris,	Cuvier, Règne An. t. ii. p. 333.
,, ,,	*Common Hake*, Flem. Brit. An. p. 195, sp. 95.
,, ,,	*The Hake*, Willughby, p. 174.
Gadus merlucius,	Linnæus. Bloch, pt. v. pl. 164.
,, ,,	*Hake*, Penn. Brit. Zool. vol. iii. p. 257.
,, ,, ,,	Don. Brit. Fish. pl. 28.

Generic Characters.—The head flattened; the body elongated; the back furnished with two dorsal fins; the first short, the second long; but one anal fin, also very long; no barbule at the chin.

THE HAKE is another of the species belonging to this large and valuable family of fishes, which has an extensive range, being found in the seas of the North of Europe, and also in the Mediterranean.

Though inhabiting the seas of the western coast of Norway, and included by Linnæus in his *Fauna Suecica*, Dr. Fleming says it is rare in Scotland; and it appears to be most abundant along the southern coast of England. Portsmouth market receives an abundant supply, which is brought by fishing-boats from the Devonshire coast; and Montagu says there is also an abundance in the market of Plymouth.

According to Mr. Couch, "The Hake is a roving fish

on the Cornish coast, without much regularity in its movements. From January to April, which is its season for spawning, it keeps near the bottom, and loses the great voracity by which it is characterised at other times, so that multitudes of them are caught in trawls, and but few with a line; but, when Pilchards approach the shores, it follows them, continuing in incalculable numbers through the winter. It rarely happens that Pilchards are taken in a sean without many Hakes being enclosed with them; and thus, when the net remains in the water for several days, they have an opportunity of glutting themselves to their heart's desire, which is to such an extent as to render them helpless, and I have seen seventeen Pilchards taken from the stomach of a Hake of ordinary size. Their digestion, however, is quick, so that they speedily get rid of their load; and fishermen observe that, when hooked, the Hake presently evacuates the contents of the stomach to facilitate its escape; so that when hundreds are taken with a line, in the midst of prey, not one will have anything in its stomach: when near the surface, however, this rejection does not take place until after they are dragged on board."

The Hake may be traced nearly all round the coast of Ireland; and is so abundant in the Bay of Galway, that, according to a recent writer, this bay is named in some ancient maps the *Bay of Hakes*. On that part of the Nymph Bank off the coast of Waterford, this fish is also so plentiful, that one thousand have been taken by six men with lines in one night. It is a voracious fish, as its systematic name of *merlucius*, Seapike, implies. It is a coarse fish, not admitted at the tables of the wealthy; but large quantities are annually preserved both by salting and drying, part of which are exported to Spain.

The Hake is very common on the northern shore of the

Mediterranean, and considerable traffic is carried on with this fish : they are packed with aromatic plants, and sent to the towns removed from the coast. The Hake is described and figured by Rondeletius, and was known to the older naturalists before him.

A Hake of three feet eight inches long in the shop of a London fishmonger, in May 1835, supplied the means of obtaining the following particulars. The length of the head, compared to the length of the body alone, as one to three ; the depth of the body not so great as the length of the head : the ventral fins are placed in advance of the pectorals ; the rays not unequally elongated : the pectoral fins commence in a line under the posterior angle of the operculum ; the rays ending with the end of the first dorsal fin : the first dorsal fin itself short and triangular in shape ; the second dorsal fin commences in a line over the vent ; the anal fin begins immediately behind the vent ; both the second dorsal fin and the anal fin terminate on the same plane, near the tail ; the rays strong and stiff : the caudal rays about three inches long, and nearly even. The fin-rays in number are—

D. 10. 29. : P. 11 : V. 7 : A. 21 : C. 19.

The head is depressed : the inside of the mouth and gill-covers black ; lower jaw the longest ; teeth slender and sharp, in a single row in each jaw : the irides yellow with a dark outer circle. The lateral line of the body straight throughout the posterior half, then gradually rising to the upper edge of the operculum ; the appearance of the lateral line is that of one white line between two dark ones : the scales large ; colour of the body dusky brown above, lighter beneath ; dorsal and caudal fins dark ; ventral and anal fins pale brown.

SUBBRACHIAL
MALACOPTERYGII.   *GADIDÆ.*

THE LING.

Lota molva,	Cuvier, Règne An. t. ii. p. 333.
Asellus longus,	Willughby, p. 175, L. 2.
Gadus molva,	Linnæus. Bloch, pt. ii. pl. 69.
,, ,,	*Ling,* Penn. Brit. Zool. vol. iii. p. 262.
,, ,,	,, Don. Brit. Fish. pl. 102.
Molva vulgaris,	*Common Ling,* Flem. Brit. An. p. 192, sp. 82.

Generic Characters.—In addition to the elongated body, with two dorsal fins and one anal fin, possessed by the species of *Merlucius* last described, may be added, chin with one or more barbules.

THE LING is a very valuable species, scarcely less so than the Coalfish or the Cod. Large quantities are taken among the Western Islands, in the Orkneys, and on the Yorkshire coast; in Cornwall, and the Scilly Islands; and may be traced nearly all round the Irish coast. The fishing for them is by hånd-lines and long-lines; and besides a portion that is consumed fresh, the fish are split from head to tail, cleaned, salted in brine, washed, and dried: but the demand generally falls short of the quantity cured, and the hardy fishermen are but poorly requited. The ports of Spain are the markets supplied; and so valuable an article of commerce was Ling considered formerly, that an act for regulating the price of Ling, Cod, &c. was passed as early as the reign of Edward the Third.

The air-bladders, popularly called Sounds, are prepared separately, and, with those of the Codfish, are sold pickled. The roes, which are of large size, are also used as food, or, preserved in brine, are sold to be employed to attract fish. Another produce of the Ling is the oil extracted from the liver, which is used by the poor to supply the cottage lamp; and as a medicine, Mr. Couch says, which those who have been able to overcome the repugnance arising from its nauseous smell and taste, have found effectual in severe cases of rheumatism, when taken in small beer in doses of from half an ounce to an ounce and a half. Formerly from fifty to sixty gallons of this oil, and that from the liver of the Codfish, were dispensed in one large establishment for this purpose, and it was found to act best when the perspiration was increased. The exudation from the skin of those to whom it was administered always became strongly tainted with it.*

In Zetland, the principal fishing for Ling is from May to August. On the Yorkshire coast, the young are called Drizzles. In Cornwall they are caught in January and February, and their favourite haunts are about the margins of the rocky valleys of the ocean. The Ling is exceedingly prolific, and of most voracious appetite, feeding on young fish, not sparing anything that has life, and the prey is swallowed whole, so that no great art is required to catch it. It is tenacious of life, and survives great injury. "I once," says Mr. Couch, "saw a Ling that had swallowed the usual large hook, shaft foremost, of which the point had fixed in the stomach, and as the line drew it, it turned round, entered the opposite side of the stomach, and fastened the organ together in complicated folds; yet having escaped by

* Memoirs of the Literary and Philosophical Society of Manchester, vol. iii; and Dr. Bardsley's Medical Reports, 8vo. 1807, p. 18.

breaking the line, it survived to swallow another hook and be taken several days after."

The most usual length of the Ling is from three to four feet; Pennant mentions having heard of one which measured seven feet; and Mr. Couch has known them weigh seventy pounds.

Not having an opportunity of describing from a specimen, I copy, by permission, the description of the Rev. Mr. Jenyns, as given in his Manual of the British Vertebrata, page 458, species 133.

" Body slender, more elongated than that of the Hake; roundish: head flat: gape large: lower jaw shorter than the upper, with a single barbule at its extremity: teeth in the upper jaw small, and very numerous; those in the lower jaw longer and larger, forming but a single row: lateral line straight: scales small, firmly adhering to the skin: two dorsal fins of equal height; the first short, commencing near the head, not pointed as in the Hake, but with most of the rays even; second long, immediately behind the first, reaching nearly to the caudal; the posterior portion the most elevated: vent in a line with the eighth or ninth ray of the second dorsal fin: anal fin immediately behind it, long, resembling the second dorsal fin, and terminating on the same line with it: caudal rounded at the extremity.

" The fin-rays are—

D. 15. 65. : P. 15 : V. 6 : A. 97 : C. 39.

" The back and sides grey, inclining to olive; sometimes cinereous, without the olivaceous tinge; belly silvery: ventrals white; dorsal and anal edged with white; caudal marked near the end with a transverse black bar; the extreme tip white."

SUBBRACHIAL
MALACOPTERYGII.   *GADIDÆ.*

THE BURBOT.

EELPOUT. BURBOLT.

Lota vulgaris, *Burbot,* Jenyns, Man. Brit. Vert. p. 448, sp. 134.
   ,,   ,,   Cuvier, Règne An. t. ii. p. 334.
   ,,   ,,   Willughby, 125.
Gadus lota,   Linnæus. Bloch, pt. ii. pl. 70.
   ,,   ,,   *Burbot,* Penn. Brit. Zool. p. 265.
   ,,   ,,   ,,   Don. Brit. Fish. pl. 92.
Molva ,,   ,,   Flem. Brit. An. p. 192, sp. 83.

THE BURBOT is the only British species of this numerous family of fishes that lives permanently in fresh water, and prefers in this country slow running rivers; but is neither so generally known, nor so much esteemed and encouraged, as from the goodness of its flesh it deserves. It is said to be found in various parts of the North of Europe, Siberia, Asia, and India. In this country it is rather local. It occurs in the Cam, and in some of the rivers of Norfolk and Lincolnshire. The Trent produces it, and Nottingham market is occasionally supplied with examples for sale. The Tame is said to contain the Burbot, and so also do several rivers in the counties of Yorkshire and Durham; as the Ouse, the

Esk, the Skern, near Mainsforth, which afterwards runs into
the Tees near Croft Bridge, and the Derwent.

The Burbot is not unlike the Eel in some of its habits,
concealing itself under stones, waiting and watching for its
prey, consisting of aquatic insects and young fish, under
arches and near eddies, into which such small and weak
animals are likely to be brought by the current of the water.
It feeds principally during the night; and, like the Eel, is
most frequently caught by trimmers and night-lines.

The Burbot is sometimes called Coney-fish, from its habit
of lurking and hiding itself in holes like a rabbit.

It spawns in February or March; is very tenacious of
life, and is said to have lived a considerable time in a
damp and cold situation, fed on small fishes and raw meat.
In this country it has been known to attain the weight of
four pounds and a half; but a more common size is about
two pounds' weight. Pennant mentions one taken in the
Trent which weighed eight pounds. In the Lake of
Geneva, into which it is stated the Burbot was introduced
from Neufchatel, it has been taken of seven pounds' weight.
The flesh is white, firm, and of good flavour, by some considered superior to that of the Eel; and as the Burbot is in
its nature extremely hardy, few difficulties present themselves in the way of their increase in quantity, while the
value of the fish would amply repay the trouble or the cost
of the experiment. It would probably thrive well and
multiply in large lakes.

Length from one to two feet: the head depressed, smooth;
jaws equal; chin with one barbule; the gape large, with
small teeth above and below; eyes of moderate size; gill-
opening large: the length of the head compared to that of
the body as one to four: the form of the body cylindrical,
compressed posteriorly. The first dorsal fin is small and

rounded ; the second elongated, reaching nearly to the tail ; both dorsal fins nearly uniform in height : ventral fins placed very forward, narrow and pointed ; the pectoral fins large and rounded ; the anal fin begins on a line behind the commencement of the second dorsal fin, but ends very nearly on the same plane : the tail oval, and slightly pointed. The fin-rays in number are—

D. 14. 68. : P. 20 : V. 6 : A. 67 : C. 36.

The colour of the body yellowish brown, clouded and spotted with darker brown, and covered with a mucous secretion ; the under parts lighter : the lateral line indistinct and straight ; scales small ; the fins partaking of the colour of the part of the body from which they emanate, those of the lower surface being much the lightest.

THE THREE-BEARDED ROCKLING.

SEA LOCHE. — WHISTLE-FISH.

Motella vulgaris,	Cuvier, Règne An. t. ii. p. 334.
,, *tricirrata,*	Nilsson, p. 48.
,, ,,	*Three-Bearded Rockling,* Jenyns, Man. Brit. Vert. p. 449, sp. 135.
,, ,,	Willughby, p. 121, H. 4, fig. 4.
Mustela marina,	*Rockling,* Ray, Syn. p. 164, sp. 9, fig. 9.
Gadus tricirratus,	Bloch, pt. v. pl. 165.
,, *mustela,*	*Three-Bearded Cod,* Penn. Brit. Zool. vol. iii. p. 267, pl. 36.
,, ,,	*Rockling,* Don. Brit. Fish. pl. 2.
,, *tricirratus,*	*Three-Bearded Gade,* Flem. Brit. An. p. 193, sp. 86.

Generic Characters.—Body elongated, cylindrical, compressed posteriorly ; the first dorsal fin very slightly elevated, delicate in structure, scarcely perceptible; second dorsal and anal fins long, continued nearly to the base of the tail.

The Three-Bearded Rockling, included by the Rev. Mr. Jago in his Catalogue of the rarer Fishes of Cornwall, and published by Ray, with a figure, at the end of his Synopsis, though not uncommon on the Devonshire and Cornish coasts, as noticed by Colonel Montagu and Mr. Couch, is more rare on our shores generally than the Five-

Bearded Rockling, of which by some it has been considered only as a variety. It frequents rocky ground that is well furnished with sea-weed, among which it threads its way with great ease and rapidity. Besides the localities mentioned, it has been taken also at Weymouth, in Belfast Bay, and in the vicinity of Carlisle, probably in the Solway Frith. The individual figured by Willughby, whose early representation of this fish is very good, was obtained by him at Chester.

Of its habits, Mr. Couch says, " It keeps in shallow water, feeds on aquatic insects, and will take a bait ; but it is not commonly used as food, because it smells unpleasantly in the course of a few hours. It is not easy to explain the use of the fringed membrane behind the head and before the dorsal fin ; it has nothing in common with the fins ; but when the fish is lying perfectly still, and all the fins are at rest, this is often in rapid motion. The barbules on the upper jaw are always extended in front, and probably serve the same purposes as the antennæ in insects."

Bloch says that it spawns in autumn; but other observers consider that it deposits its spawn in winter, like most of, if not all, those of the same family.

Pennant, in his account of the Five-Bearded Rockling, says, " The Cornish fishermen are said to whistle, and make use of the words *bod, bod, vean*, when they are desirous of taking this fish, as if by that they facilitated the capture, in the same manner as the Sicilian fishermen repeat their *Mamassu di pajanu*, &c. when they are in pursuit of the Swordfish." But this name of Whistle-fish was, according to Jago's Catalogue, attached to the Rockling with three barbules only, and even among them was but occasionally applied to the larger specimens. Pennant, it will be observed, speaks of the cause of the application of the name of

o 2

Whistle-fish on the authority of others; and on inquiry, I
find that the custom of whistling when fishing is neither
practised nor known to the Cornish fishermen of the present
day, and, in fact, that this fish is of too little value to be
an object of any solicitude. I believe, indeed, that while
preserving the sound of the name, the term has been changed,
and a very different word substituted, and that for Whistle-
fish we ought to read Weasel-fish. Both the Three and
the Five Bearded Rocklings were called *mustela* from the
days of Pliny to those of Rondeletius, and thence to the
present time.

A specimen fourteen inches long, and beautifully spotted,
was presented to the Zoological Society in 1832. The
finest examples of this species I have seen were two given
me in December 1834, by Dr. Thackeray, the Provost of
King's College, Cambridge, from the largest of which, measuring seventeen inches in length, the wood-engraving was
executed, and the following description taken.

The length of the head compared to the length of the
body alone, without the caudal rays, is as one to four; the
depth of the body equal to the length of the head : the first
dorsal fin delicate in structure ; the first ray elongated, the
rest hair-like : the second dorsal fin commencing immediately
behind the end of the first, and reaching along the back to
the tail, but ending a little short of the base of the caudal
rays : ventral fins with the first two rays elongated, the
second the most so, the two disunited ; the other five rays
nearly equal, united, and short : pectoral fins rather large and
rounded : the vent half-way between the point of the chin
and the end of the fleshy portion of the tail ; the anal fin
commences immediately behind it, is one-fourth less in
length than the second dorsal, and ends on the same plane
with it : the tail moderate in size, and rounded at the end.

The fin-rays in number are—

2nd D. 55. P. 20 : V. 7 : A. 49. : C. 18.

The head is depressed ; the mouth wide : the jaws nearly equal, but when separated the lower jaw is the longest, with one barbule at the chin ; a mixture of small and large teeth in each jaw ; the upper jaw with one barbule on each side the middle, between the lip and the nostril ; inner part of the upper lip crenate : the irides golden yellow ; the anterior portion of the body of the fish cylindrical, or slightly depressed ; the tail compressed : the general colour of the body and head is a rich yellow brown, spotted on the top of the head, along the back, the pectoral, dorsal, and caudal fins, with rich chesnut brown ; the lower part of the sides, the ventral and anal fins, pale yellow brown, approaching to white, and without spots.

Young fish of this species are of a uniform brown colour until they have acquired six or seven inches in length ; in this condition they are the *Mustela alia* of Ray.

THE FIVE-BEARDED ROCKLING.

Motella quinquecirrata,	Cuvier, Règne An. t. ii. p. 334, *note.*
,, mustela,	Five-Bearded Rockling, Jenyns, Man. Brit. Vert. p. 450, sp. 136.
Mustela vulgaris,	Willughby, p. 121.
Gadus mustela,	Linnæus.
,, ,,	Five-Bearded Cod, Penn. Brit. Zool. vol. iii. p. 268, pl. 36.
,, ,,	,, Don. Brit. Fish. pl. 14.
,, ,,	,, Gade, Flem. Brit. An. p. 193, sp. 85.

I HAVE found the Five-Bearded Rockling, when of small size, a very common fish on the Kentish coast in autumn, left by the retiring tide, in small pools among the rocks, and generally lying concealed under the tufts of sea-weed that hang over the edges of the stones into the water. I have observed this fish as far to the westward as Portland Island. Colonel Montagu considered it more rare in Devonshire than the species with three barbules at the mouth, last described: Mr. Couch observes it on the Cornish shore: it has been taken at Dublin and Belfast; and Mr. Low says it is common in Orkney, where it is found under stones among sea-weed, but seldom exceeding nine or ten inches in length.

Pennant says it attains the length of eighteen or nineteen inches. It spawns in the winter, and feeds principally on small thin-shelled crustacea and young fishes. Mr. Low says, " They are reckoned pretty good eating, but are never got in any quantity ; never caught at a hook : the only method of getting them is by shifting the stones at low water, when they are to be found with the Blennies."

In its habits it closely resembles the Three-Bearded Rockling, and several naturalists consider them only as varieties of the same species. Professor Nilsson regards them as distinct, and follows Linnæus in considering a fish with four barbules also as a distinct species.

The length of the head compared to the length of the body alone, is as one to four ; the depth of the body less than the length of the head : the shape of the body less cylindrical than that of the Three-Bearded, and the nose more pointed ; the position and elevation of the fins similar to those of the fish last described, but the first ray of the first dorsal fin is longer and more conspicuous, and the vent is nearer the head than in that species, being less than half the distance from the nose to the end of the fleshy portion of the tail. The fin-rays in number are—

2nd D. 52 : P. 14 : V. 6 : A. 40 : C. 20.

The body compressed ; the head depressed ; the mouth rather small, with a band of small teeth in each jaw, and a patch of similar teeth at the anterior part of the roof of the mouth ; the under jaw the shortest, with a single barbule at the chin ; the upper lip plain, without crenation, with two small barbules near the point of the nose, and two others, as long again, about as much before and within the nostrils as the nostrils are before and within the eyes. The eyes small, and placed near the nose. The colour of the upper part of the head, back, and sides, uniform dark brown; lower part of the sides lighter brown; under surface of the lower jaw, the ventral fins, and the belly to the vent, white; the other fins dusky brown; the course of the lateral line distinctly marked by a series of short, slender white streaks, as shown in the wood-engraving.

I have been favoured by Dr. Richardson with the following description of the appearance of a fine example of this species :—General colour of the body pale bronze, approaching to that of jewellers' gold, with streaks of purer gold colour above the lateral line in the direction of the ribs. The upper parts of the head and the gill-covers yellowish brown, blended on the cheeks with the bronze. The fins are also of a brownish orange or bronze colour, but without the metallic lustre, and their margins are blood red; the red tinge is more general on the pectorals; the irides silvery, the pupils bluish black.

Both the species last described have been called *mustela* by different authors. Linnæus attached this term to the species with five barbules : Cuvier, in the *Règne Animal,* identifies the Three-Bearded Rockling with this same word. As the number of barbules appear to be constant in each, a reference to the number in the specific name is, perhaps, the least objectionable. Linnæus, and other authors to the present time, continue, as before stated, to consider the northern species with four barbules as distinct from both.

SUBBRACHIAL
MALACOPTERYGII.   *GADIDÆ.*

THE MACKEREL MIDGE.

Motella glauca, *Mackerel Midge,* Jenyns, Man. Brit. Vert. p. 451, sp. 137.
Ciliata ,,   ,, ,, Couch, Zool. Journ. vol. i. p. 132.
 ,, ,, ,, ,, ,, Mag. Nat. Hist. vol. v. p. 15 and
              16, fig. 2, and p. 741.

Mr. Couch's MS. account of this beautiful little fish is
as follows :—" It is about one inch and a quarter in length,
moderately elongated ; head obtuse, compressed : upper jaw
the longest, with four straight barbules ; the under jaw with
one barbule ; teeth in both jaws : gill membrane with seven
rays ; eye large and bright ; a fringed membrane in a depression behind the head ; pectoral and ventral fins rather large
for the size of the fish ; dorsal and anal fins single, and reaching near to the tail ; scales deciduous ; colour on the back
bluish green ; belly and fins silvery. This seems to be one
of the species spoken of by the older naturalists under the
name of *apua;* and which, from their minute size, and the
multitudes in which they sometimes appeared, they judged
to be produced by spontaneous generation from the froth of
the sea, or the putrefaction of marine substances. The name
I have assigned to it is that in use among our fishermen,
and is descriptive of its colour and very minute size, for it is
the smallest fish with which I am acquainted."

" This fish is gregarious and migratory, never making its appearance before May, after which it is abundant from the edge of the shore to every part of the Channel. Its winter station is probably deep in the water; but in summer it keeps near the surface, and seeks the shelter of everything it finds floating ;—a circumstance that often leads to its destruction, for it is frequently hauled on board boats among the corks of nets, or with the line, or floating weeds ; and in a storm they are often thrown into boats through the breaking of the sea,—a circumstance which shows that at such seasons they must be on the crest of the wave."

" This fish dies instantly on being taken out of the water."

This small fish, with much of the appearance of being the young of a larger species, and closely allied in form to the Five-Bearded Rockling, presents in its economy some of the attributes of a species. Unlike the fish last described, which is very tenacious of life, this little fish, it is said, dies instantly on being taken out of the water: it does not appear every summer, as might be expected if it was the young of so common and local a species as the Five-Bearded Rockling ; and although present, as it is frequently said to be, during the greater part of the summer, when fry grow most rapidly, no increase is observed in its size.

SUBBRACHIAL
MALACOPTERYGII. *GADIDÆ.*

THE SILVERY GADE.

Motella argenteola, Yarrell.
Gadus argenteolus, Silvery Gade, Montagu, Mem. Wern. Soc. vol. ii. pt. 2,
p. 449.

The following is Colonel Montagu's account of this small
fish :—" There is a small species of *Gadus,* which is occasionally found on the western coast, that is nearly allied to
the Three-Bearded Cod (Rockling) in most particulars ; but
the shape of the head and the colour are essentially different.
It has very much the appearance of the fry of some larger
species, and might have been suspected to be the young of
the Ling, had it not been for a little difference in the first
dorsal fin, and the two cirri which this has before the nostrils.
If a fourth cirrus could have been discovered, suspicions
would have arisen whether it might not have been the *cim-
brius* of Gmelin. Its essential characters may stand thus :—

" With two dorsal fins, the anterior very obscure, except
the first ray, which is much the longest : cirri three, two before the nostrils, and one on the chin : upper jaw longest :
back bluish green ; sides and belly silvery.

" The head is obtuse ; eyes lateral, irides silvery : all the
fins are of a pale colour, and the whole fish is of a silvery
resplendence, except the back, which is blue, changeable
to dark green : the pectoral fin is rounded with sixteen or
eighteen rays ; ventral, six or seven, the middle ray considerably the longest, and placed much before the pectoral : first

dorsal fin commences above the gills, and the rays are very
minute and obscure, the first excepted, but more than thirty
have been counted; the second dorsal commences close to
the other, in a line with the end of the pectoral, and terminates close to the caudal; the rays are innumerable: the anal
fin begins immediately behind the vent, and terminates even
with the dorsal; the caudal fin is nearly even at the end.
Length about two inches.

"I first noticed many of these fishes thrown upon the
shore in the south of Devonshire, in the summer of 1808,
and have taken two or three since. The fishermen called it
Whitebait, but I afterwards found they had mistaken it for
the fry of Herring and Pilchard, which indiscriminately go
by that name, and are sold together in some places under
the name of Herring-Sprat.

" The Three-Bearded Cod (Rockling) is a very common
species on the western coast, and which I have taken of all
sizes, from the most minute to its full growth of sixteen or
seventeen inches, and never observed it to vary in colour,
except as it grows large it becomes more rufous and throws
out spots, which is never observed till it exceeds six or seven
inches, but is invariably rufous brown in its infant state."

It is worthy of remark, that this little fish, representing in
miniature the Three-Bearded Rockling, offers an instance
perfectly analogous to the representation in an equally diminutive size of the five-bearded species, by Mr. Couch's recent
discovery of the Mackerel Midge.

SUBBRACHIAL
MALACOPTERYGII.   *GADIDÆ.*

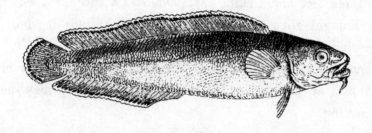

THE TORSK, OR TUSK.

Brosmius vulgaris,	Cuvier, Règne An. t. ii. p. 334.	
Brosmus ,,	*Common Tusk,* Flem. Brit. An. p. 194, sp. 90.	
Gadus brosme,	*Torsk,*	Penn. Brit. Zool. vol. iii. p. 269, pl. 37.
,, ,,	*Scotch Torsk,* Don. Brit. Fish. pl. 70.	
,, ,,	Nilsson, Prod. p. 47, sp. 14.	

Generic Characters.—Body elongated ; a single dorsal fin, extending the whole length of the back ; one barbule at the chin ; ventral fins fleshy.

The Torsk, or Tusk, is a northern species, which is only occasionally caught in the Forth, and is then brought to the Edinburgh market. It is found more frequently in the Orkney Islands, and swarms among those of Shetland, where it makes a very considerable article in their fish trade. It is caught with lines and hooks when fishing for Ling and Cod, and is salted and dried in the same manner. When eaten fresh, it is very firm and rather tough ; which makes most people prefer it dry. It is one of the best fishes when cured, swells much in boiling, and parts into very thick flakes. I observed three examples of this fish, each about sixteen inches in length, in the London market, during the

month of January 1831. These were brought from the
North in the lobster-boats. The length assigned to this
species by M. Nilsson is from eighteen inches to two feet,
rarely three feet. Mr. Low says the largest he had heard of
was three feet and a half. Mr. Donovan's specimen, which
was brought alive to London in the well of a fishing-boat,
measured twenty-five inches.

But little being known in the South of the habits of
this fish, an abridgment of Faber's account of it may be
interesting.

" A northern fish, scarcely occurring below 60° or above
73°; not migrating regularly, and therefore rarely seen by
the ichthyologists of the South. Plentiful on the coasts of
Norway as far as Finmark, of the Faroë Islands, and the
west and south coasts of Iceland ; rare on the north and east
coasts of Iceland. It must be uncommon in Greenland, as
Fabricius only knew it from the report of the natives.
Just touches the most northern point of Denmark, at Skagen
in Jutland, where it is sometimes taken ; not at all in the
south. Approaches the land early in the year in shoals, that
of Iceland in January ; remains there in company with the
Five-Bearded, and goes away again late in summer. Lives in
deep water, and is therefore seldom taken, even when it is
most abundant. Prefers a rocky bottom, on which sea-
weeds grow. Never found anything in its stomach; and this
has probably given rise to the saying, that it lives on the
juice of sea-weeds. Spawns in April and May among the
fuci along the coast. Is rarely taken with the Cod hooks,
more frequently at the smaller lines. Sometimes taken by
the Norwegian fishermen among the Holibuts. It must
have less power of resisting the violence of the sea than its
congeners, as it is thrown up dead in incredible numbers on
the coasts of the Faroë Islands and the south coast of Ice-

land after a storm. Its flesh is hard, but well flavoured.
In Iceland seldom dried, but eaten fresh. Jan Olsen says
that the fresh flesh is badly tasted, but when dried it is the
best food. In Norway it is treated like the Stockfish, but
forms no branch of merchandise. The hard roe, according
to Pontoppidan, has a good flavour. Its enemies are the
larger species of Cod. It is much infested by a worm which
forms a *nidus* in its skin, and produces rounded swellings."

The description of this fish by Mr. Low is here adopted,
with slight modification. The measurements of the specimen
from which this description was taken were the following :—
" The whole length twenty inches and a half: the greatest
breadth four and a half, which was taken at the end of the
pectoral fin ; at the vent four inches ; something more than
half-way from the vent to the tail, two inches ; at the tail,
one inch and a quarter : the length of the head four inches ;
from the point of the nose to the commencement of the dorsal
fin, six inches ; length of the dorsal fin thirteen inches ;
from the point of the lower jaw to the vent, eleven inches ;
length of the anal fin, eight inches ; tail something more
than two inches."

" The head small in proportion to the fish, with a single
barbule under the chin : the upper jaw very little longer than
the lower ; in the jaws there are great numbers of very small
teeth, and in the roof of the mouth a rough or toothed bone,
much in the shape of a horse-shoe ; a pretty broad furrow
runs from the nape to the commencement of the dorsal fin,
which runs the whole length of the back to within about an
inch of the tail ; the tail is rounded ; the anal fin begins at
the vent and ends at the tail, but is not joined with it ; the
rays of the dorsal and anal fins are numerous, but the softness of these and the thickness of the investing skin hinder
them from being counted with exactness : the edges of the

dorsal, anal fin, and tail, are white; the rest dusky: the pectoral fins are rounded, broad, and of a brown colour; the ventrals small, thick, and fleshy, ending in points; the body to the vent is roundish; the belly from the throat growing suddenly very prominent, continuing so to the vent, where it becomes smaller to the tail; behind the vent the body is pretty much compressed: the colour of the head is dusky; the back and sides yellow, which becoming lighter by degrees, is lost in the white of the belly; the lateral line is scarcely discernible, but runs nearer the back than the belly, till towards the middle of the fish, in its passage backwards, it curves a little downwards, and runs straight to the tail."

The fin-rays, according to Mr. Donovan, are—

D. 49 : P. 21 : V. 5 : A. 37 : C. 35.

The vignette represents a fishing-boat of Cadiz Bay.

SUBBRACHIAL
MALACOPTERYGII.   *GADIDÆ.*

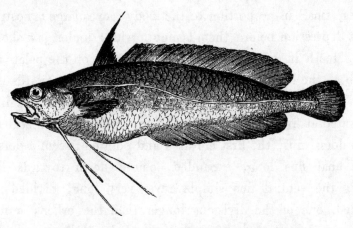

THE GREAT FORKED BEARD.

FORKED HAKE.   HAKE'S DAME, *Cornwall.*

Phycis furcatus, Common Fork Beard, FLEM. Brit. An. p. 193, sp. 84.
    ,,   ,,   CUVIER, Règne An. t. ii. p. 335.
Barbus major,   *Great Forked Beard,* RAY, Syn. 163, fig. 7.
Blennius physis, Forked Hake, PENN. Brit. Zool. vol. iii. p. 259, pl. 35.

Generic Characters.—Body elongated; two dorsal fins, the first short, the second long; ventral fins with a single ray only at the base, afterwards divided; chin with one barbule.

THE GREAT FORKED BEARD was first discovered on the Cornish coast by Mr. Jago, and inserted by Ray, with a figure, in his Synopsis, as referred to. Pennant's fish was taken on the coast of Flintshire. A specimen appeared in Carlisle market in December 1833, which was caught near Bowness; communicated to me by T. C. Heysham, Esq.: and this fish has also occurred at St. Andrews in Scotland, as noticed in the sixth volume of the Memoirs of the Wernerian Natural History Society, page 569. It is obtained occasionally in Cornwall.

VOL. II.   P

The figure here given is taken from a drawing by Mr. Couch, whose MS. contains the following notice of this species :—" The head flat on· the top, compressed at the sides, small in proportion to the body : eyes large ; nostrils in a depression before them : mouth wide: under jaw short-est ; teeth in both fine ; some larger teeth on the palate : a barb at the lower jaw : body compressed, slender towards the tail, which is small in proportion ; belly tumid ; lateral line elevated at first, afterwards low ; body and head with scales : two dorsal fins, the first elevated and pointed ; second dorsal and anal fins long, expanded, bound down towards the tail ; the ventral fins simple rays, very long, divided or forked, one of the divisions longer than the other ; a few spines before the anal fin ; tail rounded, all the rays soft. Colour of the sides and back dusky brown ; on the gill-covers sometimes greenish ; fins dusky purple, except the ventrals ; belly whitish.

This fish grows to the length of two feet : in a specimen of this size the longest portion of the ventral ray was eight inches, the shortest five inches and a half.

" Hake's Dame is the name by which alone this fish is known to our fishermen. It is not uncommon in Cornwall ; but I have never seen it except in winter, when it seems to come into shallow water to spawn. It takes a bait, and is used as food, but is not much esteemed."

The number of fin-rays, according to Dr. Fleming, are—

1st D. 10 : 2nd D. 62 : P. 12 : V. 1 : A. 56.

It is desirable to notice the specific characters of this fish, in order to distinguish between it and a Mediterranean species of the same genus, which, according to Cuvier, is the true *Blennius phycis* of Linnæus, and not the British fish, as supposed by Pennant and others. The British fish has the first dorsal fin triangular, much higher than the second, the anterior rays produced ; the ventral rays twice as long as the head. The Mediterranean fish, of which I possess a specimen, has the first dorsal fin low and rounded, very similar in character to that of the Burbot, as figured at page 183 of this volume, with the ventral rays much shorter. A description and figure of this fish is given by Willughby, page 205, pl. N. 12, fig. 3.

I have not seen a specimen of the British Great Forked Beard.

P 2

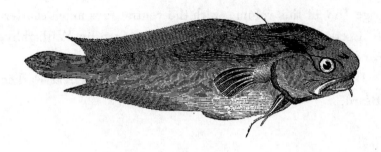

THE LESSER FORKED BEARD.

TRIFURCATED HAKE.   TADPOLE FISH.

Raniceps trifurcatus,	*Trifurcated Hake,* Flem. Brit. An. p. 194, sp. 88.
,, *Jago,*	,, ,, ,, ,, ,, ,, 89.
,, ,,	Cuvier, Règne An. t. ii. p. 336.
Barbus minor,	*Lesser Forked Beard,* Ray, Syn. p. 164, sp. 8, fig. 8.
,, ,,	*Forked Hake,* Penn. Brit. Zool. vol. iii. p. 261.
Batrachoides trifurcatus,	*Trifurcated Tadpole Fish,* Penn. Brit. Zool. vol. iii. p. 272, pl. 38.

Generic Characters.—Head depressed, body compressed ; two dorsal fins, the first very small ; the second dorsal and the anal fins elongated ; ventral fins small, the first two rays lengthened and separated.

Dr. George Johnston, of Berwick, in his address to the members of the Berwickshire Naturalist's Club, read at the first anniversary meeting in September 1832,* when referring to the various species of fishes which had occurred to him during the previous twelvemonths, remarks at page 7 : " Of the Tadpole Fish, which is one of the rarest British

* See also Mr. Loudon's Mag. Nat. Hist. vol. vi. page 11.

species, and previously known only as an inhabitant of the shores of Cornwall, I had the pleasure of exhibiting to you a living specimen, which had been captured in Berwick Bay. When alive, and when recently dead, the body appeared everywhere smooth and even ; but after having lain three days on a plate and become a little shrivelled, there appeared an obscure row of tubercles, running backwards from the pectoral fins,—and these pea-like tubercles could be more readily distinguished by drawing the finger over the skin. I would call attention to this fact, because the only good distinction between the *Raniceps trifurcatus* and *R. Jago* of Dr. Fleming is derived from the presence of these tubercles ; in the former, the lateral line is said to be tuberculated above the pectoral fins, in the latter it is said to be smooth : but here we have a specimen which when alive exhibits the character of *Jago*,—when dead, that of the *trifurcatus* ; and hence I am induced to think that both are the same animal, having the tubercles more or less prominent and obvious according to the leanness or other conditions of the body."

The difficulty of deciding the point without possessing a specimen, which the rarity of the fish rendered almost hopeless, probably induced Dr. Fleming to follow Pennant in giving both names a place in his History of British Animals. The description of Mr. Couch is quoted by Dr. Fleming as belonging to the Cornish fish and the Lesser Forked Beard of Jago ; and Cuvier, in a note at the foot of page 336 of the second volume of the *Règne Animal*, quotes the *Gadus trifurcatus* of Pennant as belonging to his genus *Raniceps*.

The advantages of equal communication and assistance on this point from Mr. Couch and Dr. Johnston enable me to carry the comparison of the two fishes still further.

Mr. Couch has favoured me with a drawing and a description of a specimen taken in Cornwall. The description is already given by Dr. Fleming, and the drawing has supplied the means of giving the representation at the head of this article. Dr. Johnston has also furnished me with a coloured drawing, a penciled sketch, and a description. A copy of the sketch, carefully reduced in size, forms the vignette at the end. These two compared together, these again compared with the double representations in the last two octavo editions of Pennant's British Zoology, and each with the figure of Jago's fish in Ray's Synopsis, will, I think, leave little doubt that all are intended to represent the same fish.

Dr. Johnston's description is as follows:—

" The comparison implied in the name Tadpole Fish is very expressive of its general form and colour; for when alive it was entirely black, and the anterior parts are large and tumid, while the hinder are much compressed. The extreme length of our Berwickshire specimen was eleven inches; and its greatest circumference, which is immediately before the pectoral fins, was seven inches, whence it tapered rapidly to the tail. The head is very large, obtuse, and flattened on the crown, where there is a slight depression between the eyes, which are an inch distant from each other, lateral, prominent, round, and black. The mouth is wide; and under the chin there is a small conical barb or feeler: the lips are rounded and white; the inferior jaw armed with two close rows of sharp teeth, and the upper, which is moveable, with similar teeth, but more numerous, and not distinctly rowed. On the palate, behind the jaw, there is a semilunar cartilaginous prominence or tubercle roughened with small teeth; and the wide entrance into the œsophagus is guarded with four similar tubercles, but of a roundish figure, two above, and two smaller below. The branchial rays are few in number, and on the inner side of each of them there are two rows of minutely spinous tufts. The first dorsal fin is very minute, but is terminated by a rather long ray: the second dorsal fin commences just behind it, or one-third of the whole length from the head, and extends nearly to the tail; it is half an inch broad, equal throughout, the rays ending in free single points. The anal fin is like the dorsal: the pectorals are oblong wedge-shaped, one inch and a half long: the ventral fins are small, and their two anterior rays are very long, white, and detached; the foremost one-half the length of the second, which measures little less than two inches. Tail wedge-shaped. The scales are small, and lie close to the body: they have an oblong square form, marked with parallel lines or striæ, which on the exposed part of each scale run in a transverse, and on the covered parts in a longitudinal direction."

The numbers of the different fin-rays, according to Pennant, are—

1st D. 3 : 2nd D. 62 : P. 23 : V. 6 : A. 59 : C. 36.

Mr. Couch says this fish is too rare for us to be much acquainted with its history. The only specimen he ever saw was taken with a line in rocky ground, in the month of April; at which time its roe was small. The remains of an echinus were in its intestines.

The following note appears at the end of Mr. Couch's account of this fish :—

" Mr. Jago, whose name occurs at the head of a list of fishes at the end of Ray's *Synopsis Piscium*, was a native of Cornwall, and a minister of the Church of England. When Bishop Trelawney, so well known as one of the six bishops committed to the Tower by James the Second, endowed the Chapel of Ease at East Looe, and thereby obtained the consent of the Rector of St. Martin to name the curate, he appointed his friend Mr. Jago to the curacy ; and the latter embraced the favourable opportunity thus placed within his reach to make collections for an intended History of Cornish Fishes, which, however, he never perfected. Never having been married, his MS. and drawings at his decease came into the possession of his friend Mr. Dyer, by whom they were delivered to Dr. Borlase, the author of the History of Cornwall."

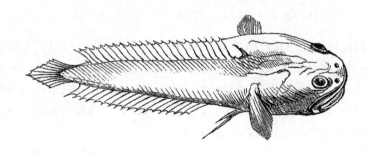

SUBBRACHIAL
MALACOPTERYGII.   *PLEURONECTIDÆ.**

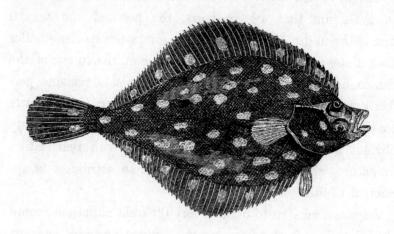

THE PLAICE.

Platessa vulgaris,   *Plaise,* Flem. Brit. An. p. 198, sp. 103.
  ,,   ,,   Cuvier, Règne An. t. ii. p. 338.
  ,,   ,,   *Plaise,* Willughby, p. 96, F. 4.
Pleuronectes platessa, Linnæus. Bloch, pt. ii. pl. 42.
  ,,   ,,   *Plaise,* Penn. Brit. Zool. vol. iii. p. 304.
  ,,   ,,   ,,   Don. Brit. Fish. pl. 6.

Generic Characters.—Body rhomboidal, depressed ; both eyes on the right side of the head, one above the other ; a row of teeth in each jaw, with others on the pharyngeal bones ; dorsal fin commencing over the upper eye, that fin and the anal fin extending nearly the whole length of the body, but neither of them joined to the tail ; branchiostegous rays 6.

The character and appearance of the various species of *Pleuronectidæ,* or Flatfish, as they are popularly called, are so peculiar and so unique among vertebrated animals as to claim particular notice.

The want of symmetry in the form of the head ; both eyes placed on the same side, one higher than the other,

* The family of the Flounders, popularly called Flatfish.

frequently not in the same vertical line, and often unequal
in size; the position of the mouth; the inequality of the
two sides of the head, and the frequent want of uniformity
in those fins that are in pairs, the pectoral and ventral
fins of the under or white side being in some species smaller
than those of the upper; and the whole of the colour of the
fish confined to one side, while the other side remains perfectly white,—produce a grotesque appearance: yet a little
consideration will prove that these various and seemingly
obvious anomalies are perfectly in harmony with that station
in nature which an animal bearing these attributes is appointed to fill.

As birds are seen to occupy very different situations, some
obtaining their food on the ground, others on trees, and not
a few at various degrees of elevation in the air, so are fishes
destined to reside in different situations in the water: the
Flatfishes and the various species of Skate are, by their depressed form of body, admirably adapted to inhabit the
lowest position, and where they occupy the least space,
among their kindred fishes.

Preferring sandy or muddy shores, and unprovided with
swimming-bladders, their place is close to the ground, where,
hiding their bodies horizontally in the loose soil at the bottom, with the head only slightly elevated, an eye on the
under side of the head would be useless; but both eyes
placed on the upper surface affords them an extensive range
of view in those various directions in which they may either
endeavour to find suitable food, or avoid dangerous enemies.
Light, one great cause of colour, strikes on the upper surface
only; the under surface, like that of most other fishes, remains perfectly colourless. Having little or no means of
defence, had their colour been placed only above the lateral
line on each side, in whatever position they moved, their

piebald appearance would have rendered them conspicuous objects to all their enemies. When near the ground, they swim slowly, maintaining their horizontal position; and the smaller pectoral and ventral fins on the under side are advantageous where there is so much less room for their action, than with the larger fins that are above. When suddenly disturbed, they sometimes make a rapid shoot, changing their position from horizontal to vertical: if the observer happens to be opposite the white side, they may be seen to pass with the rapidity and flash of a meteor; but they soon sink down, resuming their previous motionless, horizontal position, and are then distinguished with difficulty, owing to their great similarity in colour to the surface on which they rest.

Though the appearance and situation of the eyes and mouth seem to indicate a degree of deformity, yet the head contains modifications of all the bones that are found in a symmetrically-formed head. The vent is situated very far forward between the ventral fins and the commencement of the anal fin; but the abdominal cavity, though circumscribed, extends backwards to a considerable distance, the intestine returning by a convolution.

Most of the Flatfishes are deservedly in great request as articles of food. The number of species diminishes as the degrees of northern latitude increase. In this country we have sixteen species; at the parallel of Jutland, Denmark, and the islands at the mouth of the Baltic, there are thirteen; on the coast of Norway they are reduced to ten species; at Iceland the number is but five, and at Greenland only three.

The Plaice is described and figured by Rondeletius, and was known to the older naturalists long before his time. It inhabits sandy banks and muddy grounds in the sea; and among the Orkney islands is caught by lines and hooks;

but as it is not of large size there, it is not much sought after : it is common, however, in the Edinburgh market, where the small ones are called Fleuks.   On the English coast the Plaice is taken in abundance generally wherever either lines or trawl-nets can be used ; and in Ireland, this fish is recorded to be taken from the shores of the county of Cork on the south, round by the eastern coast to the county of Donegal on the north-west.

The Plaice spawns in February or March, and is considered to be in the finest condition for the table at the end of May. Diamond Plaice is a name attached to those which are caught at a particular fishing-station off the Sussex coast, which is called the Diamond ground.   The fish are remarkable for the purity of the brown colour and the brilliancy of the spots.

Plaice feed on the soft-bodied animals generally, with young fish and small crustacea, and have been known to attain the weight of fifteen pounds ; but one of seven or eight pounds' weight is considered a Plaice of large size.   It is taken sometimes in almost incredible numbers.   So great a glut of Plaice occurred once in Billingsgate market, that, although crowded with dealers, hundreds of bushels remained unsold.   Great quantities of Plaice, averaging three pounds' weight each, were sold at one penny per dozen.   One salesman, having in vain endeavoured to sell an hundred bushels at the rate of fifty Plaice for four-pence, left them with Mr. Goldham, the clerk of the market, requesting him to sell them for anything he could get.   Unable to dispose of them otherwise, Mr. Goldham, by direction of the Lord Mayor, divided them among the poor.

In some parts of the North of Europe, where from the rocky nature of the soil the sea is remarkably transparent, Plaice and some other Flatfish of large size are taken by dropping down upon them, from a boat, a doubly-barbed short spear, heavily leaded to carry it with velocity to the bottom, with a line attached to it, by which the fish when transfixed is hauled up.

In East Friesland the Plaice has been transferred to fresh-water ponds, where it is established and thrives well.

Like other ground-fish, all the *Pleuronectidæ* are very tenacious of life.

The length of the head compared to the whole length of the head, body, and tail, is as two to nine ; the depth of the solid part of the body, without including the dorsal or anal fins, rather more than one-third of the whole length ; the form subrhomboidal ; the mouth and teeth rather small ; the upper eye the largest, and placed rather more backward than the lower eye, with a strong and prominent bony ridge between the orbits, and several tubercles forming a curved line from the posterior part of the ridge to the commencement of the lateral line : the preoperculum is in a vertical line over the origin of the ventral fin ; the operculum terminates in an angle upon the base of the pectoral fin ; the lateral line prominent, commencing at the upper margin of the operculum, arched over the pectoral fin, then straight along the middle of the fleshy portion of the tail, and extending over the membrane connecting the central caudal rays. The dorsal fin commences over the upper eye ; the longest rays rather behind the middle of its whole length : the anal fin, preceded by a spine, begins in a line under the origin of the pectoral fin ; the longest rays rather before the middle : both dorsal and anal fins end on the same plane, and short of the end of the fleshy portion of the tail, which, as well as the caudal rays, is narrow and elongated ; the tail rounded.

The fin-rays in number are—

D. 73 : P. 11 : V. 6 : A. 55 : C. 16.

The body is smooth on both sides, the scales small; the colour of the upper or right side a rich brown, with a row of bright orange red spots along the dorsal and anal fins, and other spots of the same colour dispersed over the body; the under side entirely white. Young Plaice have frequently a dark spot in the centre of the red one.

The fishes of this first division of the *Pleuronectidæ* with the eyes and the colour on the right side of the body are further distinguished by the term *dextral* fishes.

SUBBRACHIAL
MALACOPTERYGII.   PLEURONECTIDÆ.

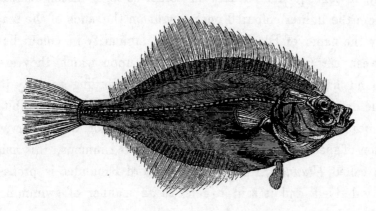

THE FLOUNDER.

FLOOK, *Merret.*—MAYOCK FLEUKE, *Edinb.*—BUTT.

Platessa flesus,   Flounder, FLEM. Brit. An. p. 198, sp. 104.
  ,,   ,,   Le Flet, CUVIER, Règne An. t. ii. p. 339.
Pleuronectes fluviatilis, Fluke, WILLUGHBY, p. 97, F. 4.
  ,,   *flesus,*   ,, LINNÆUS. BLOCH, pt. ii. pl. 44 & 50.
  ,,   ,,   Flounder, PENN. Brit. Zool. vol. iii. p. 305.
  ,,   ,,   ,,   DON. Brit. Fish. pl. 94.

THE FLOUNDER is one of the most common of the Flatfish, and is found in the sea and near the mouths of large
rivers all round our coast, being more particularly abundant
where the bottom is soft, whether of sand, clay, or mud.
All the bays, creeks, and inlets of Orkney produce it, according to Mr. Low; and it is taken in abundance in different
parts of Scotland, where it is called Fluke and Mayock
Fleuke,—a term having reference to the flattened form of
the fish.   It is common at Berwick and Yarmouth, at which

latter place it is called a Butt—a northern term ; and those
Flounders that are caught in the extensive backwaters behind
Yarmouth, where there is a considerable deposit of mud,
are in consequence so dark in colour as to be distinguished,
from the lighter coloured ones caught on the sands of the sea,
by the name of Black Butts. This similarity in colour between certain fishes and the bottom upon which they are
found has been already referred to as affording security to
the defenceless from the attacks of their enemies, and exhibits
a beautiful instance of the design employed for the preservation of species. In Sweden, according to Linnæus, this fish
is called *Flundra,* from which our word Flounder is probably derived, and is said to refer to its manner of swimming
when close to the ground.

The Flounder lives and thrives whether stationary in the
sea, the brackish water, or the fresh water. In the Thames
it is taken as high up as Teddington and Sunbury: Mr.
Jesse mentions having seen the Flounder pursue Minnows
with great eagerness into the shallows where the Mole runs
into the Thames at Hampton Court. This species is caught
in considerable quantities from Deptford to Richmond by
Thames fishermen, who, with the assistance of an apprentice,
use a net of a particular sort, called a tuck-net, or tuck-sean.
One end of this net is fixed for a short time by an anchor or
grapple, and its situation marked by a floating buoy ; the
boat is then rowed, or rather sculled, by the apprentice in
a circle, the fisherman near the stern handing out and clearing the net : when the circle is completed and a space inclosed, the net is hauled in near the starting point in a
direction across the fixed end.

Flounders ascend rivers generally. Colonel Montagu says
they are found up the Avon within three miles of Bath.
They have been successfully transferred to fresh-water ponds:

being long-lived out of water, the carriage from one place to another is a matter of very little difficulty. Along our southern shore the Flounder is very common; and it occurs on the Irish coast from Cork up the eastern side to Antrim, and thence northward and westward to Donegal.

The Flounder feeds upon aquatic insects, worms, and small fishes, and has been known to acquire the weight of four pounds, but is not usually seen near so large. It spawns in February or March, and the young Flounders may be seen alive by the end of April. Varieties of the Flounder occur much more commonly than those of any other species of Flatfish. I have before me, while now writing, specimens without any colour on either side; specimens coloured on both sides; and specimens with both eyes and the whole of the colour on the left side instead of the right. Those without any dark colour on either side are albino varieties, through the transparent skins of which the colour of the bloodvessels and muscles has suggested the trivial names of *rosea* and *carnaria* to the authors who considered them species. The *Pl. passer*, figured by Bloch, pt. ii. pl. 50, is certainly only a reversed Flounder, having the eyes and the colour on the left side;—a variety so common, that it is scarcely possible to examine a peck-measure of Flounders without finding one or more reversed specimens. One of the most remarkable specific distinctions of the Flounder, the series of denticulated tubercles placed between the rays of the fins along the dorsal and abdominal lines, is distinctly figured in both Bloch's plates as quoted, pt. ii. plates 44 and 50.

The length of the head is to the whole length of the fish as one to four; the greatest width of the body, without the fins, is to the whole length of the fish as one to three: the mouth small; the teeth in one row in each jaw, small and numerous ; the upper eye nearly over the lower ; the lateral line but very slightly curved over the pectoral fin, and marked with numerous rough stellated tubercles at its commencement, some more of which are arranged in two lines, one above, the other below the lateral line throughout its course : the body smooth ; the scales small ; the dorsal and abdominal lines armed with a series of denticulated tubercles, one in each space, between the rays, and alternating with them ; the dorsal fin extends from the eye almost to the tail ; the ventral fin is placed a little farther back than in the Plaice, under the margin of the operculum ; the anal fin, preceded by a spine directed forwards, also commences farther back ; both dorsal and anal fins terminate on the same plane ; the fleshy portion of the tail narrow, its rays elongated, and almost square at the end. The fin-rays in number are—

D. 55 : P. 11 : V. 6 : A. 42 : C. 14.

The colour of this species is variable, the shades of brown depending on the nature of the ground from which the fish was taken, but generally mottled with darker brown ; the fins light brown, occasionally varied with patches of darker brown, but generally lighter than the body. Examples sometimes occur with a few indistinct reddish spots on the upper surface ; but the roughness of the lateral line in the Flounder, and its smoothness in the Plaice, is a distinguishing character in these two species, however similar they may happen to be in colour or size.

SUBBRACHIAL MALACOPTERYGII.   *PLEURONECTIDÆ.*

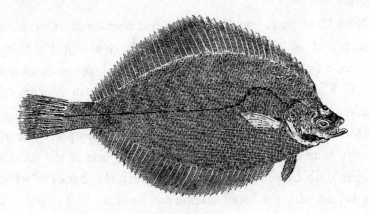

COMMON DAB.

SALTIE, AND SALT-WATER FLEUK, *Edinburgh.*

Platessa limanda,	Dab,	FLEM. Brit. An. p. 198, sp. 105.
,, ,,	*La Limande,*	CUVIER, Règne An. t. ii. pp. 339 & 340.
Passer asper,	Dab,	WILLUGHBY, p. 97, F. 4.
Pleuronectes limanda,	LINNÆUS.	BLOCH, pt. ii. pl. 46.
,, ,,	Dab,	PENN. Brit. Zool. vol. iii. p. 308.
,, ,, ,,	DON. Brit. Fish. pl. 44.

THE DAB is common to all the sandy parts of the coast, and is usually caught along with Plaice and Flounder; but is immediately distinguished from either by its more uniform and lighter brown colour, its more curved lateral line, and the roughness of the scaly surface, from which latter circumstance it has been called in Latin, *limanda,* from *lima,* a file. Dr. Neill reports it as common in the market of Edinburgh, where it is called Saltie, or Salt-water Fleuk. I have received it from Berwick, and it is taken at Yarmouth. It is very common in the London market, and Colonel Montagu says it is caught along the Devonshire coast with the seine-net; many are also taken by trawling.

Q 2

It occurs in Cornwall, and is recorded as found on several parts of the east coast of Ireland between Waterford and Belfast.

The Dab feeds on small fish, crustacea, and marine insects; and is in best condition for the table in February, March, and April. Its flesh is considered superior to that of the Plaice or Flounder: Cuvier says it is in higher estimation in Paris than the Flounder, because it bears carriage better. It spawns in May or June; inhabits deeper water generally than the Flounder; and on some parts of the coast is caught both by sea-lines and hand-lines, the hooks of which are baited with the usual marine sand-worm, or a portion of the body of some of the testaceous mollusca. The size of the Dab is commonly about eight or nine inches in length, and seldom exceeds twelve inches.

The form of the body is like that of the Flounder: the length of the head is to that of the body as one to five; the greatest breadth compared to the whole length is as two to five: the mouth and teeth small, the latter separated; the eyes rather large, the orbits divided, but the bony ridge is not very prominent; the length of the pectoral fin nearly two-thirds the length of the head; ventral fins small, in a line under the origin of the pectoral fin: the dorsal and anal fins extending along the body nearly to the tail, both ending on the same plane; the longest rays of both are placed behind the centre: tail slender, elongated, and slightly rounded. The fin-rays in number are—

D. 76 : P. 11 : V. 6 : A. 59 : C. 14.

The form of the body is subrhomboidal; the scales rough, their margins ciliated; the lateral line arched high over the pectoral fin, the remainder to the tail straight; the rays of the dorsal and anal fins scaled; the colour of the fish a uniform pale brown, with the under surface white.

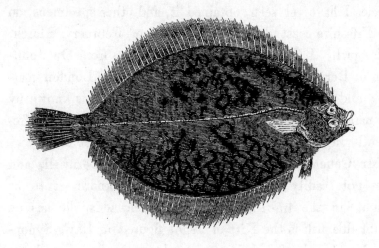

LEMON DAB. SMOOTH DAB.

SMEAR DAB.—SANDFLEUK, *Edinburgh.*

TOWN-DAB, *Hastings.*—MARY-SOLE, *Devonshire.*

Platessa microcephalus, Smear Dab, FLEM. Brit. An. p. 198, sp. 106.
     ,,   *microcephala, Lemon Dab,* JENYNS, Man. Brit.Vert. p. 457, sp. 144.
Pleuronectes lævis, Smear Dab, PENN. Brit. Zool. vol. iii. p. 309, pl. 47.
     ,,   *microcephalus, Small-headed Dab,* DON. Brit. Fish. pl. 42.

THE LEMON DAB, or SMOOTH DAB, is not of such frequent occurrence as the common Rough Dab; and is, on account of a mixture of various shades of reddish brown and yellow colours, a prettier fish to look at. It approaches to a rhomboid in form, even more so than any of the species of this genus as now restricted to a certain portion only of those Flatfishes that have the eyes on the right side.

In tracing the occurrence of the Smooth Dab round the coast, Dr. Neill of Edinburgh says it is taken off Seton

Sands and Aberlady Bay. Dr. Richard Parnell, who has
devoted great attention to the fishes of the Forth, and to
whose kindness I am indebted for the largest example of this
species I have yet seen, obtained it, and other specimens, on
the Fifeshire coast during the months of February, March,
and April. I have received specimens also from Dr. Johnston of Berwick. It is not uncommon in the London market; and is taken on the Sussex coast, where it is known by
the name of Town-Dab. Colonel Montagu observed it frequently in Devonshire, where it is called Mary-Sole.

Mr. Couch says it is rather a rare fish in Cornwall, and
does not readily take a bait; but he has known it to be
caught in the trawl-nets. Mr. Couch adds, he has no
doubt this fish is the Kitt of Jago, figured in Ray's Synopsis, No. 1; the fish being reversed on the paper, and appearing with its eyes and colour on the left side, as in the
sinistral fishes.

The flesh of the Smooth Dab is considered equal to that
of the Common Dab, and the substance of the body is much
thicker. This species spawns in May.

Duhamel was well acquainted with the Smooth Dab; but
considered it a rare fish on the coast of France, and on some
parts of it entirely unknown.

The form of the body rhomboidal; small-sized specimens
are more elongated: the length of the head is to that of the
head and body, without the caudal rays, as one to five and a
half; the depth of the body, including the dorsal or anal fin,
only just equal to half the whole length of the fish: the
mouth small; lips tumid; the jaws equal in length; teeth in
an even, close, regular row in each jaw, but extending further
back on the white under side of the fish than on the upper;
nostrils double: the eyes exactly over each other; the orbits
separated by a strong, prominent, bony ridge, but without

tubercles : the head small ; the pectoral fin but little more than half the length of the head ; ventral fin small ; the dorsal and anal fins reaching near to the tail, but distinct; the tail rounded.   The fin-rays in number are—

D. 86 : P. 10 : V. 5 : A. 70 : C. 16.   Vertebræ 46.

The general colour of the upper surface of the body is a mixture of pale reddish brown and yellow, with small dark brown specks ; the lips are orange, as is also the posterior edge of the operculum, and the anterior edge of the body immediately behind it : the body smooth, and covered with a mucous secretion ; the lateral line but little arched over the pectoral fin ; the under parts white.

The vignette below represents a Peter-boat as used by the Thames fishermen between and above the different bridges.

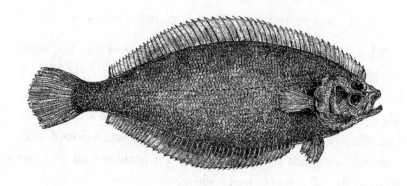

LONG ROUGH DAB.

SANDNECKER.—SAND FLEUK, AND LONG FLEUK, *Edinb.*

Platessa limandoïdes, Sandnecker,   JENYNS, Man. Brit. Vert. p. 459, sp. 146.
Pleuronectes ,,   ,,   PARNELL, Edinb. Phil. Journ. July 1835,
                                     p. 210.
   ,,   ,,   *Long Rough Dab,* BLOCH, pt. vi. pl. 186.

THE LONG ROUGH DAB is a very recent addition to
the catalogue of British Fishes.   The first notice I had
of the occurrence of this species on our coast was in the
autumn of 1833, from Dr. John Harwood, of St. Leonard's,
near Hastings, who had seen a specimen that was taken
on the Sussex coast.   In the summer of 1834 I received
two specimens from Dr. George Johnston of Berwick, which
had been taken in that vicinity.   In May 1835 I was
favoured by Dr. Parnell with the largest specimen I have
yet seen, measuring fifteen inches; who, at the same time,
sent me word they were to be had frequently in the Frith

of Forth in May, June, and July. Soon afterwards I learned by a letter from George T. Fox, Esq. of Durham, that a specimen of *Pl. limandoïdes* of Bloch had been taken some years before on the coast of Sunderland, and was still preserved in the possession of Thomas Wilkinson, Esq. of Bishop Wearmouth. The first recorded notice of this fish as British, that I am aware of, is that by Dr. Parnell, in the Edinburgh New Philosophical Journal, already quoted, where, by an error of the press, the fish is called *Pl. limandanus*.

Bloch received his specimen from Hamburgh, and states that this fish is caught by the hook in the vicinity of Heligoland. He says it feeds on young crabs and young lobsters, and that its flesh is white and good.

The length of the head compared to the whole length of the fish is as one to five ; the breadth of the body, not including the dorsal or anal fins, is equal to one-third of the whole length ; with the dorsal and anal fins it is equal to half the distance from the point of the nose to the end of the fleshy portion of the tail : the form of the body is an elongated oval, almost equally pointed at both ends ; the parts of the mouth capable of some protrusion ; teeth in a single row in each jaw, separate, conical, and curving slightly inwards : eyes rather large ; the upper one a little before the line of the other ; the orbits separated by a bony ridge : pectoral and ventral fins small ; the former only half the length of the head : dorsal and anal fins extending nearly to the tail ; both fins ending on the same plane : the tail slightly rounded.

The cheeks, operculum, and body, covered with harsh, ciliated scales, the surface exceedingly rough to the touch ; a row of ciliated scales along each ray of the dorsal and anal fins ; the lateral line straight, or very slightly inclining upwards as it approaches the operculum ; the head and body one uniform pale brown ; the fins lighter ; the under surface of the body rough and white.

The fin-rays in number are——

D. 76 : P. 10 : V. 5 : A. 64 : C. 16.

The vignette represents a Folkstone fisherman selling his fish by auction on the beach after landing. This is done, according to the Dutch fashion, by lowering the price demanded for the lot till a bid is made, when the bargain is struck by dropping the shingle, which is held, as represented, between the fore-finger and thumb.

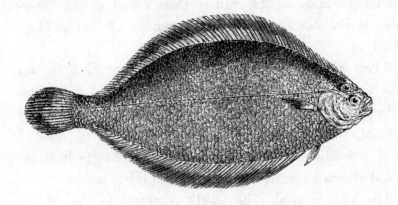

THE POLE, OR CRAIG FLUKE.

Platessa Pola, La Pole, Cuvier, Règne An. t. iii. p. 339.
,, ,, The Pole, Jenyns, Man. Brit. Vert. p. 458, sp. 145.
,, ,, Craig Fluke, Parnell, Edinb. Phil. Journ. July 1835, p. 210.
,, ., La Vraie Limandelle, Duhamel, sect. ix. pl. 6, figs. 3 & 4.
Pleuronectes Pola, La Pole, Lacepede, vol. iv. p. 368. New 8vo. Edition,
vol. x. p. 74.

This second addition to the British Fishes in the genus *Platessa* is still more rare on our coast than the *Pl. limandoïdes* last described. In the month of May 1833 I observed a specimen twelve inches long in the shop of Mr. Groves of Bond-street ; and on pointing out the differences between this and other Flatfishes by comparison with several species among which it was placed for sale, and mentioning its rarity, Mr. Groves immediately sent it to the Zoological Society for preservation, observing that he had not noticed the difference, and did not recollect that he had ever seen that species before. From this specimen the woodcut was executed. In May 1835, Dr. Parnell very kindly sent to me from Edinburgh, for examination, a skin of this species nineteen inches in length, with several other preserved skins of fishes taken in the Frith of Forth, where the Craig Fluke, as this fish is there called by the fishermen, is occasionally taken in the months of April, May, and June.

These are the only examples of this fish taken in our seas that I am acquainted with. Of its habits but little, I believe, is known. Baron Cuvier states, in his *Règne Animal,* that in France the flesh of this species is in as great estimation as that of the Sole; and Lacépède states that it attains a length of twenty-four to thirty inches.

The head is small; its length compared to that of the whole fish is as one to six: the greatest breadth of body, dorsal and anal fins excluded, is to the whole length rather more than a third; including the dorsal and anal fins, rather less than a half: the form of the body an elongated oval, pointed at each end: the mouth small; the lips thin; a single row of teeth in each jaw, close set, smooth, incisorlike, with thin and even-cutting edges: the eyes rather large; the upper one ranging vertically behind the line of the lower, with the usual intervening bony ridge; irides orange: pectoral and ventral fins small: dorsal and anal fins extending very nearly the whole length of the body; both ending on the same plane; the rays about the middle of each the longest; those at the extreme ends, before and behind, very short: tail rather long and rounded.

The fin-rays in number in one specimen were—

D. 109 : P. 11 : V. 7 : A. 93 : C. 19.

The body is quite smooth; the scales rather large, deciduous, but neither ciliated nor roughened in any way beyond a few radiating striæ; the head smooth, without tubercles; lateral line straight, and extending, as in all the other species of *Platessa*, to the end of the membrane connecting the caudal rays.   The colour of the body uniform yellowish brown; the edges of all the fins darker: when dried for preservation, the colour of the skin of the body becomes clove brown; that of the fins, brocoli brown.

The vignette below represents the Thames Peter-boat rigged with a fore-sail and main-sail, as used by the fishermen about Greenwich, and from thence along the course of the river downwards.

THE HOLIBUT.

Hippoglossus vulgaris,	*Holibut,*	Flem. Brit. An. p. 199, sp. 108.
,, ,,	*Fletan,*	Cuvier, Règne An. t. ii. p. 340.
,, ,, ,,	Willughby, p. 99, F. 6.
Pleuronectes hippoglossus, ,,	Linnæus. Bloch, pt. ii. pl. 47.
,, ,,	*Holibut,* Penn. Brit. Zool. vol. iii. p. 302.
,, ,, ,,	Don. Brit. Fish. pl. 75.

Generic Characters.—With both eyes and the colour on the right side in the British specimen of this genus, and with fins similar to those of the species of the genus *Platessa* — the jaws and the pharynx are armed with teeth that are sharper and stronger, and the form of the body is more elongated.

The Holibut is one of the largest species of the *Pleuronectidæ*, but its capture is principally confined to the Northern fisheries: it is noticed by Pennant in his Arctic Zoology, and is well known on the coasts of Norway, Iceland, and Greenland. It is usually caught with lines and hooks. The Greenlanders eat the flesh of this fish both fresh and dried, for which latter purpose it is cut into long slips and exposed to the air. They are fished for successfully by the natives of the Orkneys, who ply their lines in the slack water and various eddies produced by the different islands, out of the race of the tides ; these quiet places being more particularly the haunts of the Holibut and Flatfish generally. A large quantity of oil is obtained from them.

In the Northern seas, Holibuts weighing near five hundred pounds are said to have been obtained; and examples of large size have occasionally occurred nearer home. In April 1828, a Holibut seven feet six inches in length, three feet six inches in breadth, and weighing three hundred and twenty pounds, was taken off the Isle of Man and sent to Edinburgh market. It was said to have been the largest specimen ever exhibited there.

The Holibut has occurred, as might be expected, on the northern coast of Ireland, from Belfast Bay to the shores of the county of Donegal ; but in consequence, probably, of the word *maximus* having been applied by some authors as a specific term to the Holibut, and by others to the Turbot, some confusion has arisen by a misappropriation of the English names. I am in doubt, therefore, whether some localities on the south coast of Ireland claimed for the Holibut do not in reality refer to the Turbot, as neither Colonel Montagu nor Mr. Couch mentions the occurrence of the Holibut either in Devonshire or Cornwall, but the Turbot is well known to be common to both. In London this fish is occasionally seen in the months of March and April: here, from its large size, it is sold in slices at a low price by the pound weight. The flesh, though white and firm, is dry, the muscular fibre coarse, with but little flavour : the head and fins are said to be the best parts. Specimens only two feet long are occasionally seen, but in general the examples are much larger.

The Holibut feeds close to the ground, on the smaller species of Flatfish and various crustacea. It spawns in spring: the roe is of a pale red colour, and the ova in the female very numerous. A specimen five feet two inches long, in the shop of a London fishmonger, supplied the means of obtaining the following description :—

The length of the head compared to the whole length of the fish without the caudal rays, is as one to four ; the greatest breadth one third of the whole length, dorsal, anal, and caudal fin-rays all excluded : the head small, but the mouth large ; teeth in two rows in each jaw, small for the size of the fish, conical, pointed, and separated ; the irides yellow, the pupils black : the pectoral fin on the coloured or dextral side one-fourth larger than that on the white or under side ; the dorsal fin commences in a line over the eye, the rays longest over the widest part of the body ; the anal fin of similar character : both dorsal and anal fins terminate on the same plane, and distinct from the caudal fin, the posterior margin of which is concave ; the ventral fins are small, the white ventral fin of the under side the smaller of the two : immediately in advance of the commencement of the anal fin are two apertures ; the anterior opening large, and evidently connected with the intestines ; the posterior opening smaller, and apparently the outlet from the urinary bladder and sexual organs.   The fin-rays in number were—

D. 104 : P. 16 : V. 6 : A. 81 : C. 16.

The form of the body elongated ; the surface smooth, covered with small oval-shaped soft scales ; the lateral line arched over the pectoral fin ; the colour composed of different shades varying from light brown to dusky brown ; the surface of the under side perfectly smooth and white.

SUBBRACHIAL
MALACOPTERYGII.   PLEURONECTIDÆ.

THE TURBOT.

RAWN FLEUK, AND BANNOCK FLEUK, *Scotland.*

Rhombus maximus,	*Le Turbot,*	Cuvier, Règne An. t. ii. p. 340.
,, ,,	*Turbot,*	Willughby, p. 94, F. 2.
Pleuronectes ,,	Linnæus.	Bloch, pt. ii. pl. 49.
,, ,,	*Turbot,*	Penn. Brit. Zool. vol. iii. p. 315, pl. 49.
,, ,,	,,	Don. Brit. Fish. pl. 46.
,, ,,	,,	Flem. Brit. An. p. 196, sp. 96.

Generic Characters.—Colour and eyes on the left side; teeth in the jaws and pharynx; dorsal fin commencing anterior to the upper eye; dorsal and anal fins extending very nearly to the tail.

THE TURBOT, so well known, so highly and so justly esteemed, is considered the best, as it is also one of the largest, of our Flatfishes; and, like the Salmon, notwithstanding its great excellence, and the immense numbers that are caught in various ways, it is still in great abundance, but not equally so on all parts of the coast.

VOL. II.   R

According to Mr. Low, it is rare in Orkney; but the numbers taken increase on coming southward; and in the market of Edinburgh, according to Dr. Neill, it is commonly denominated Rawn Fleuk, from its being thought best for the table when in rawn,* or roe; it is sometimes also called Bannock† Fleuk, on account of its shape.

On the coasts of Durham and Yorkshire, a considerable fishery for Turbot is carried on by the fishermen of Hartlepool and Scarborough with long lines, the mode of using which was described when speaking of the common Codfish. A large proportion of the Turbot produced in the English market is taken on or near the various sandbanks between the long line of our eastern shore and the coast of Holland. The writer of the article " Fisheries," in the edition of the Encyclopædia Britannica now in course of publication, says, " The only fishery, perhaps, which neither the Scotch nor the English follow up with the same success as the Dutch, is that of the Turbot; the finest of which are supposed to be taken upon the Flemish banks. The Turbot fishery begins about the end of March, when the Dutch fishermen assemble a few leagues to the south of Scheveling. As the warm weather approaches, the fish gradually advance to the northward, and during the months of April and May they are found in great shoals on the banks called the Broad Forties. Early in June they have proceeded to the banks which surround the small island of Heligoland, off the mouth of the Elbe, where the fishery continues to the middle of August, when it terminates for the year. The mode of taking Turbot is as follows :—At the beginning of the season the trawl-net is used; which being drawn along the banks, brings up

* In the West of England a different meaning is attached to this word : see the bottom of page 170.

† *Bannock*, a round flat cake.

various kinds of Flatfish, as Soles, Plaice, Thornbacks, and Turbots ; but when the warm weather has driven the fish into deeper water, and upon banks of a rougher surface, where trawling is no longer practicable, the fishermen have then recourse to their many-hooked lines. The hooks are baited with the common Smelt, and a small fish resembling an Eel, called the Gorebill.* Though very considerable quantities of this fish are now taken on various parts of our own coasts, from the Orkneys to the Land's End, yet a preference is given in the London market to those caught by the Dutch, who are supposed to have drawn not less than 80,000*l.* a year for the supply of this market alone ; and the Danes from 12,000*l.* to 15,000*l.* a year for sauce to this luxury of the table, extracted from one million of lobsters, taken on the rocky shores of Norway,—though our own shores are in many parts plentifully supplied with this marine insect, equal in goodness to those of Norway."

About one-fourth of the whole supply of Turbot to the London market is furnished by Dutch fishermen, who pay a duty of 6*l.* per boat, each boat bringing from one hundred to one hundred and fifty Turbot. A very considerable quantity is also purchased of the Dutch fishermen at sea on the fishing stations near their own shore by English fishermen, and is brought by them to our market in their own boats, paying no duty.

Along our southern coast many Turbot are caught by the trawling vessels, and long-line fishing at particular seasons on the Varne and on the Ridge,—two extensive banks of sand, the first about seven miles, and the second about twelve miles, from Dover, towards the French coast. On these banks French fishermen also lay their long-lines ; and when they do not succeed in selling their Turbot at sea, which

* Vol. i. page 391.

R 2

suits them best, they freight one or more of their own boats
with them, and send them into Dover harbour for sale, paying the usual duty.   They are not, however, allowed to sell
any fish but Turbot, except under particular circumstances.
If in want of provision, or their boat has suffered damage
from bad weather, they are then permitted, by certificate
from a magistrate, to sell as much fish as will procure them
food, or pay the cost of repairs.

Along the Devonshire coast, where trawling on an extensive scale is practised, a portion of the Turbot and Dory is
forwarded during the season to Bath and Exeter; the remainder is sent to the London market by land-carriage.   It
is observed that the Turbot of the northern part of our own
coast, and those bought of or brought by the Dutch fishermen, are darker in colour than those from the south-western
shores of England.

Mr. Couch says, " The Turbot keeps in sandy ground,
and is a great wanderer, usually in companies; and though
its proper habitation is close to the bottom, it sometimes
mounts aloft, and I have known it upon the surface over a
depth of thirty fathoms : I have been informed also of its
pursuing to the surface a companion that was drawn up by
the line, when both were taken together."

The Turbot, though a voracious fish, is particular as to
the quality of his food : the bait used for him should be very
fresh ; if it happens to be in the least degree tainted, the
Turbot will not touch it.   The most enticing baits to use
are those small fishes which are either very bright in colour
or very tenacious of life ; the Atherine, and the two common
species of the genus *Cottus*, the Sea-Scorpion and Father-
Lasher, are most frequently used : the first attracts by its
shining silvery appearance, and the others by living a long
time on the hook, and showing themselves in their struggles

to get free.   The River Lampern was formerly used in large quantities by the Dutch, and was a great favourite with them as baits for Turbot, on account of the facility with which they could be kept alive while the boats were at sea, and combining bright silvery colour with great power of resisting the usual effect of mutilation.   The principal food of the Turbot is small fish, crustacea, and mollusca.   It spawns about August, but rapidly recovers its condition and firmness.

Turbot are recorded as having been taken on the south coast of Ireland; I have seen one that was caught on the coast of Londonderry in the north; and there is little doubt but this valuable species occurs at many intermediate localities.

" The Turbot was known to the Athenians, and has been ever since a worthy object of gastronomical worship."   The most common size varies from five to ten pounds' weight; occasionally this fish attains to twenty pounds, and sometimes thirty pounds.   Mr. Couch notices, in his MS. a record of one taken in the year 1730, at Cawsand, near Plymouth, which weighed seventy pounds.   On the 18th of February 1832, an unusually large Turbot was caught at Staiths, near Whitby, which weighed thirteen stone eight pounds (one hundred and ninety pounds), and measured six feet across.   Rondeletius, however, states that he had seen a Turbot five cubits in length, four in breadth, and a foot in thickness.   The Turbot is considered to have been the *Rhombus* of the ancient Romans, of which a specimen of enormous size is said to have been taken in the reign of Domitian, who ordained a *Senatus Consultum* to devise the best mode of bringing it to table.*

> " No vessel they find fit to hold such a fish,
> And the senate's convoked to decree a new dish."

* Juven. Sat. iv.

Quin, of epicurean notoriety, is said to have given it as his opinion that the flesh on the dark-coloured side of the Turbot was the best meat ; and as examples occasionally occur that are dark-coloured on both sides, some London fishmongers, from experience in their good qualities, recommend such fish as deserving particular attention. Reversed Turbots, as they are called,—that is, Turbots having the eyes and dark colour on the right side instead of the left,—are also occasionally brought to market : I have seen two or three such ; but they have exhibited a slight degree of malformation in the form of a notch or depression on the top of the head. The *Pleuronectes cyclops* of Mr. Donovan, plate 90, I believe to be an example of the young fry of the Turbot, the head of which is not perfectly formed.

The number of Turbot brought to Billingsgate market within twelve months, up to a recent period, was 87,958 ; and the number of lobsters within the same period 1,904,000.

The form of the Turbot, exclusive of the caudal rays, is nearly round : the length of the head compared to the length of the head and body alone is as one to three ; the depth of the body, including both dorsal and anal fins, is equal to the length from the nose to the end of the fleshy portion of the tail : the mouth is large, the direction of the opening obliquely upwards; the teeth small and numerous in both jaws; the eyes in a vertical line one directly over the other ; the whole surface of the cheeks, and all the parts of the gill-cover on the upper or coloured side, studded with numerous tubercles ; the operculum ending in an angle directed backwards and over the base of the pectoral fin ; the gill-openings large; the pectoral fin small ; the dorsal fin, commencing by short rays immediately over the nostril and anterior to the upper eye, extends very nearly to the end of the fleshy portion of the tail, where the rays are again short, the longest rays being over the middle of the body; the ventral fins broad, placed very far forward, appearing like the commencement of the anal fin, and only separated from it by a narrow space; the anal fin ending by short rays near the tail, and on the same vertical plane as the dorsal: the caudal rays moderately long, and slightly rounded. The fin-rays in number are—

D. 64 : P. 12 : V. 6 : A. 48 : C. 15 : Vertebræ 30.

The whole of the upper or coloured side of the body studded with hard roundish tubercles, the surface otherwise smooth; the scales small, the prevailing colour varying shades of brown, the fins a little lighter; the lateral line arched high over the pectoral fin, then straight to the tail; the under surface of the body is smooth, and generally perfectly white.

The vignette represents a Dutch boat.

THE BRILL.

PEARL, KITE, BRETT, BONNET-FLEUK.

Rhombus vulgaris,   *La Barbue,* Cuvier, Règne An. t. ii. p. 341.
  ,, *non aculeatus,* Willughby, p. 95, pl. F. 1.
Pleuronectes rhombus, Linnæus. Bloch, pt. ii. pl. 43.
  ,,   ,,   *Pearl,* Penn. Brit. Zool. vol. iii. p. 321, pl. 50.
  ,,   ,,   *Brill,* Don. Brit. Fish. pl. 97.
  ,,   ,,   ,,   Flem. Brit. An. p. 196, sp. 97.

THE BRILL is a well-known fish, brought in abundance
to the London market, and procured from the same localities
and by the same modes as the Turbot ; but is not held in
equal estimation, being considered by some as inferior to
the Sole, but very superior to the Plaice.

Dr. Neill says it is found in Aberlady Bay, where it is
called Bonnet-Fleuk ; it is taken also at Yarmouth, and
other places along our eastern coast.   It is abundant on our
southern coast, inhabiting sandy bays as well as deep water,

from whence the principal part of the supply for the London market is derived. It has been taken also at Belfast. In its food, as well as in its season of spawning, it is similar to the Turbot, but does not usually appear so large, seldom exceeding eight pounds in weight. It should be borne in mind, that the Kite of the Devonshire and Cornish coasts is the same as the Brill; but that the Kit of Jago is the smooth or small-headed Dab, figured and described in this volume at page 221.

The writer of the supplementary part to the Class Fishes, in Mr. Griffith's edition of Cuvier's Animal Kingdom, says that the enormous fish presented to the Roman Emperor Domitian was a Brill, *Rhombus vulgaris* of Cuvier, and not the Turbot; but the authority or the reasons for this opinion are not given. Bloch, in his account of the Brill, makes a similar statement.

The length of the head from the point of the lower jaw to the edge of the operculum is, when compared to the length of the body alone without the head or caudal rays, as one to two; the breadth of the body, dorsal and anal fins excluded, equal to half the whole length of the entire fish; the whole breadth, dorsal and anal fins included, is to the whole length as two to three: the form of the body rhomboidal; the surface perfectly smooth; the position and extent of the fins very similar to those of the Turbot last described; a few of the most anterior rays of the dorsal fin extend beyond the connecting membrane; the tail rounded.

The fin-rays in number are—

D. 76 : P. 10 : V. 6 : A. 59 : C. 16.   Vertebræ 35.

The mouth is large, deeply cleft; under jaw the longest; teeth numerous, small, pointed, and sharp: the upper eye behind the lower one in a vertical line; irides yellow: cheek and operculum smooth, without tubercles; basal and ascending marginal lines of the preoperculum forming nearly a right angle; lateral line arched over the pectoral fin, then straight to the end of the tail: the scales are nearly round, small, and smooth; the colours of the body a reddish sandy brown, varied with darker brown, and sprinkled over with white pearl-like specks, whence, probably, one of the names bestowed on this fish has originated: the under surface is smooth and white.

The young are of a pale reddish brown, marked with very dark brown or black spots.

The vignette represents the outline of the anterior part of a Brill with a malformed head. For the fish from which this sketch was taken, I am indebted to the kindness of Mrs. Nelson of Devonport. It was taken in that vicinity in June 1835, and was brought on shore alive.

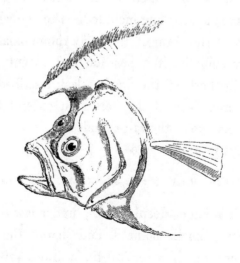

SUBBRACHIAL
MALACOPTERYGII.   *PLEURONECTIDÆ.*

MULLER'S TOPKNOT.

Rhombus hirtus,   *Muller's Topknot,* Yarrell.
Pleuronectes hirtus, Muller, Zool. Dan. vol. iii. p. 36, pl. 103.
    ,,   *punctatus, Topknot,* Penn. Brit. Zool. vol. iii. p. 322, pl. 51, and
                       edit. 1776, pl. 41, but named by mistake
                       Smear Dab.
Le Gros Plie ou Targeur, Duhamel, sect. ix. pl. 5, fig. 4.
Pleuronectes hirtus,   *Muller's Topknot,* Jenyns, Man. Brit. Vert. p. 463,
                       sp. 151.

Several modern authors have confounded the present
fish with the species next to be described ; and Cuvier,* as
well as Professor Nilsson,† have brought together the *Pleu-*
ronectes hirtus of Muller, and the *Pl. punctatus* of Bloch,
apparently considering them as the same fish. Muller
doubted whether the *punctatus* of Bloch was the same as
his fish, and notices the points in which they differ. The
opportunity of examining some specimens very recently,

* Règne An. t. ii. p. 341.   † Prod. Ichth. Scand. p. 59, sp. 11.

confirms the existence of two distinct species on our shores. Though somewhat similar in the form of the body, the colouring, and the spots, there are still the following well-marked distinguishing specific characters.

The *hirtus* of Muller, and those included in the synonymes here given, have the eye or coloured side only of the body rough; the under side smooth; the eyes and mouth small: the first ray of the dorsal fin not longer than the succeeding rays; the ventral and anal fins united; the dorsal and anal fins also connected to the tail by a membrane; the tail short and rounded; the scales of the body when detached higher than wide.

The *punctatus* of Bloch, and the fishes included under the synonymes given with the next species, have both sides of the body rough; the eyes large and prominent; the mouth larger than in *hirtus*, and not placed so obliquely; the first ray of the dorsal fin elongated; the ventral and anal fins separated; the tail rather long; the scales of the body when detached wider than high.

The *hirtus* of Muller appears to be the most common species of the two; but neither occur very frequently. I have received a specimen from Dr. George Johnston, which was taken near Berwick Bay; and I am indebted to Professor Henslow of Cambridge for a drawing of one taken in the Medway. Dr. John Harwood, of St. Leonard's, possesses a specimen taken on the Sussex coast; and both Colonel Montagu and Mr. Hanmer obtained specimens in Devonshire.

Mr. Couch considers it not an uncommon fish in the West of England, and has furnished me with two examples, from one of which the figure was taken. It appears to keep among rocks, where it is not readily distinguished, on account of the similarity in its colour to the seaweed; and it is chiefly taken in the nets which are set for Red Mullet. In winter the boys find small ones, not larger than a half-crown piece, in the pools left by the tide. This species of Flatfish does not probably attain any great size; the largest examples I have seen not exceeding seven or eight inches in length.

Mr. Baker, of Bridgewater, sent me a specimen, beautifully preserved, that had been taken in the Bristol Channel: and I have a record of one that was caught on the coast of the county of Down in Ireland.

The whole length of the specimen described is five inches and one quarter; the length of the head compared to that of the body, without the caudal rays, is as one to two and a half; the breadth of the body, not including the dorsal and anal fins, half of the whole length: the form of the body, including both these fins, is rhomboid: the dorsal fin commences immediately over the upper lip, the rays lengthening by degrees, and being longest over the posterior third part of the body; the pectoral fin small; the ventral fins placed in a vertical line under the middle of the head, and attached to the commencement of the anal fin by a membrane: this latter-named fin commences under the line of the ascending posterior margin of the preoperculum: both dorsal and anal fin end on the same plane, and are connected to the fleshy portion of the tail by a membrane; the tail small and rounded. The fin-rays in number are—

D. 90 : P. 11 : V. 6 : A. 70 : C. 14. Vertebræ 33.

The mouth is small, the position almost vertical; the teeth distinct, small, conical, and sharp: the diameter of the eye equal to one-fourth of the length of the head; the upper eye placed behind the line of the lower to the distance of nearly one-half its width: the basal and ascending marginal lines of the preoperculum form an obtuse angle; the cheeks, operculum, and body, covered with denticulated scales, which in shape, when detached, are longest in their vertical diameter.

The colour of the body is a reddish brown, mottled and spotted with very dark brown or black ; a large, conspicuous dark spot behind, but above the ends of the pectoral fin-rays; the lateral line curved over the pectoral fin, then descending and intersecting the lower portion of the large dark spot, afterwards passing straight to the tail ; the fins paler brown than the body ; all the rays of the dorsal and anal fins with a line or row of denticulated scales along their upper surface; the under side of the body smooth and white.

The vignette represents a fishwoman at Scheveling.

SUBBRACHIAL
MALACOPTERYGII.   *PLEURONECTIDÆ.*

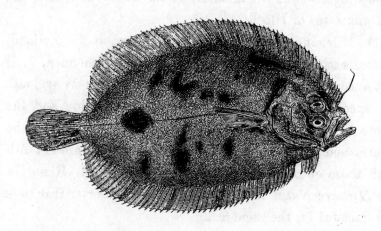

BLOCH'S TOPKNOT.

Rhombus punctatus,   *Bloch's Topknot,* YARRELL.
Pleuronectes ,,   BLOCH, pt. vi. pl. 189.
  ,,   ,,   FLEM. Wern. Mem. vol. ii. p. 241.
  ,,   ,,   ,,   Phil. Zool. pl. 3, fig. 2.
  ,,   ,,   ,,   Brit. An. p. 196, sp. 99.
Rhombus unimaculatus, RISSO, Hist. tom. iii. p. 252, fig. 35.
Pleuronectes punctatus, Bloch's Topknot, JENYNS, Man. Brit. Vert. p. 462,
                          sp. 150.

By an oversight, the use of a looking-glass was omitted when making the drawing of this fish on the wood from the specimen, and the figure therefore appears with the eyes and the colours on the right side, like a *Platessa*, instead of on the left, as in a true *Rhombus*. The fish ought to have been represented with the head placed in the same direction as in the species last described.

The *Pleuronectes punctatus* of Bloch, or, as it is here called to preserve the appropriation, Bloch's Topknot, is much more rare than the Topknot of Muller ; but appears, like it, to have an extended range. Professor Nilsson includes but one species in his Fishes of Scandinavia, but brings together the trivial names of the *hirtus* of Muller and the *punctatus* of Bloch.

Dr. Fleming procured the true *punctatus* in Zetland, where, according to the testimony of the fishermen, it is not uncommon. Professor Henslow obtained at Weymouth the specimen from which Mr. Jenyns' description and the figure here inserted were taken ; and a comparison of the figures and descriptions referred to under the present fish with those of the *Rhombus unimaculatus* of M. Risso, in his *Histoire Naturelle,* will convince the observer that they are intended for the same fish.

Bloch, if he has correctly figured his species, was, I think, mistaken in supposing his fish to be the same as *Le Gros Plie ou Targeur* of Duhamel ; as the separation between the ventral and the anal fins, and the want of connexion between the ends of both dorsal and anal fins with the tail, will demonstrate on comparing the two figures : but the character and disposition of the spots are something like those of Muller's fish. The figure by Dr. Fleming, in his Philosophy of Zoology, wants only the greater elongation of the first ray of the dorsal fin, perhaps a sexual distinction, to render it identical with the figure here given, and that by M. Risso.

I avail myself, by permission, of the very full description of this fish given by Mr. Jenyns in his Manual, taken from the specimen in the collection of the Philosophical Society of Cambridge.

" Length five inches and a half. Form roundish oval ; the dorsal and ventral lines equally convex : greatest breadth, fins excluded, just half the length : head a little less than one-third of the same : profile notched immediately before the eyes : mouth of moderate size, very protractile ; jaws nearly equal ; the lower one a very little the longest, and ascending obliquely at an angle of rather more than forty-five degrees : teeth so fine as to be scarcely visible : eyes large, remarkably full and prominent ; their diameter about one-fourth the length of the head ; placed on the left side ; approximating ; the lower one rather more advanced than the upper ; between them a projecting ridge : basal and posterior margins of the preopercle meeting at a very obtuse angle, the former rising obliquely to meet the latter ; lateral line commencing at the upper part of the opercle, at first very much arched, but afterwards straight : both sides of the body, but more especially the upper, extremely rough : scales minute ; those on the upper side having their free margins set with from four to six denticles ; those beneath having the denticles finer and more numerous : dorsal fin commencing immediately in advance of the upper eye, and extending very nearly to the caudal, at the same time passing underneath the tail, where the rays become very delicate ; greatest elevation of the fin near its retral extremity ; first ray very much produced, nearly three times the length of those which follow ; most of the rays divided at their tips ; some of the last in the fin branched from the bottom : anal fin commencing in a line with the posterior angle of the pre-opercle, answering to the dorsal, and terminating in the same manner beneath the tail ; greatest elevation corresponding : caudal oblong, the extremity rounded : pectorals inserted behind the posterior lobe of the opercle, a little below the middle ; the first ray very short, the next three or four longest, the succeeding ones nearly as long ; pectoral on the eye side rather larger than that on the side opposite : ventral fins immediately before the anal, and appearing like a con-

VOL. II.   S

tinuation of that fin, but not connected with it, as in the other species : vent situated between the two last pairs of rays : the rays of all the fins covered with rough scales nearly to their tips.   The numbers of the fin-rays are—

D. 87 : P. left side 12 : right side 11 : V. 6 : A. 68 : C. 16.

The colour above brown, or reddish brown, mottled and spotted with black ; a large round spot, more conspicuous than the others, in the middle of the side towards the posterior part of the body ; fins spotted : under side plain white.

SUBBRACHIAL
MALACOPTERYGII. PLEURONECTIDÆ.

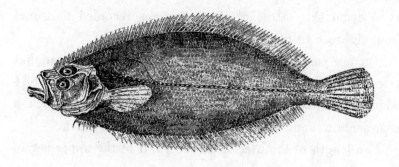

THE WHIFF.

THE CARTER, *Cornwall.*

Rhombus megastoma,	*Whiff,*	YARRELL.
,, ,,	*La Cardine,*	CUVIER, Règne An. t. ii. p. 341.
Passer Cornubiensis,	*Whiff,*	RAY, Syn. p. 163, fig. 2.
Pleuronectes pseudopalus,	,,	PENN. Brit. Zool. vol. iii. p. 324, pl. 52.
,, *megastoma,*	,,	DON. Brit. Fish. pl. 51.
,, ,,	,,	FLEM. Brit. An. p. 196, sp. 98.

THE WHIFF appears to have been first described and figured by Ray from Mr. Jago's Catalogue of Cornish Fishes, which is introduced, with short notices and representations, in Ray's Synopsis. This fish seems to occur more frequently in Cornwall than on any other part of our coast. Mr. Couch says, "This species is well known to the Cornish fishermen, who apply the name of Carter to it. It keeps on sandy ground, at no great distance from land, and takes a bait, so that it is caught as often as any of the salt-water Flatfishes; but it is not highly esteemed for the table, chiefly from being so thin."

s 2

Mr. Donovan found it in Wales; Colonel Montagu obtained two on the Devonshire coast; and Mr. Jenyns has described it in his valuable Manual of the British Vertebrate Animals, from a specimen obtained by Professor Henslow at Weymouth. Most of the specimens recorded measured from eighteen to twenty-one inches in length.

But few particulars are known of this fish. It appears but seldom in the London market: I obtained one in June 1834 which measured seventeen inches in length, from which a representation and the following description are taken.

The length of the head from the point of the upper jaw to the posterior edge of the operculum, compared to the length of the body alone, without the head or caudal rays, is as one to three; the breadth of the body, dorsal and anal fins excluded, is to the whole length of the fish rather less than one third: the dorsal fin commences half-way between the point of the nose and the anterior edge of the upper orbit, and extends to within three-quarters of an inch of the end of the fleshy portion of the tail and the base of the caudal rays; the pectoral fin on the under or white side is considerably smaller in size, and contains two rays less, than that on the upper side; the ventral fins are of some extent at the base, as in the preceding species of the genus *Rhombus;* the anal fin commences in a line under the origin of the pectoral fin, extends along the whole length of the abdominal line, and ends near the tail on the same plane as the dorsal fin; the fleshy portion of the tail is narrow; the caudal rays three inches long, and slightly rounded.

The fin-rays in number are—

D. 89 : P. 11 : V. 6 : A. 71 : C. 13. Vertebræ 41.

The mouth is large; the lower jaw the longest, with a rounded projection under the symphysis; the teeth on both jaws numerous, pointed and sharp : the eyes large ; the upper one the most so, and placed farther back than the lower ; the orbits separated by a prominent bony ridge : the lateral line conspicuous, elevated, and double over the pectoral fin, one portion being a continuation of the prominent straight line along the body, the other rather less conspicuous, taking a high curve over the lower and the pectoral fin ; both lines ultimately approaching each other again at the upper angle of the operculum, as shown in the woodcut : the form of the body is an elongated oval ; the surface rough ; the scales rather large ; the colour a uniform yellow brown ; the fins rather lighter ; the under side smooth and white.   A specimen in the British Museum exhibits faint indications of various spots, as shown in Mr. Donovan's coloured figure.

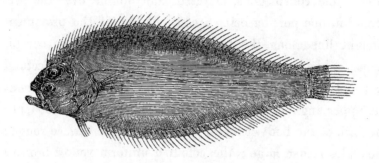

THE SCALDFISH.

MEGRIM, *Cornwall.* SMOOTH SOLE.

Rhombus Arnoglossus,	*Scaldfish,*	YARRELL.
,, ,,	,,	CUVIER, Règne An. t. ii. p. 342.
Arnoglossus lævis,	,,	WILLUGHBY, p. 102, F. 8, fig. 7.
Pleuronectes casurus,	,,	PENN. Brit. Zool. vol. iii. p. 325, pl. 53.
,, *Arnoglossus,*	,,	FLEM. Brit. An. p. 197, sp. 100.
Rhombus nudus,	,,	RISSO, Hist. t. iii. p. 251, sp. 141.

THE SCALDFISH, or MEGRIM, as it is called in Cornwall, appears, like the species last described, to be in this country, as far as we yet know, exclusively confined to the southern coast, and is only at present recorded as having been taken between Weymouth and the Land's End.

Mr. Couch says, " he has never known it take a bait, and its diminutive size prevents its being an object of attention to fishermen ; but they say it is much preyed upon by Congers and other large fishes, in the stomachs of which they often find it : it follows from this that it keeps in deep water." It seldom exceeds four or five inches in length ; but Mr. Couch has seen one that measured six inches and a half. M. Risso says the females are very prolific.

The length of the head is to that of the body as one to three, caudal rays excluded ; the depth of the body, without the dorsal or anal fins, equal to one-third of the whole length ; the dorsal fin commences over the upper eye, and reaches very nearly to the end of the fleshy portion of the tail; the pectoral fin long and narrow, but shorter and smaller on the under side ; ventral fins under the gill-cover ; the anal fin commencing in a line under the pectoral, and ending near the tail on the same plane as the dorsal fin ; caudal rays of moderate length, and slightly rounded ; but the rays of all the fins in both the specimens before me, from which the description is taken, extend considerably beyond the connecting membranes of each, as shown in the woodcut.

The fin-rays in number are—

D. 87 : P. 6 : V. 10 : A. 60 : C. 18.

The mouth is large, with small teeth in both jaws ; lower jaw the longest when separated : eyes rather large ; pupils blue ; irides yellow ; orbits separated by a bony ridge ; upper eye larger than the lower, and placed more backward in a vertical line : body in shape an elongated oval, narrowed towards the tail ; the scales large, round, thin, and transparent, almost all wanting, so easily are they removed on the slightest touch ; the body of the fish appears naked. I am indebted to the kindness of Mr. Couch for a Cornish specimen : I also possess one from the Mediterranean, which enables me to say that our fish is the *Rhombus nudus* of M. Risso, as quoted. The lateral line after its commencement at the posterior edge of the operculum rises slightly over the pectoral fin ; then descending gradually, deviates but little from a straight line throughout the remainder of its course to the tail. The colour of both specimens is alike, a uniform pale yellow brown.

SUBBRACHIAL
MALACOPTERYGII.   *PLEURONECTIDÆ.*

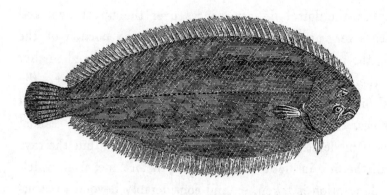

THE SOLE.

Solea vulgaris,	*La Sole,*	CUVIER, Règne An. t. ii. p. 342.
,, ,,	*A Sole,*	MERRETT, Pinax, p. 187.
Buglossus seu Solea,	*Sole,*	WILLUGHBY, p. 100, F. 7.
Pleuronectes Solea,	,,	LINNÆUS. BLOCH, pt. ii. pl. 45.
,, ,,	,,	PENN. Brit. Zool. vol. iii. p. 311.
,, ,,	*Common Sole,*	DON. Brit. Fish. pl. 62.
Solea vulgaris,	*Sole,*	FLEM. Brit. An. p. 197, sp. 101.

Generic Characters.—Both eyes and colour on the right side ; the mouth distorted on the side opposite the eyes ; small teeth in both jaws, but confined to the under side only, none on the same side as the eyes ; form of the body oblong ; dorsal and anal fins extend to the tail.

THE common Sole is so universally known as to require only a particular notice of those points in its economy that are the least obvious. It inhabits the sandy shore all round our coast, where it keeps close to the bottom, feeding on the smaller testaceous animals, and the spawn and fry of other fishes. It is taken among the Orkneys, and along the north-east coast ; but it is of small comparative size : the Soles of the south and west are much larger, and considered otherwise superior to those of the north and east.

The Sole is found northward as far as the Baltic and the

seas of Scandinavia; and southward, along the shores of Spain, Portugal, and the Mediterranean. It was first described by Bellon.

Soles—and of these an enormous quantity—are caught almost entirely by trawling; they seldom take any bait. It is usual to send them to market in baskets, within which the Soles of small size, called Slips, are arranged nearest the wicker-work forming the outside of the basket: the larger Soles, being more valuable, are packed in the middle. Eighty-six thousand bushels of Soles were received at Billingsgate market only within the last twelvemonths.

The Sole is found full of roe at the latter end of February. They are then for a few weeks soft and watery; but they soon recover, and throughout a great portion of the year are deservedly in high estimation: the flesh is white, firm, and of excellent flavour; those from deep water are the finest in quality.

The principal trawling-ground in England is along the south coast from Sussex to Devonshire: the Sole has also been taken on the shores of various counties in Ireland, viz. Cork, Waterford, Antrim, Londonderry, and Donegal. On the Devonshire coast there are two great fishing stations, Brixham and Torbay; the boats from which, using large trawling-nets from thirty to thirty-six feet in beam, produce a continual supply. Soles of very large size are occasionally taken. I have a record of one pair taken in Torbay which measured twenty-three inches in length each, and weighed together ten pounds; but for the particulars of the largest I have heard of, I am indebted to the kindness of the Rev. W. F. Cornish, of Totness. This specimen, a remarkably fine-grown fish, and very thick, was twenty-six inches long, eleven inches and a half wide, and weighed nine pounds.— Totness market, June 21st, 1826.

Soles appear to thrive well in fresh water. Dr. M'Culloch, in his papers on " Changing the Residence of certain Fishes from salt water to fresh,"* says, he was informed that a Sole had been kept in a fresh-water pond in a garden for many years ; and adds, that in Mr. Arnold's pond at Guernsey, which has been before referred to, the Sole becomes twice as thick as a fish of the same length from the sea. A letter from a gentleman residing on the banks of the Arun contains the following statement :—" I succeeded yesterday in seeing the person who caught the Soles about which you inquire, and who has been in the constant habit of trawling for them with a ten-feet beam trawl in this river for the last forty years. The season for taking them is from May till November. They breed in the river (Arun), frequenting it from the mouth five miles upwards,† which is nearly to the town of Arundel, and remain in it the whole year, burying themselves in the sand during the cold months. The fisherman has occasionally taken them of large size, two pounds' weight each, but frequently of one pound ; and they are thicker in proportion than the Soles usually caught at sea. In other respects, precisely the same ; and it is evident they breed in great numbers in the river from the quantity of small ones about two inches long that are constantly brought on shore when drawing the net for Grey Mullet."

Reversed Soles—that is, having the eyes and the brown colour on the left side instead of the right—are not uncommon : and I possess a specimen that is of the usual dark colour, with rough ciliated scales on both sides.

The length of the head is to the whole length of the entire fish as one to six ; the breadth of the body, dorsal and anal

* Royal Institution Quarterly Journal, No. xxxiv. July 1824, and No. xxxviii. July 1825.

† For a view of this part of the Arun, see vol. i. page 209.

fins excluded, compared to the whole length, as one to three : the nose is rounded and produced, projecting beyond the mouth : the upper jaw the longest ; both jaws furnished with minute teeth on the under or white side of the fish only : the eyes small ; the lower eye over the angle of the mouth ; the upper eye placed more forward in a vertical line ; the irides yellow ; the pupils blue : the space between the eyes, the cheek, and operculum, flat, and covered with small rough ciliated scales : the pectoral and ventral fins small ; the dorsal fin begins at the point of the nose, the anal fin under the line of the edge of the gill-cover ; both extend the whole length of the body, ending on the same plane, near the base of the caudal rays ; and both these fins have a series of small, rough, ciliated scales, extending along the line of each ray : the tail rather long, and slightly rounded.

The fin-rays in number are—

D. 84 : P. 7 : V. 5 : A. 67 : C. 17.   Vertebræ 47.

The form of the body is a long oval, widest at a short distance behind the head, becoming narrower and rather pointed towards the tail; the colour on the upper side almost a uniform dark brown ; the scales small, each ciliated at the edge, and rough to the touch ; the lateral line running straight from the tail forward to the operculum, then rising and ending on a line with the superior edge of the upper orbit ; the pectoral fin tipped with black.   On the under side the colour is white : about the nostril and mouth are numerous soft papillæ.

SUBBRACHIAL
MALACOPTERYGII.   *PLEURONECTIDÆ.*

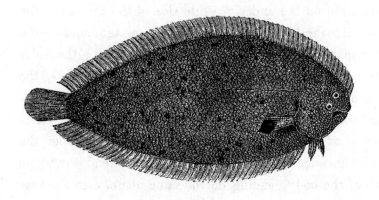

THE LEMON SOLE.

THE FRENCH SOLE, *Sussex coast.*

Solea pegusa, Lemon Sole, Yarrell, Zool. Journ. vol. iv. p. 467, pl. 16.
  ,,   ,,   ,,   ,,   Jenyns, Man. Brit. Vert. p. 467, sp. 155.

During a short visit to Brighton in the last week of February 1829, I obtained a single example of this species of Sole, which appeared to have been previously unnoticed as occurring on our shores. Since it was described in the Zoological Journal, as above quoted, I have obtained, but at considerable intervals, two or three other specimens of this fish in the London market, and have now deposited examples in the British Museum and the collection of the Zoological Society.

This species is occasionally taken with the common Sole when trawling over a clear bottom of soft sand, about sixteen miles from Brighton in a direction towards the coast of France; from which circumstance this fish is known to some of our fishermen by the name of French Sole ; others call it by that of Lemon Sole, in reference to its prevailing yellow-ish colour.

In shape the Lemon Sole is wider in proportion to its whole length than the common Sole, and it is also somewhat thicker; the head is smaller, being in proportion to the whole length of the fish rather less than as one to seven; the breadth of the body, dorsal and anal fins included, exactly half the whole length : the arrangement of the fins is nearly the same as in the common Sole; but the fin-rays and the number of vertebræ differ.

D. 81 : P. 8 : V. 5 : A. 69 : C. 17. Vertebræ 43.

The prevailing colour is a mixture of orange and light brown, freckled over with numerous small round spots of dark nutmeg brown, giving a mottled appearance to the whole upper surface. The scales differ in character; the lateral line is straight, but not so prominent or strongly marked; the tail is narrower than in the common Sole, though containing the same number of rays; the end of the pectoral fin spotted with black. On the under side the appearance is still more characteristic of the distinction of the species. The under surface of the head is almost smooth, without any of the papillary eminences so numerous and remarkable in the common Sole, and the nostril is pierced in a prominent tubular projection, which is wanting in the other : the under surface is white, with the appearance of the scales more strongly marked than upon the upper.

SUBBRACHIAL
MALACOPTERYGII.   *PLEURONECTIDÆ.*

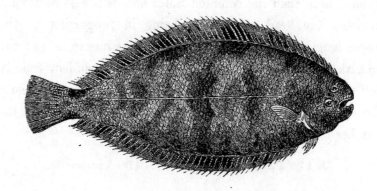

THE VARIEGATED SOLE.

Monochirus linguatulus,	Cuvier, Règne An. t. ii. p. 343.
Solea parva sive lingula,	Rondeletius.
,, ,, ,, ,,	Willughby, p. 102, F. 8, fig. 1.
Pole panachée,	Duhamel, sect. ix. pl. 2.
Pleuronectes lingula,	*Redbacked Flounder,* Penn. Brit. Zool. vol. iii. p. 313, pl. 49.
,, *variegatus,*	*Variegated Sole,* Don. Brit. Fish. pl. 117.
Solea variegata,	,, ,, Flem. Brit. An. p. 197, sp. 102.

Generic Characters.—The pectoral fin on the upper or eye side small ; that on the under side minute, almost imperceptible, or entirely wanting : in other respects like *Solea.*

The Variegated Sole appears, like the Lemon Sole last described, to be a rare species, and but few specimens are to be found in collections, though it seems to have a wide range. According to Professor Reinhardt, it is found on the shores of Scandinavia. In the Magazine of Natural History, conducted by Mr. Loudon, a notice appears, vol. vi. page 530, that it has been taken at Rothsay. Mr. Donovan obtained one seven inches long in the London market ; and Mr. Jenyns has described this fish in his Manual from a specimen procured at Weymouth. Colonel Montagu mentions that Dr. Leach bought three in Plymouth market in August 1808, and gave him one of them, measuring nine inches in length, from which his notes of the species were recorded in his MS.; and Mr. Couch has observed it in Cornwall, very kindly sending me a specimen, from which the woodcut was executed. But little is known of the habits of this species ; but it is stated in Pennant that it appears about Plymouth in the spring.

It is immediately distinguished from either of the Soles previously described here, by its variegated colour ; by its scales, which are larger ; by its pectoral fins, which are much smaller, that on the under side being very minute ; and by the dorsal and anal fins, as shown in the cut, ending considerably short of the tail.

The whole length of the specimen described was five inches ; the breadth without the fins, one inch and three-eighths ; the length of the head compared to that of the body alone, as one to four : the dorsal and anal fins ending on the same plane, but not reaching the base of the caudal rays, and both having the numerous rays projecting beyond the edges of the connecting membranes, as shown in the cut ; the right pectoral fin small, that on the under side consisting of only two unequal, slender, and short rays.

The fin-rays in number are—

D. 67 : P. right side 4 : left side 2 : V. 5 : A. 52 : C. 16.

The body is thicker in proportion than either of the Soles previously described ; the scales larger, the divisions strongly marked, the edges ciliated, rough to the touch ; the lateral line straight : the colour of the upper side reddish brown, clouded both on the body and fins with darker brown ; the under surface white ; scales also ciliated and harsh to the touch.

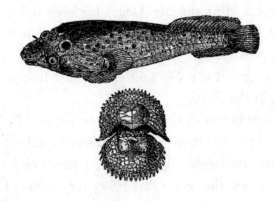

THE CORNISH SUCKER.

Lepidogaster Cornubiensis,　Cornish Sucker,　FLEM. Brit. An. p. 189, sp. 71.
Cyclopterus Lepidogaster,　Jura Sucker,　　PENN. Brit. Zool. vol. iii. p. 181,
　　　　　　　　　　　　　　　　　　　　　　　pl. 25.
　　,,　　ocellatus,　　Ocellated Sucker, DON. Brit. Fish. pl. 76.
Lepidogaster biciliatus,　　　　,,　　　,,　　RISSO, Hist. tom. iii. p. 272,
　　　　　　　　　　　　　　　　　　　　　　　sp. 163.
　　,,　　Cornubiensis,　Cornish Sucker, JENYNS, Man. Brit. Vert. p. 469,
　　　　　　　　　　　　　　　　　　　　　　　sp. 157.

Generic Characters.—Body smooth, without scales; dorsal and anal fins opposite, and near the tail; pectoral fins large, descending to the inferior surface of the body, and by an extension of the membrane surrounding an oval disk; ventral fins united by a membrane which extends circularly under the belly, forming a second concave disk.

BARON CUVIER has called the third family of the Subbrachial Malacopterygii, *Discoboles,* on account of the disk formed by the union of the ventral fins. The term CYCLOPTERIDÆ, derived from an original Linnæan generic name, is here adopted in order to preserve uniformity in the names of

* The family of the Sucking-fishes.

the families. The pectoral fins in these fishes are large,
descending to the inferior surface of the body, where they are
joined by four strong rays, and, united by a membrane to a
similar structure on the opposite side, form the boundary of
an adhesive disk. In the species of the first genus, a second
disk is formed by the union of the ventral fins.

The few species belonging to this small family are very
remarkable for the power they possess of attaching themselves to stones, rocks, or other substances, by means of the
adhesive apparatus on the under surface of their bodies, apparently deriving some degree of protection and support from
the contact.

The two British species belonging to the first genus are
small, defenceless, their bodies smooth, without scales ; and
the power of attaching themselves to stones, &c. which they
are seen to exercise, may be useful by enabling them to resist
the action of strong currents or dashing waves, and is perhaps
applicable with them to other uses, with which naturalists are
not yet acquainted.

The first prettily-marked species of Sucking-fish was discovered by Dr. Borlase, who found it on the coast of Cornwall, and described it under the name of the Lesser-Sucking-
fish, in his Natural History of that county. Pennant
afterwards found it at Jura, in the Hebrides, and called it in
consequence the Jura Sucker ; but if any name indicative of
a peculiar geographical locality is admissible, it ought to have
been that only in which it was first discovered ; and I have
therefore followed Dr. Fleming and Mr. Jenyns in calling it
the Cornish Sucker, although this name is not entirely free from
objection, two other species of fishes, provided with suckers,
being found in Cornwall. Mr. Couch says, however, that this
fish is there called pre-eminently the Sucking-fish by fishermen, from the readiness with which it adheres to any substance,

VOL. II.   T and even to the hand that seizes it,—a circumstance which has also been noticed by Colonel Montagu. " It is sluggish in its habits ; but seems to wander, since it is sometimes abundant, and at others rare. Its usual haunts are about low-water mark, where it is often left by the tide, concealed beneath a stone. I find it," says Mr. Couch, " large with spawn in March. Its food is crustaceous animals and marine insects, which it swallows entire."

The whole length of the specimen described was two inches and a half ; the distance from the point of the nose to the end of the gill-cover was equal to one-third of the whole length of the fish : the head depressed ; mouth produced ; very much flattened ; narrower than the head ; has been aptly called spatula-like ; gape elongated : numerous small teeth in both jaws, forming a band in each : under surface of the head very flat ; the first disk before the line of the opening of the gill-cover ; the second disk behind it : upper surface of the head smooth ; before the inner corner of each eye a small flattened filament, about equal in length to the diameter of the eye itself ; behind this a second, but much shorter ; both of a bright carmine colour ; behind the eyes, which are widely separated, are two distinct, red, eye-like spots : the dorsal fin commences about half-way between the eyes and the end of the tail ; the anal fin begins still nearer the tail, and both are joined to it by a membrane ; the tail rounded ; the posterior part of the body compressed. The pectoral fin large, with an extension underneath of four stronger rays, which with the connecting membrane form the sides of the most anterior disk of the two ; an extension of the membrane only, without rays, being continued along the front. Immediately behind the broad swimming portion of the pectoral fin on each side, a membrane arises in the same vertical position, which joining the united ventral fins forms the free edge of the second disk, the rays of the two ventrals occupying the posterior portion, and the continuation of the connecting membrane making the circle entire.

The fin-rays in number are—

D. 18 : P. 19 : A. 10 : C. 18.

The general tint a pale flesh colour with spots and patches of carmine about the upper and under surface of the jaws, around the eyes, on the top of the head, sides of the body and abdomen. The description was taken from the largest of five specimens, on three of which the spots behind the eyes were conspicuous, but wanting in the other two.

The appearance of the surface of the disk is shown in the woodcuts of some of the more closely allied species, to assist in determining specific distinction.

The vignette below represents a man fishing for prawns on a rocky coast. The fisherman deposits around him eight or ten hoop-nets, each baited with a piece of stale fish : a large bung by way of a buoy is attached to each hoop. The man, with a long forked stick raises the nets in succession, by putting the fork of the stick under the bung, and deposits them again after examination.

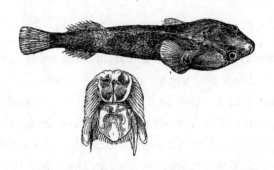

THE BIMACULATED SUCKER.

Lepidogaster bimaculatus, Bimaculated Sucker, Flem. Brit. An. p. 190, sp. 72.
Cyclopterus   ,,   ,,   ,,   Penn. Brit. Zool. vol. iii. p. 182,
                                                pl. 25.
    ,,   ,,   ,,   ,,   Don. Brit. Fish. pl. 78.
    ,,   ,,   ,,   ,,   Montagu, Linn. Trans. vol. vii.
                                                  p. 293.
Lepidogaster   ,,   ,,   ,,   Jenyns, Man. Brit. Vert. p.
                                                  470, sp. 158.

THIS very distinct species was first described by Pennant from a specimen sent him by the Duchess of Portland, which was taken at Weymouth. It has since been taken by Mr. Donovan on the coast of Kent; by Professor Henslow at Weymouth; by Colonel Montagu in Devonshire, and at two different localities in Cornwall, Polperro and Penzance. It has also been taken by Mr. William Thompson of Belfast, when dredging for shells on the coast of the county of Down in Ireland.

Colonel Montagu obtained it by deep dredging at Torcross, adhering to stones and old shells, and kept some specimens alive for a day or two in a glass of sea-water.

" In this situation they always adhered to the sides of the glass by the apparatus termed the sucker, and frequently remained fixed till they died; and even after death the power of adhesion continues; the wet finger being applied to the part, the fish becomes suspended: when alive they instantly attach themselves to the hand if taken out of the water."

Mr. Couch says it keeps in deeper water than the preceding species; but is occasionally found under stones at low-water mark.

In this species, of which I possess several examples, varying in length from three-quarters of an inch to one inch and three-quarters in length, the head is depressed; the posterior portion of the body compressed; the head is shorter, compared to the whole length, than in the preceding species: the mouth wider; but the jaws not so much produced; the teeth similar; no filaments before the eyes; the irides pink and gold; the pupils blue: the additional rays at the inferior part of the pectoral fin, and the connecting membrane on each side, making up the lateral portions of the anterior disk, are much longer: the ventral fins form the sides of the second or posterior disk, and are also elongated; the dorsal and anal fins of equal size, opposite, short, placed far back; commencing and ending on the same planes: not connected with the tail, between which and the two fins just named there is a considerable space: tail rather elongated.

The fin-rays in number are—

D. 6 : P. 19 : A. 6 : C. 10.

The general colour carmine red; pale flesh colour underneath, with a light-coloured patch between the eyes, and otherwise liable to some variation in the markings; the two spots on the sides not always very obvious; young specimens are without these lateral markings.

THE LUMP SUCKER.

SEA-OWL, *and* COCK PADDLE.

Cyclopterus lumpus,	LINNÆUS. BLOCH, pt. iii. pl. 90.	
,, ,,	CUVIER, Règne An. t. ii. p. 346.	
Lumpus Anglorum,	WILLUGHBY, p. 208, N. 11.	
Cyclopterus lumpus,	*Lump Sucker,* PENN. Brit. Zool. vol. iii. p. 176, pl. 24.	
,, ,, ·	,, ,, DON. Brit. Fish. pl. 10.	
,, ,,	*Lump-fish,* FLEM. Brit. An. p. 190, sp. 75.	

Generic Characters.—Head and body deep, thick, and short ; back with an elevated ridge, the investing skin enclosing simple rays ; pectoral fins uniting under the throat, and with the ventrals forming a single disk.

THE LUMP SUCKER is remarkable for its very grotesque form, while from the large size of its body, both as to depth and thickness in reference to its length, and the comparatively small size of its fins, it appears calculated to make but slow progress through the water.

It is more plentiful northward than on our southern coast, and beyond this country has a most extensive range. Pennant includes it in his Arctic Zoology. It is caught on the coast of Greenland, where it is eaten. Professors Nilsson and Reinhardt include it among the fishes of Scandinavia; and Mr. Low considers it common in the Orkneys. Dr. Neill says that in the spring months it is caught on the sands of Portobello, and sent for sale to the Edinburgh market, where it is purchased for table, and the male fish considered superior to the female. Along our eastern and southern coasts it is also taken more exclusively during spring, when it approaches the shore for the purpose of depositing its spawn, which happens in April or the beginning of May. This species has also been taken at Belfast; and the Lump Sucker of the North American shores is apparently identical with our own.

Some of our fishermen consider that we have on our coast two species of Lump-fish, which they distinguish by the names of Red-Lump and Blue-Lump, considering the first only as eatable; but the difference in colour, and also in the quality of the flesh, is only the effect of season; the fine external colour, as well as the firmness of the flesh, being lost to the fish for a time by the exhausting process of spawning; it is then by them considered as the worthless Blue-Lump. The ova forming the hard roe are of large size, and of a fine orange colour. The young are four inches and a half long, and three inches in height, by the end of November. Shaw's specimen, of six inches in length, to which he attached the specific name of *Pavonina*, is only a young fish of our common species, which for want of sufficient age had not attained its perfect colour. As the Lump-fish is retentive of life, its power of adhesion is sometimes made the subject of experiment. Pennant says, " That on placing a fish of this species, just caught, into a pail of water, it fixed itself so firmly to the bottom, that on taking it by the tail, the whole pail by that means was lifted, though it held some gallons, and that without removing the fish from its hold."

The Lump-Sucker feeds principally on young fish, of which it devours a large quantity. Mr. Couch says that it sometimes takes a bait, and he has found in its stomach various *onisci*.

In the month of March the colours of the Lump-fish are in the highest perfection, combining various shades of blue, purple, and rich orange ; it is then frequently to be seen in the shops of London fishmongers, suspended by the middle of the back, attracting attention from the combination of singular form and brilliant colours.

A specimen sixteen inches long is usually about eight inches deep, and four inches wide : the length of the head is about one-fourth of the whole length of the fish ; the descending line of the profile of the head is abrupt ; the back highly arched and somewhat compressed, forming a ridge, with a row of tubercles along the upper edge ; on cutting through the integument, the ridge is found to be supported by several rays, which sometimes from abrasion of the hard skin appear externally, and have been considered as bearing some resemblance to an anterior dorsal fin. Behind this central ridge, and over the last third portion of the curve of the dorsal line, is the true dorsal fin, the length of the base of which is about equal to the length of the longest of its rays ; the pectoral fins descend low on the sides, and passing forwards enclose the adhesive apparatus which extends anteriorly to the edge of the membrane connecting the branchiostegous rays, and backwards as far in a vertical line as the posterior angle of the operculum : the union of the ventral fins complete the single disk of the only species of this genus that inhabits our seas. The anal fin is under or opposed to the dorsal, and of nearly the same size and shape : the tail moderate.

The fin-rays in number are—

D. 11 : P. 20 : A. 9 : C. 10.

Each of the rays with a row of hard tubercles along a considerable portion of their length. The whole surface of the head and body is covered with small bony tubercles, most of which are more or less stellated in form. Along several parts of the body are rows of larger and more prominent tubercles, with surfaces minutely granulated ; one row occupies the central ridge of a portion of the back ; two or three tubercles are placed on each side just in advance of the dorsal fin ; one long row extends from the upper angle of the operculum in a straight line to the upper part of the end of the fleshy portion of the tail ; a second long row reaches from the space above the pectoral fin to the lower part of the fleshy portion of the tail ; another row of large size extends along the abdomen on each side as far as the commencement of the anal fin.

The mouth is wide ; the lips fleshy ; the lower jaw the longest : a band of short and small teeth in each jaw ; a small patch of rounded teeth on the root of the tongue, with others at the pharynx : the irides a fine red ; the colour of the sides of the head and body, and all the upper parts, varying shades of dark blue, lighter blue, and purple ; the lips, under surface of the head and body, fine rich orange ; all the fins tinged with the same colour. After the season of spawning is over, much of the brilliant colouring is lost for a time.

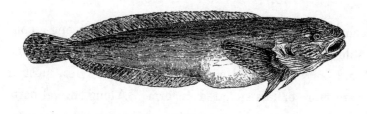

THE UNCTUOUS SUCKER,

OR, SEA-SNAIL.

Liparis vulgaris,   *Sea Snail,* FLEM. Brit. An. p. 190, sp. 73.
    ,,   ,,   CUVIER, Règne An. t. ii. p. 346.
    ,,   *nostras, Sea Snail,* WILLUGHBY, App. p. 17, H. 6, fig. 1.
Cyclopterus liparis, LINNÆUS.   BLOCH, pt. iv. pl. 123, fig. 3.
    ,,   ,,   *Unctuous Sucker,* PENN. Brit. Zool. vol. iii. p. 179, pl. 24.
    ,,   ,,   ,,   ,,   DON. Brit. Fish. pl. 47.

Generic Characters.—Body without scales, smooth, elongated, compressed posteriorly ; a single dorsal fin rather lengthened : ventral fins united to the pectorals, and surrounding a single disk.

THE UNCTUOUS SUCKER, or SEA-SNAIL, so called from the soft and slimy surface of its body, appears to be much more common in the northern parts of the British Islands than in the southern. Mr. Scoresby, and other observers, have even found it as far north as Greenland ; and specimens of it were taken in the trawl-net on the west coast of Davis's Straights during the first Arctic voyage of Captain Sir Edward Parry ; yet it does not appear to be mentioned by Professor Nilsson or Reinhardt in their accounts of the fishes of the Scandinavian shores ; nor is it included by Linnæus in his *Fauna Suecica.*

Mr. Low says, " The Sea-Snail is found under the stones of many places of Orkney ; but no place more frequent than that at the point of the Ness of Stromness, where they may be picked up by dozens."

Mr. Donovan obtained a specimen from among a parcel of Sprats at Billingsgate fish-market ; and those who recollect the wholesale mode of fishing for Sprats practised by the Stow-boatmen, as described at page 123, will not be surprised that many rare and curious fishes of small size are caught with the Sprats. It is also obtained on the southern coast, under stones, and in small pools of water left by the ebbing tide. Dr. Mac Culloch says this species ascends rivers from the sea to deposit its spawn, and it is frequently found near the mouths of rivers. Pennant says it is full of spawn in January, and the matured ova are said to be very large. It feeds on aquatic insects, testaceous animals, and very small fishes.

The whole length of the specimen described was four inches, which is the common size of the adult of this species ; but it is said to grow much larger in the Northern Seas : the head is about one-fourth of the whole length of the fish ; the eyes widely separated, the space between them depressed ; the nose blunt ; the lips thick and fleshy ; the mouth wide, but not deeply divided. Mr. Low says it has no teeth ; but this is an oversight ; the teeth are very numerous, and small, with minutely recurved points, forming a broad rasp-like band in each jaw ; the tongue also broad, covered with prominent papillæ ; the lower jaw rather the longest ; the gill-opening placed high up ; the form from the shoulder is compressed, and tapering all the way to the tail ; the body invested with a thin semi-transparent membrane, which encloses it like a bag, the fixed points being the lines of the dorsal and anal fins ; the pectoral fins are large, and the lower portions descending the side are attached to additional rays like ventral fins, which extending far forward are situated exterior to the sides of the adhesive disk ; the belly tumid ; the dorsal fin begins much nearer the head than the anal fin, and both end close to the tail ; the caudal rays rather long and narrow.   The fin-rays in number are—

D. 36 : P. & V. 32 : A. 26 : C. 12.

The colour of the body is a pale brown, irregularly striped with lines of a darker colour, which take different directions, and give a variegated appearance to the head, back, and sides ; these lines are confined to the outer thin skin, and do not appear upon the more solid surface underneath ; in this state some authors have called this species *lineatus ;* but these markings are not constant, and many examples are without any streaks or lines, the edges of the dorsal and anal fins only being edged with a darker colour ; the tail, and sometimes the pectoral fins, slightly barred and spotted. When kept in diluted spirit of wine, the coloured lines and characters of the species may be easily preserved ; but this fish loses both markings and size if allowed to become dry.

SUBBRACHIAL
MALACOPTERYGII.   *CYCLOPTERIDÆ.*

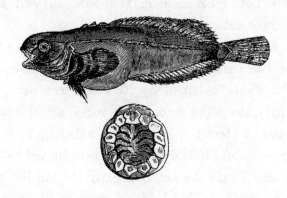

MONTAGU'S SUCKING-FISH.

DIMINUTIVE SUCKER.

Liparis Montagui,   *Montagu's Sucker,*   FLEM. Brit. An. p. 190, sp. 74.
   ,,   ,,   CUVIER, Règne An. t. ii. p. 346, note 2.
Cyclopterus Montagui, Diminutive Sucker, MONTAGU, Wern. Mem. vol. i.
                                        p. 91, pl. 5.
   ,,   ,,   ,,   ,,   DON. Brit. Fish. pl. 68.
   ,,   ,,   *Montagu's Sucker,* PENN. Brit. Zool. vol. iii. p. 183.

THIS species of Sucking-fish, smaller in size than the one
last described, was first discovered by Colonel Montagu. A
drawing of it was sent by that excellent observer to Mr.
Donovan, who was then publishing his Natural History of
British Fishes, and with whom the specific name, referring to
Colonel Montagu, originated. The first specimen obtained
was of very diminutive size. Subsequently Colonel Montagu
having acquired various other larger and adult specimens,
published a description and figure of this species himself in
the Memoirs of the Wernerian Natural History Society, as
already quoted.

This fish has since that period been found on various parts of the coast. Dr. George Johnston has obtained it in Berwick Bay; Mr. Thompson has taken it in Belfast Bay; and it is not uncommon in Cornwall, as well as on the Devonshire coast.

Colonel Montagu says this species inhabits only the rocky parts of the coast, and of course is rarely taken with the dredge. Those obtained by its discoverer were found at exceedingly low tides among the rocks at Milton, on the south coast of Devon. When it is adhering to a rock the posterior part of the body is frequently turned to one side, nearly parallel with the anterior part, the tail being brought close to the head. This habit of curving its body has been observed by all those who have found this species.

Mr. Couch's notice of it in his MS. is as follows:—" This is a common species in the West of England, where, however, it seems to wander, since at certain times it is much more rare than at others. It possesses considerable activity; and when the tide has ebbed it is often found concealed beneath a stone, where when at rest it usually throws the tail forwards towards the head. I have never seen it adhere to any fixed substance. The young come to life in September."

Montagu's Sucker, in the adult state, is from two inches and a half to three inches long: the body is rounded as far as the vent; the posterior end somewhat compressed; head broad, a little depressed, and inflated about the gills; mouth moderately large; both jaws armed with several rows of minute teeth: eyes small, and placed high; irides golden; pupils dark blue, with a single blue line descending from the eye to the angle of the mouth: the operculum angular; the branchiostegous membrane transparent; the pectoral and ventral fins unite; the first is rounded; in the last, four or five rays on each side invest the adhesive disk, which is single, small, and circular : an enlarged representation of the Sucker is here added to assist in affording the means of determining the species : the belly is very tumid ; the vent far removed behind the sucker. The dorsal fin commences farther from the head than in the last species ; the most anterior rays short, but gradually increasing in length form a broad fin towards the tail, where it is rounded : the anal fin shorter than the dorsal. The fin-rays in number are—

D. 26 : P. & V. 29 : A. 24 : C. 12.

This description is partly obtained from Montagu's paper.

The prevailing colour is a dull orange, varied with occasional bluish tints ; the fins brighter orange red ; the lateral line perceptible by a lighter-coloured streak ; the under parts of the body, and about the throat and sucker, white, tinged with flesh colour.

THE COMMON REMORA.

Echeneis remora,	*Sucking-fish,*	Turton, Brit. Faun. p. 94, sp. 38.
,,	,,	*Le Remora,*	Bloch, pt. v. pl. 172.
,,	,,	,,	Cuvier, Règne An. t. ii. p. 347.
,,	,,	*Mediterranean Remora,*	Penn. Brit. Zool. vol. iii. App. p. 524.
,,	,,	*Common Remora,*	Jenyns, Man. Brit. Vert. p. 473, sp. 162.

Generic Characters.—Body elongated, covered with very small scales; a single dorsal fin placed opposite the anal; the head very flat, covered with an oval disk formed by numerous transverse cartilaginous plates, the edges of which are directed backward; the mouth wide, with numerous small recurved teeth on both jaws, the tongue, and the vomer.

Dr. Turton in his British Fauna includes this species of Sucking-fish, having taken a specimen himself at Swansea from the back of a Codfish in the summer of 1806.

The species of this singular family are not numerous: Cuvier enumerates but four that are as yet made known, and another large West Indian one has been more recently described. They are immediately recognized by the flattened, oval, adhesive disk, on the top of the head, by means of which they are able to attach themselves firmly to the surface of other fishes, or the bottoms of vessels ; but whether for protection or conveyance, or both, is a question which has not been satisfactorily ascertained.

The Greeks and Romans were well acquainted with the Mediterranean species, which is the fish under present consideration.

The length of the head, from the end of the upper jaw, which is much the shortest, to the end of the operculum, is nearly one-fifth of the whole length of the fish ; the depth of the body about one-seventh of the whole length : the form of the head is flattened, very much depressed ; the body about the middle nearly round in form, the posterior half compressed : the mouth is wide ; the opening nearly horizontal, with two bands of minute teeth in the elongated lower jaw, a single band on the upper jaw, with others on the tongue and vomer, all curving inwards : the eye placed about half-way between the point of the upper jaw and the rounded end of the operculum ; the gill-aperture very large ; the adhesive disk in this species contains seventeen or eighteen transverse laminæ, divided by a longitudinal mesial ridge ; the disk commences just behind and above the upper lip, and extends nearly as far back as the line of the ends of the pectoral fin-rays : all the fins are covered with a dense membrane, which imparts to them the consistence of leather ; the pectoral fins are rather small and rounded ; the ventrals narrow, very close together, the inner ray of each attached to the central line of the belly by a membrane ; the dorsal and anal fins are both placed behind the mid-length of the fish, beginning and ending on the same plane ; the end of the caudal rays crescent-shaped.

VOL. II.   U

The fin-rays in number are—

D. 21 : P. 22 : V. 4 : A. 20 : C. 20.

The colour is dusky brown ; the under part of the body rather lighter than the back ; the fins darker in colour than the body.

The disk of the adhesive apparatus in the specimen now described with seventeen transverse laminæ was one-third of the whole length of the fish, not including the caudal rays ; the breadth one inch and one quarter. The figure on the left side of the vignette represents the outer surface of the anterior half of the disk : the margin is free, flexible, and of considerable breadth, to secure perfect contact with the surface to which it is opposed ; the parallel laminæ are represented as only slightly elevated ; the degree of adhesion is in proportion to the power used to raise the inner surface of the disk in a direction perpendicular to the plane of contact. The figure on the right side of the vignette represents the inner surface of the posterior half of the disk. The vertical direction of the moveable laminæ is effected by sets of muscles going off obliquely right and left from two elongated bony processes, one on each half of each of these moveable divisions. The contraction of these muscles, acting upon these levers, raises the external edges of the parallel divisions, increasing the area of the vacuum ; and it will be observed that the points of the moveable transverse divisions to which the muscles are attached are nearer the middle line than the outer edge, by which the chance of interfering with the perfect continuity of the free margin, and thereby destroying the vacuum, is diminished. All the bony laminæ, the outer edges of which are furnished with rows of minute tooth-like projections, are moved simultaneously, like the thin vertical divisions of our common wooden window-blinds by means of the mechanical contrivance on the frame-work. The longer muscles placed nearer the outer oval edge are probably instrumental in preserving the contact of the more flexible margin, and the serrated external edges of the parallel laminæ help to preserve the degree of elevation obtained : the adhesive power is in proportion to the area of the vacuum.

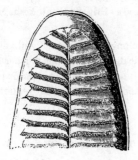 

u 2

APODAL
MALACOPTERYGII.   *MURÆNIDÆ.*

SHARP-NOSED EEL.

Anguilla acutirostris,   Sharp-nosed Eel,   YARRELL, Proceed. Zool. Soc. 1831,
pp. 133 and 159.   Zool. Journ. vol.
iv. p. 469.
,,   *omnium autorum,*   WILLUGHBY, p. 109, G. 5.
,,   *acutirostris,   Sharp-nosed Eel,*   JENYNS, Man. Brit. Vert. p. 474,
sp. 163.

Muræna anguilla,   *L'Anguille,*   LINNÆUS. BLOCH, pt. iii. pl. 73.
,,   ,,   *Common Eel,*   PENN. Brit. Zool. vol. iii. p. 191.
Anguilla vulgaris,   ,,   ,,   FLEM. Brit. An. p. 199, sp. 109.
,,   ,,   *Long-bec,*   CUVIER, Règne An. t. ii. p. 349.
,,   ,,   *Common Eel,*   BOWDICH, Brit. Fr. Wat. Fish. No. 7.

Generic Characters.—Body cylindrical, elongated, covered with a thick
and smooth skin ; the scales very small ; lubricated with copious mucous secretion ; mouth with a row of teeth in each jaw, and a few on the anterior part of
the vomer ; pectoral fins close to a small branchial aperture ; no ventral fins ;
dorsal fin, anal fin, and caudal fin united.

BARON CUVIER, in this family of the *Murænidæ,* or Eel-
shaped Fishes, which includes several genera forming his
fourth order, has brought together those fishes with soft
fins which have an elongated form of body : they are also
destitute of ventral fins, and are in consequence called *Apodal.* The genus *Anguilla,* including our common Eels, is
the first of this order.

The general appearance of the Eel is so well known, and so unlike that of most other fishes, as to require but a slight description ; yet it was not till a period of very modern date that naturalists became acquainted with the fact that the fresh waters of several countries produce three or four distinct species which had previously been confounded together. Thus the first edition of the *Règne Animal*, published in 1817, included but one species of common fresh-water Eel as well known : the second edition, published in 1829, contains a short notice of four different species ; three of which, if not all four, are found in this country.

The form of the Eel, resembling that of the serpent, has long excited a prejudice against it, which exists in some countries even to the present time ; and its similarity to snakes has even been repeated by those, who, from the advantages of education, and their acquirements in natural history, might have been supposed capable of drawing more accurate conclusions. There is but little similarity in the snake and the Eel except in the external form of the body : the important internal organs of the two animals, and the character of the skeleton, are most decidedly different.

Eels are in reality a valuable description of fish : their flesh is excellent as food ; they are very numerous, very prolific, and are found in almost every part of the world. The various species are hardy, tenacious of life, and very easily preserved. In this country they inhabit almost all our rivers, lakes, and ponds ; they are in great esteem for the table, and the consumption in our large cities is very considerable. The London market is principally supplied from Holland by Dutch fishermen. There are two companies in Holland, having five vessels each : their vessels are built with a capacious well, in which large quantities of Eels are preserved alive till wanted. One or more of these vessels may be constantly seen lying off Billingsgate ; the others go to Holland for fresh supplies, each bringing a cargo of 15,000 to 20,000 pounds' weight of live Eels, for which the Dutch merchant pays a duty of 13*l.* per cargo for his permission to sell. Eels and Salmon are the only fish sold by the pound weight in the London market.

Eels are not only numerous, but they are also in great request, in many other countries. Ellis, in his Polynesian Researches, vol. ii. page 286, says : " In Otaheite, Eels are great favourites, and are tamed and fed until they attain an enormous size. These pets are kept in large holes, two or three feet deep, partially filled with water. On the sides of these pits they generally remained, excepting when called by the person who fed them. I have been several times with the young chief, when he has sat down by the side of the hole, and, by giving a shrill sort of whistle, has brought out an enormous Eel, which has moved about the surface of the water, and eaten with confidence out of its master's hand."

" Most of the writers on the habits of the Eel have described them as making two migrations in each year : one in the autumn *to* the sea ; the other in spring, or at the beginning of summer, *from* the sea. The autumn migration is performed by adult Eels, and is believed to be for the purpose of depositing their spawn ; it is also said that these parent fish never return up the rivers. The spring migration is commonly supposed to be confined to very small Eels, not more than three inches in length, and in reference to the fry alone, it is too well known, and too often recorded, to be matter of doubt. The passage of countless hundreds of young Eels has been seen and described as occurring in the Thames,* the Severn, the Parrett, the Dee, and the Ban.

* See an excellent account by Dr. William Roots, of Kingston, published in the second series of Gleanings in Natural History, by Edward Jesse, Esq. p. 50.

I am, however, of opinion, that the passage of adult Eels to the sea, or rather to the brackish water of the estuary, is an exercise of choice, and not a matter of necessity ; and that the parent Eels return up the river as well as the fry."

" All authors agree that Eels are extremely averse to cold. There are no Eels in the arctic regions,—none in the rivers of Siberia, the Wolga, the Danube, or any of its tributary streams ; yet the rivers of the southern parts of Europe produce four species. There is no doubt that fishes in general, and Eels in particular, are able to appreciate even minute alterations in the temperature of the water they inhabit. The mixed water they seek to remain in during the colder months of the year is of a higher temperature than the pure fresh water of the river, or that of the sea. It is a well-known law in chemistry, that when two fluids of different densities come in contact, the temperature of the mixture is elevated for a time in proportion to the difference in density of the two fluids, from the mutual penetration and condensation. Such a mixture is constantly taking place at the mouths of rivers that run into the sea, and the mixed water maintains a temperature two degrees warmer than that of the river or the sea. This elevation in the temperature of the water of estuaries and the mouths of rivers is, I have no doubt, one reason why they in general abound in young fish."

In a tideway river the descent of the Eels towards the brackish water takes place during the autumn, and various devices are employed in different streams to intercept them in their progress. The vignette at the bottom of the next page represents the form of an apparatus used in various parts of the Thames, called an Eelbuck, consisting of a framework of wood supporting various wicker-baskets of a particular form. The large open end of each basket is opposed to the stream, and by the peculiar structure of the inside, any fish once within the body of the basket, cannot escape.

During the cold months of the year Eels remain imbedded in mud ; and large quantities are frequently taken by Eel-spears in the soft soils of harbours and banks of rivers, from which the tide recedes, and leaves the surface exposed for several hours every day. The Eels bury themselves twelve or sixteen inches deep, near the edge of the navigable channel, and generally near some of the many land-drains, the water of which continues to run in its course over the mud into the channel during the whole time the tide is out. In Somersetshire the people know how to find the holes in the banks of rivers in which Eels are laid up, by the hoar-frost not lying over them as it does elsewhere, and dig them out in heaps. The practice of searching for Eels in mud in cold weather is not confined to this country ; Dr. Mitchill, in his paper on the Fishes of New York, published in the Transactions of the Literary and Philosophical Society of that city, says, " In the winter Eels lie concealed in the mud, and are taken in great numbers by spears." Thus

imbedded in mud, in a state of torpidity, the Eel indicates a low degree of respiration. Dr. Marshall Hall has shown that the quantity of respiration is inversely as the degree of irritability. With a high degree of irritability and a low respiration, co-exist—1st. The power of sustaining the privation of air and of food ; 2nd. A low animal temperature ; 3rd. Little activity ; 4th. Great tenacity of life. All these peculiarities Eels are well known to possess. The high degree of irritability of the muscular fibre explains the restless motions of Eels during thunder-storms, and helps to account for the enormous captures made in some rivers by the use of gratings, boxes, and Eel pots or baskets, which imprison all that enter. The power of enduring the effects of a low temperature is shown by the fact, that Eels exposed on the ground till frozen, then buried in snow, and at the end of four days put into water, and so thawed slowly, discovered gradually signs of life, and soon perfectly recovered.

The mode by which young Eels are produced appears to have long been a subject of inquiry, and the notions of the ancients as well as of some of the moderns were numerous and fanciful. Aristotle believed that they sprang from the mud ; Pliny, from fragments which were separated from their bodies by rubbing against rocks ; others supposed that they proceeded from the carcases of animals; Helmont believed that they came from May-dew, and might be obtained by the following process :—" Cut up two turfs covered with May-dew, and lay one upon the other, the grassy sides inwards, and thus expose them to the heat of the sun ; in a few hours there will spring from them an infinite quantity of Eels." Horse-hair from the tail of a stallion, when deposited in water, was formerly believed to be a never-failing source of a supply of young Eels. It was long considered certain that they were viviparous : this belief had its origin probably in the numerous worms that are frequently to be found in various parts of the bodies of Eels, sometimes in the serous cavities, at others in the intestinal canal. Rudolphi has enumerated eight different species of entozoa common to fresh-water Eels. The enormous number of young known to be produced by Eels is a good negative proof that they are oviparous; viviparous fishes producing, on the contrary, but few young at a time, and these too of considerable size when first excluded. Having devoted time and attention to the close examination of numbers of Eels for many months in succession, the further details of which will be found in Mr. Jesse's second series of Gleanings in Natural History, I need only here repeat my belief that Eels are oviparous, producing their young like other true bony fishes.

" The sexual organ consists of two long narrow sacs extending one on each side of the air-bladder throughout the whole length of the abdominal cavity, and continued for two inches posterior to the vent. The membranes forming this tubular sac, secreting on the inner surface the milt of the male, and affording attachment for the ova in the female, are puckered or gathered along the line of junction to the peritoneal covering of the spine, and the free or loose floating edge is therefore thrown into creases or plaits like a frill. It is probably from this folded or convoluted appearance the sexual organs of the Eel have frequently been called fringes. By the kindness of my friends Mr. Clift and Mr. Owen, of the Royal College of Surgeons, I have had the pleasure of seeing some drawings belonging to the collection of John Hunter, in which these peculiarities of the sexual organs in the Eel are beautifully exhibited in various magnified representations."

Dr. Mitchill of New York, whose paper on Fishes has been already referred to, says, " the roes or ovaria of Eels may be seen by those who will look for them in the proper season, like those of other fishes."

Eels that have lain in brackish water all the winter under the constant influence of the higher temperature of that locality, probably deposit their spawn earlier in the spring than those which have passed the winter in places from which there existed for them no possible egress. In the Mole, the Wey, the Longford river, and in some large ponds, the Eels in the spring of 1833 did not deposit their spawn till near the end of April; but in two Eels from Sheerness received and examined on the 18th of May, the internal appearances induced me to believe that the roes had been passed some time. How long the ova remain deposited before the young Eel is produced, is, I believe, unknown. The duration of this interval is very variable in different fishes. The roe of the Herring, deposited at the end of October or the beginning of November, is said to become living fry within three weeks: the ova of Eels, the produce of which is very small, do not probably require a longer period. Both the parent Eels and the fry occupying the brackish water appear to have the power of going either to the salt water or to the fresh without inconvenience, from the previous preparation which the respiratory organs have undergone, and many of both are found in pure sea water: the great bulk of the young, however, certainly ascend the stream of the river, and their annual appearance in certain places is looked for with some interest. The passage of young Eels up the Thames at Kingston in the year 1832 commenced on the 30th of April, and lasted till the 4th of May; but I believe I am correct in stating that few young Eels were observed to pass up the Thames either in the year 1834 or 1835. Some notion may be formed of the quantity of young Eels, each about three inches long, that pass up

the Thames in the spring, and in other rivers the beginning
of summer, from the circumstance that it was calculated by
two observers of the progress of the young Eels at Kingston
in 1832, that from sixteen to eighteen hundred passed a
given point in the space of one minute of time. This passage of young Eels is called Eel-*fare* on the banks of the
Thames,—the Saxon word signifying to go, to pass, to travel ;* and I have very little doubt that the term *Elver*, in
common use on the banks of the Severn for a young Eel,
is a modification or corruption of Eel-fare.

" When the Elvers appear in the Severn, they are taken
in great quantities with sieves of hair-cloth, or even with a
common basket, and, after being scoured and boiled, are
offered for sale. They are either fried in cakes or stewed,
and are accounted very delicious."

There is no doubt that Eels occasionally quit the water,
and when grass meadows are wet from dew, or other causes,
travel during the night over the moist surface in search of
frogs and other suitable food, or to change their situation.
Some ponds continually produce Eels, though the owners of
these ponds are most desirous of keeping the water free
from Eels, from a knowledge of their destructive habits towards the spawn and fry of other fishes. Other ponds into
which Eels have been constantly introduced are obnoxious to
them from some quality in the water ; and they are known to
leave such places during the night, and have been found on
their passage to other retreats. Dr. Hastings, in his Illustrations of the Natural History of Worcestershire, says at
page 134 : " I will here mention a curious confirmation of
the opinion in favour of the overland migration of Eels. A

---

* A pedestrian on the road is called " a way-faring man ;" and hence, also,
the price for travelling by a conveyance is called " the fare." We have also
" thoroughfare," &c.

relative of the late Mr. Perrott was out in his park with his keeper near a large piece of water, on a very beautiful evening, when the keeper drew his attention to a fine Eel quietly ascending the bank of the pool, and with an undulating motion making its way through the long grass : on further observation he perceived a considerable number of Eels quietly proceeding to a range of stews, nearly the distance of a quarter of a mile from the large piece of water from whence they started. The stews were supplied by a rapid brook, and in all probability the instinct of the fish led them in that direction as a means of finding their way to some large river from whence their ultimate destination, the sea, might be obtained. This circumstance took place at Sandford Park, near Enstone."

That Eels breed also in the fresh water of inland rivers and lakes from which they are unable to visit the sea, is, I believe, certain. A constant supply for the table is obtained throughout the winter in these localities, as well as at other seasons, by gamekeepers and fishermen, who have charge of waters thus situated ; and no doubt exists in their minds that these Eels are bred in the places from which they are obtained, and of which the great variation that occurs in the size is an additional proof.

The Eel is a voracious feeder during certain months of the year. In winter the stomachs of those which I examined were empty : by the middle of March I found the stomachs of others distended with the larvæ of various insects, and the bones of small fishes. They are known to consume a large quantity of spawn, and will attack large Carp, seizing them by the fins, though without the power of doing them further injury. Occasionally they eat vegetable substances, and have been seen swimming about the surface of water, cropping the leaves of small aquatic plants. By means of a long and capacious air-bladder, Eels rise to various elevations in the water with great ease, and sometimes swim very high even in deep water. When Whitebait-fishing in the Thames, I once caught an Eel in the net in twenty-six feet depth of water, though the Whitebait-net does not dip more than about three feet below the surface.

Eels appear to be slow of growth, not attaining greater length than twelve inches during the first year, and do not mature roe till the second or third year. The sharp-nosed species, however, acquires a large size. I saw at Cambridge the preserved skins of two which weighed together fifty pounds; the heaviest twenty-seven pounds, the second twenty-three pounds. They were taken on draining a fen-dyke at Wisbeach.

Ely is said to have been so named from rents being formerly paid in Eels: the lords of manors in the isle were annually entitled to more than 100,000 Eels. A stich or stick of Eels was twenty-five; and the practice of stringing Eels on tough slender willow-twigs, put in at the gill-aperture and out at the mouth, still prevails in Dorsetshire among those who carry Eels about for sale from house to house; one, two, or three pounds' weight being thus strung on a stick, to suit different customers. Elmore on the Severn obtained its name from the immense number of Eels which are taken there.

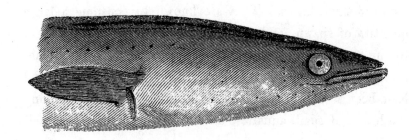

In a sharp-nosed Eel of twenty-two inches in length, three distances taken from the point of the lower jaw are to the whole length of the Eel as follows :—to the upper part of the base of the pectoral fin, as two to seventeen ; to the commencement of the dorsal fin, as two to seven ; and to the commencement of the anal fin, as nine to twenty-two. In a sharp-nosed Eel of twenty inches in length, the pectoral fin will be almost one inch, and the vent more than an inch, nearer the head than the same parts in a broad-nosed Eel of the same length.

The head is compressed, the top convex, depressed as it slopes forward : the eyes small, placed immediately over the angles of the mouth ; irides reddish yellow : the jaws very narrow, slightly rounded at the end ; the lower jaw the longest : nostrils with two openings on each side, one tubular, the other a simple orifice ; both jaws furnished with a narrow band of small teeth ; gape small ; various mucous pores about the mouth and other parts of the head ; gill-opening a small aperture immediately before and rather below the origin of the pectoral fin ; the scales on the body rather small : dorsal fin extending over more than two-thirds of the whole length of the fish ; anal fin occupying more than half of the whole length ; both united at the end, forming a tail ; the number of rays in the fins not easily ascertained, from the thickness of the skin ; the lateral line exhibits a long series of mucous orifices ; vertebræ 113. The vent includes four distinct openings, the most anterior of which leads upwards to the intestine, the posterior to the urinary bladder in a direction backwards, and one elongated lateral opening on each side communicating with the cavity of the abdomen, as in other bony fishes.

The cranium on the right hand of the three, figured at page 303, is that of the sharp-nosed Eel.

The prevailing colour of all the upper surface is a dark olivaceous green; the sides lighter; the belly white. When the fish are obtained from pure streams, the colours are clear and bright, and it is called a Silver Eel; when taken from water over a muddy bottom, the colours are brown and dusky.

Dr. Marshall Hall, in 1831, while pursuing some physiological investigations on the circulation of the blood in various reptiles and fishes, observed a pulsating sac near the tail of the Eel. The form, action, and connexions of this sac are best seen under the microscope. A young Eel of six or seven inches in length, if rolled up in a strip of linen cloth, leaving out a small portion only of the tail, will remain quiet when placed on a long slip of glass, or may be tied to it with thread. The pulsation observed in this sac is entirely independent of the action or influence of the heart, and the number of beats more than double in the same period of time; they also continue after the heart has been removed. Some Continental physiologists have ascertained that these pulsating sacs, which are found in the frog, the toad, the salamander, and the green lizard,* contain lymph, and direct its motion, and they have accordingly called them lymphatic hearts. They are only observed in connexion with veins. " Such is," says Dr. Muller, " the pulsating organ discovered by Dr. Marshall Hall at the end of the vena caudalis of the Eel, where that organ receives the venous branches of the extremity of the tail, and conducts its blood into the vena caudalis. But organs of pulsation in the lymphatic system have hitherto been altogether unknown; it is not probable that they should exist only in amphibia, and important discoveries of a like nature in the higher animals, such as birds

---

* See a paper in the Philosophical Transactions for 1833, by Dr. John Muller, Professor of Physiology in the University of Bonn.

and mammalia, may be expected ; my researches, as regards these, have however been hitherto unsuccessful." In another part of his paper, Dr. Muller observes, " I have never discovered a trace of motion in the cysterna chyli and ductus thoracicus of mammalia."

In a conversation with Mr. Owen on this subject, he suggested, that as the valves of the lymphatic vessels are very few and imperfect in reptiles and fishes, especially in the latter, these pulsating sacs would seem to be superadded as a compensating power in the absence of that mechanism which impresses a definite direction and an unintermitting flow upon the currents of the lymph in the higher vertebrata, especially mammalia.

I am indebted to the kindness of Dr. Marshall Hall for permission to copy the excellent illustration of this structure in the tail of the Eel, from his very interesting critical and experimental essay on the circulation of the blood.

In the vignette the arrow-heads indicate the direction of the currents.

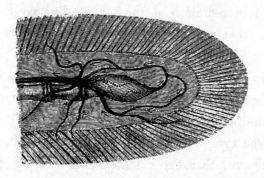

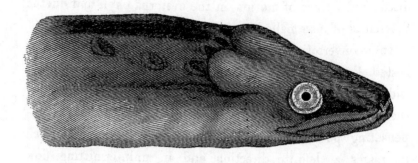

THE BROAD-NOSED EEL.

Anguilla latirostris, Broad-nosed Eel, Yarrell, Proceed. Zool. Soc. 1831,
  pp. 133 and 159. Zool. Journ.
  vol. iv. p. 469.
  ,,   ,,   ,,   ,, Jenyns, Man. Brit. Vert. p. 476,
  sp. 164.
  ,,   ,,   *A. pimperneaux,* Cuvier, Règne An. t. ii. p. 349.
  ,,   ,,   *Glut Eel,* Bowdich, Brit. Fr. Wat. Fish, No. 22.

The Broad-nosed Eel is almost as common a species
as the Sharp-nosed Eel, but is immediately distinguished
from it by the much greater comparative breadth of the
head; the representation at the top of the page is therefore
confined to that part of the fish which exhibits the best
distinctions; and the vignette to the Snig Eel, page 303,
represents in the left-hand figure of the three heads the cranium of the Broad-nosed Eel, to show this character as it
exists in the bone. This Eel is the Grig or Glut Eel of
Pennant, who says, " They have a larger head, blunter nose,
and thicker skin than the common sort." It is, probably,
also the Frog-mouthed Eel of the Severn, referred to by
Dr. Hastings, in his Natural History of Worcestershire,

page 135, and so called by the fishermen from the extraordinary width of the mouth.

In its habits the Broad-nosed Eel has not been distinguished by any peculiarity that I am aware of from the other common Eel; but it does not appear to attain so large a size, the largest I have seen not exceeding five pounds in weight. It exists in many of the waters which produce the Sharpnosed Eel, is much thicker in the body in proportion to its length, and fishermen can distinguish this species readily when fishing in the dark by its more soft and unctuous feel in the hand.

The term Grig is, however, in and about London, applied to a particular Eel of small size, of which the figure here introduced represents the head. This Eel is the *An-*

guille plat-bec of Cuvier, *Règne Animal*, tom. ii. p. 349, who considers it a distinct species. It is the Grig Eel also of Mrs. Bowdich's British Fresh Water Fishes, No. 28, in which work the three Eels already spoken of here are well figured; and the species were considered by Cuvier as identical with those of the *Règne Animal*.

The name Grig is also applied by Thames fishermen to any small-sized Eel of any species when not longer than eight or nine inches, and of which eight or ten are required to make up a pound weight.

In a Broad-nosed Eel of twenty-two inches in length, three distances taken from the point of the lower jaw are to the whole length as follows:—to the upper part of the base

x 2

of the pectoral fin, as two to thirteen; to the commencement
of the dorsal fin, as one to three; and to the commencement
of the anal fin, as ten to twenty-two.

The Broad-nosed Eel has the head rounded at the back
part, and flattened from the eyes forward; both jaws broad
and blunt; the lower jaw the widest, and longer than the
upper: nostrils double, one tubular, the other a plain orifice;
the gape large; lips fleshy: teeth more numerous than in
either of the other British fresh-water species, larger, stronger, and forming a much broader band in each jaw: the eyes
large, placed before the line of the gape; irides golden yellow: the gill-openings, pectoral fins, the commencement of
the dorsal fin, and the vent, placed farther back than in the
Sharp-nosed Eel; dorsal and anal fins also much deeper and
thicker; the tail broad and rounded; the body of the fish
thicker for the same length than in other Eels: the number
of vertebræ 115.

The colour of the upper surface of the body is a dark-
greenish brown, subject to some variation, depending on locality, soil, and the quality of the water.

THE SNIG.

Anguilla mediorostris, Snig Eel, YARRELL. JESSE, Glean. Nat. Hist. 2nd
Series, pp. 75 and 76.

    ,,   ,,   ,, ,, JENYNS, Man. Brit.Vert. p. 477, sp. 165.

I AM indebted to the kindness of Mr. Jesse, and his
friend, Francis Mills, Esq. for the only specimens of this
Eel I have yet seen ; and from some differences in its external characters, in its habits, and also in the comparative size of
the head, as well as some peculiarity in the five cervical vertebræ that are nearest the head, I believe it to be a different
species from either of those previously described in this work.

The specimens I have had were from the Avon in Hampshire, where this Eel, rather remarkable for its yellow colour,
is called the Snig, and is considered distinct from the other
well-known and more common Eels.

Dr. Hastings, in the Appendix to his Illustrations of the
Natural History of Worcestershire, page 135, says, that
besides an Eel called the Frog-mouthed Eel by the fishermen, from the extraordinary width of the mouth,—identical,
probably, with the Broad-nosed Eel of this work,—" there are
two distinct kinds of Eels in the Worcestershire Avon, the
Silver and Yellow Eel," which last may be similar to the Snig
of the Avon of Hampshire.

The term Snig, it should however be stated, is in some counties a general name for any sort of Eel; and a particular mode of fishing for Eels, which is described in most of the works on Angling, is called Sniggling.

The Hampshire Snig differs from our other Eels in its habit of roving and feeding during the day, which other Eels do not. It is considered excellent as an article of food, and of a superior flavour to other Eels: it does not however attain a large size, seldom exceeding half a pound in weight.

The fishermen make a certain difference in the mode of placing their eel-pots when they are desirous of catching Snigs; finding by long experience that the Snigs get into those pots the mouths of which are set in the opposite direction, in reference to the stream, to others in which the common Eels are taken.

In the comparative breadth of the nose, the Snig is intermediate in reference to the Sharp and Broad-nosed Eels, but rather more resembles that with the sharp nose; it has a slight but elongated depression extending from the anterior edge of the upper jaw to the upper and back part of the head; the tubular openings of the nostrils are longer, and the mucous pores about the lips larger and more conspicuous; both jaws rounded at their extremities, the lower one the longest; teeth longer and stronger than in the common sharp-nosed species; gape large; the angle and the posterior edge of the eye on the same vertical line; the pectoral fins, the commencement of the dorsal fin, and the vent, are each placed nearer the head than in either of our fresh-water Eels.

The general colour olive green above, passing by a lighter green to yellowish white below.

Desirous of obtaining internal characters of distinction among our fresh-water Eels, I prepared skeletons of each species, selecting three examples that measured exactly the same length, in order to afford a more just comparison. The vignette at the bottom of the page represents correctly the relative size and power of bone in each species. The cranium on the left is that of the Broad-nosed Eel; that in the middle is from the Snig; the head on the right hand is from the Sharp-nosed Eel. It is obvious that each is able to overcome a larger and more powerful victim as food than the other. It will also be seen, that independent of some difference in the length and form of some of the bones, as well as in the size of the head in the middle, belonging to the Snig, as compared with that on either side, there is a characteristic distinction in the form of the bones of the vertebral column. The first five cervical vertebræ are smooth and round, entirely destitute of superior or lateral spinous processes, both of which are possessed by the other two, of a size corresponding to the character of the vertebral bone itself to which it belongs. With this exception, the skeleton of the Snig most resembles that of the Sharp-nosed Eel; but is somewhat stronger, and particularly so in the processes of the other vertebræ generally.

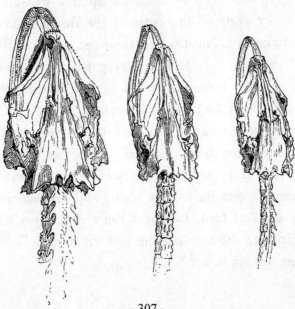

THE CONGER.

Conger vulgaris,	*Le Congre,* Cuvier, Règne An. t. ii. p. 350.
	Conger, Willughby, p. 111, G. 6.
Muræna Conger,	,, Linnæus. Bloch, pt. v. pl. 155.
,, ,,	*Conger Eel,* Penn. Brit. Zool. vol. iii. p. 196.
,, ,,	,, Don. Brit. Fish. pl. 119.
,, ,,	,, Flem. Brit. An. p. 200, sp. 110.

Generic Characters.—The dorsal fin commences much nearer the head than in the fresh-water Eels; the upper jaw the longest; in other respects resembling the genus *Anguilla.*

THE CONGER EEL is a marine species well-known on all the rocky parts of the coast of the British Islands, but nowhere more abundant than on the coast of Cornwall.

Mr. Low says, " It is found very frequently round the Orkney Islands: some are caught at the fishermen's lines; but the otter is by far the most successful in killing Congers. He brings them ashore, and eats but a very small part, leaving the rest for the next comer; and where his haunts are known, the country people are very careful every morning to search for the remains of the night, and are seldom disappointed, but find Cod, Ling sometimes, but especially Congers, which are oftener seen amongst the deep hollows of the rocks than farther to sea."

Dr. Neill says this species is sometimes, but not very often, found in the Forth ; and, in proof of great tenacity of life, mentions having seen one alive and vigorous in the Edinburgh market, although it had been ten hours out of water.

The Conger is frequently caught at various rocky parts of our eastern coast, and I have known specimens of large size taken in winter about the mouth of the Thames. Congers are caught by bulters, or long-lines, and hand-lines—modes of fishing already described, and the most esteemed bait is the sandlaunce. " So well assured," says Colonel Montagu, " are the French fishermen of the advantage derived from the use of this little fish, that the fishing-boats in times of peace run over from the coast about Dieppe to Slapton Bay, on the south coast of Devon, on purpose to purchase launce ; and for that purpose alone do some of our fishermen keep fine nets for the purpose of supplying bait to these foreigners, for which they obtain about twenty-pence the bushel. Some principal Conger banks lie off the French coast, from which a prodigious quantity are taken to feed the poorer classes on maigre days.

The principal fishery for Congers in this country is on the Cornish coast; where, according to Mr. Couch, it is not uncommon for a boat with three men to bring on shore from five hundred weight to two tons, the fishing being performed during the night ; for this fish will not readily take a bait by day, and even on moonlight nights it is more shy than when in the dark, except in deep water. The most usual bait with the Cornish fishermen is a Pilchard. The Congers that keep among rocks hide themselves in crevices, where they are not unfrequently left by the retiring tide ; but in situations free from rocks, Congers hide themselves by burrowing in the ground.

The flesh is not in much estimation, but meets a ready sale at a low price among the lower classes. Formerly a very considerable quantity was prepared by drying in a particular manner, and exported to Spain : Bayonne also received a part. When thus dried, the flesh was ground or grated to powder, and in this state was used to thicken soup.

Congers spawn in December or January; and the distinction of the sexes is obvious on the examination of the roe during the cold months. Small ones, about the size of a man's finger, are found among rocks, close to land, during the summer. The small Eels which ascend the Severn in such numbers in the spring, and were considered by Willughby and Pennant as the young of the Conger, are in reality the young of fresh-water Eels.

The adult fish is most voracious, not sparing even those of its own species. From the stomach of a specimen weighing twenty-five pounds, I took three common Dabs, and a young Conger of three feet in length. The power of the jaws in this fish is very great : in the stomach of small specimens examined on the coast, I have found the strong testaceous coverings of our shell-fish comminuted to fragments. They are often tempted by the crustacea entrapped in the lobster-pots to enter those decoys in order to feed on them, and are thus frequently captured.

Congers acquire a very large size. Specimens weighing eighty-six pounds, one hundred and four pounds, and even one hundred and thirty pounds, have been recorded, some of them measuring more than ten feet long, and eighteen inches in circumference. They possess great strength, and often prove very formidable antagonists if assailed among rocks, or when drawn into a boat on a line.

Three measurements taken from the point of the nose, as in the fresh-water species, give the following proportions in reference to the whole length :—the distance to the origin of the pectoral fin is as two to thirteen ; to the commencement of the dorsal fin, as one to five ; and to the vent, as two to five.

The head is long and depressed : the upper jaw the longest ; both jaws furnished with strong teeth, forming a broad band in each : the lips fleshy : the nostrils double ; the most anterior near the edge of the lip, and tubular ; the other a simple orifice : numerous mucous pores about the parts of the mouth and head : the mouth deeply divided, making the gape long ; the angle forming a tangent with the posterior edge of the pupil : the eyes large ; body nearly cylindrical ; dorsal fin commencing but little behind the pectorals, extending along four-fifths of the whole length of the body ; anal fin commencing immediately behind the vent, and extending along three-fifths of the whole, and joining the dorsal fin, forms a pointed tail.

The colour of the upper surface of the body is a uniform pale brown, becoming lighter on the lower part of the sides, and passing into dull white underneath ; the dorsal and anal fins whitish, edged with black ; lateral line almost white.

The notion entertained by some, that river Eels on going to the sea remain there and become Congers, scarcely requires a serious remark. No one who looks for specific distinctions can fail to observe them when comparing either of our fresh-water Eels with the Conger. These differences, which extend to colour, form of body, and situation of fins, receive further confirmation on examining their internal structure : independent of comparative difference of relative position in some of the most important of the viscera, the greatest number of vertebræ found in our fresh-water Eels is 116, those of the Conger amount to 156.

APODAL
MALACOPTERYGII.   *MURÆNIDÆ.*

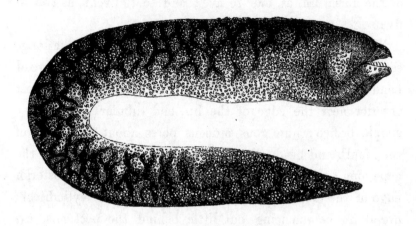

THE MURÆNA.

Muræna Helena,   Linnæus. Bloch, pt. v. pl. 152.
   ,,   ,,   *La Murène,* Cuvier, Règne An. t. ii. p. 352.
   ,,   ,,   *The Muræne,* Couch, MS.

Generic Characters.—Body elongated ; no pectoral fins ; branchial opening a minute orifice on each side ; a single row of teeth in each jaw ; dorsal and anal fin very low, united.

Mr. Couch is the only British naturalist I have heard of who has obtained an example of this beautifully marked species on the English coast. The following is Mr. Couch's account, copied from his MS. :

A specimen, the first on record as a British fish, was caught by a fisherman of Polperro, October 8th, 1834.

Its length was four feet four inches ; body very flaccid, rounded anteriorly, compressed and tapering towards the tail : the whole body seemed plump. Before the eyes it is slender and sharp ; jaws equal ; gape moderately large ; teeth long, incurved, sharp, separate, in one row, a row on the palate ; tongue adherent, scarcely perceptible ; a nasal barb on each

side of the end of the snout, another a short distance above
each eye, and a probe passed down the latter found its way
out at the former ; large mucous orifices encircle both jaws
at equal distances, four on each row. Eye rather small, one
inch and one quarter from the snout ; irides light bluish
grey, having a lively look : cheeks tumid ; an extensive depression at the side of the thorax, in which is the simple
orifice of the gills ; the external appearance of the branchial
aperture very much resembles that belonging to the Lamprey : from the snout to the branchial opening, six inches ;
from the part above the eye the head is much elevated ; the
skin wrinkled ; thorax remarkably protuberant ; the distance
from the top of the head to the thorax five inches and three-
quarters. The vent is exactly half-way between the two
ends of the body, from whence proceeds a line to the end of
the tail parallel to the anal fin, and half an inch from its
base : this line must be the lateral line, since there is no
appearance of any other. The dorsal fin begins five inches
and a half from the snout, and proceeds round the extremity
of the body to join the anal, which begins at the vent ; but
these fins are thick and fleshy, and not readily distinguished
from the margin of the body.

The ground colour of the anterior part of the body is a
fine lively yellow, the hinder part a fine purple ; but the
whole, including the fins, is divided into segments, forming
irregularly shaped spots, which yet have a tendency to regular distribution ; towards the tail the yellow spots more
resemble irregular rings, with larger spaces between them ;
the whole is interspersed with innumerable spots of whitish
and deep yellow, golden, brown, and purple, forming a most
beautiful arrangement : under the thorax and to the gill-
opening are a few lines marked in the skin as if to facilitate
motion, though the skin is exceedingly smooth and soft ;

it is strong also, and the colours were remarkably slow to fade, contrary to what is observed in most fishes. This specimen was taken with a line, and manifested great strength after it was taken on board the boat.

Of this singular and beautifully marked fish Mr. Couch very kindly sent me for my use a coloured drawing made from the fresh specimen, from which the figure on the preceding page, carefully reduced in size, was drawn and engraved.

This Muræna is considered very common in almost every part of the Mediterranean. It was a great favourite with the ancient Romans, who preserved large quantities of them in their numerous vivaria, where they were fed with great care. On the celebration of one of his triumphs, Cæsar distributed six thousand specimens of this Muræna among his friends.

The flesh is said to be delicately white, and very agreeable eating. In the Mediterranean it is fished for with lines. It is very voracious, and its bite is very severe, which, from the nature of the teeth, and the large size of the muscles about the head, might be expected.

This fish is said to live with equal facility in fresh or salt water, though generally found at sea.

The vignette represents a Venetian pleasure-boat.

APODAL
MALACOPTERYGII.   *MURÆNIDÆ.*

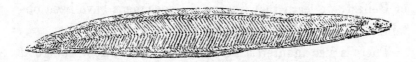

THE ANGLESEY MORRIS.

Leptocephalus Morrisii, Anglesey Morris, PENN. Brit. Zool. vol. iii. p. 212,
pl. 28.

      ,,   ,,   ,,   ,,   MONTAGU, Wern. Mem. vol. ii. p.
436, pl. 22, f. 1.

      ,,   ,,   ,,   ,,   FLEM. Brit. An. p. 200, sp. 111.

      ,,   ,,   *Le Leptocephale,* CUVIER, Règne An. t. ii. p. 358.

Generic Characters.—Head small and short ; teeth numerous ; pectoral fins and gill-opening very small ; body compressed and very thin, tape-like ; dorsal and anal fins small, united at the tail, forming a point.

THIS species was discovered in the sea near Holyhead by Mr. William Morris, who sent the specimen to Pennant, by whom it was named after his friend. Pennant subsequently sent the same specimen to Gronovius, who described it under the generic name of *Leptocephalus,* in reference to the small size of the head.

Any doubts which might formerly have been raised as to the real existence of such a species, to which Colonel Montagu has alluded, must have ceased to exist, as this fish has now been taken and recognised in various localities. Pennant in his first description, perhaps from the state of his specimen, was not aware of all the characters this delicate fish possesses; but Colonel Montagu has well described and given a figure of it in the Wernerian Memoirs, as quoted.　More than twenty specimens have within a few years been taken at different parts of the coast of England, Wales, and Ireland. By·the kindness of Mr. Couch, I possess three examples that were taken in Cornwall ; and from Mr. William Thompson, of Belfast, we learn that five or six specimens have been obtained by him and his collecting friends.

There is also an interesting account of this fish, with a good figure, in the sixth volume of Mr. Loudon's Magazine of Natural History, page 330, by H. V. Deere, Esq. who states that his specimen, to all appearance dead, was brought to him by a Devonshire fisherman, who had carried it in his pocket, wrapped in brown paper, for three hours.　After this gentleman had held the fish in his hand for about a minute, examining it, symptoms of life appeared, and then the little animal was placed in a tumbler of salt and water, where it survived its incarceration in brown paper for several hours.　Its appearance is described as most pleasing, from its semitransparent and silvery hue, its prominent eye, and graceful motions.　It is usually found among seaweed.

I carefully dissected off the whole of one side from one of the three specimens sent me by Mr. Couch, laying bare the vertebral column and the intestinal canal.　The bones forming the vertebræ have no spinous processes whatever, either superior or inferior ; the angles of the ascending and descending oblique indented striæ, visible on the external surface of the skin, mark the points of union of the different vertebræ ; the oblique muscles between the striæ are attached to the bodies of the bones forming the column ; the margin all round each vertebral bone is opaque, but the centre or body of each is transparent.

The intestine is a single straight canal of small calibre, reaching from the head to the vent ; after passing from the posterior part of the head, it descends to the abdominal line, which it traverses without convolution to the vent. This canal may be distinctly seen in the perfect fish when placed flat on a slip of glass, and looked at against a good light, particularly the descending portion from the head to the level of the abdominal line.

The head is small, short, and rather blunt : the eyes large; irides silvery, the pupil dark : the lower jaw slender ; teeth in both jaws, numerous and minute : gill-openings and pectoral fins very small; the body behind the head becomes deeper, very much compressed, as thin as tape, and when rendered opaque by the effect of a mixture of spirit of wine and water, which is the best mode of preserving them, this fish very much resembles a piece of a tape-worm.

The dorsal fin commences rather before the middle of the whole length of the fish ; the anal fin rather behind it ; and both extend to the tail, where they are united, and end in a point. These fin-like appendages have the appearance of an extension of the skin, and are so delicate that it is not always easy to decide where they do begin, or may be called fin ; the dorsal and abdominal margins, as well as the lateral line, exhibit a series of small black specks : the obliquely striated appearance of the sides has been already referred to. The general colour is most like that of opal.

I have had opportunities of examining specimens from the Mediterranean which were identical with those from Cornwall, as well as those described and figured in the English works already referred to. M. Risso includes but one species in his fishes of Southern Europe and the Environs of Nice, which he has named *Leptocephalus Spallanzani*, tom. iii. p. 205 ; but the description so exactly accords with English specimens, that I have no doubt it is the fish I have seen, and the same as that on our own shores.

VOL. II.   Y

APODAL
MALACOPTERYGII.   *MURÆNIDÆ.*

THE BEARDLESS OPHIDIUM.

Ophidium imberbe, Linnæus.
    ,,   ,,   *Beardless Ophidium,* Penn. Brit. Zool. vol. iii. p. 208, pl. 29.
    ,,   ,,   ,,   ,,   Montagu, Wern. Mem. vol. i. p. 95,
            pl. 4, f. 2.
    ,,   ,,   ,,   ,,   Flem. Brit. An. p. 201, sp. 112.

Generic Characters.—Head smooth; body elongated, compressed; teeth in both jaws, the palate, and pharynx; gill-aperture rather large; dorsal, anal, and caudal fin united.

The Beardless Ophidium was first added to the ca-talogue of British Fishes by Pennant, to whom it was communicated by the Duchess of Portland: the specimen was found near Weymouth. Pennant gave a figure of his fish in the Appendix to the fourth volume of the British Zoology, edition of 1777, but no description. Colonel Montagu afterwards obtained a specimen on the south coast of Devon, which is figured and described in the first volume of the Wernerian Memoirs, as quoted. The editor of the edition of Pennant's British Zoology, published in 1812, left out the figure of the Beardless Ophidium, given in the previous edition, but copied the figure and description of Colonel Montagu.

Never having seen a specimen of this fish, Colonel Montagu's figure and description are here given, with some additions to be hereafter explained.

" Length about three inches ; depth about a quarter of an inch. The head is very obtuse, and rounded in front : eyes large, placed forward and lateral ; irides dark, with a circle of silver round the pupil : mouth, when closed, inclines obliquely upwards ; the lips are marginated : the gill-membranes inflated beneath. The body is ensiform, considerably compressed towards the tail, and in shape is not unlike that of *Cepola rubescens*, vol. i. page 195, of this work ; the lateral line is nearly in the middle, originating at the angle of the operculum to the gills, but rather obscure : vent nearly in the middle : the pectoral fin is rounded : the dorsal fin commences immediately above the base of the pectoral, and is at first not so broad, and usually not so erect, as the other part : the anal fin commences at the vent, and, together with the dorsal, unites with the caudal fin, which is cuneiform, but obtusely pointed. The colour is purplish brown, disposed in minute speckles ; and along the base of the anal fin are about ten small bluish-white spots regularly placed, but scarcely discernible without a lens, and possibly peculiar to younger fishes : all the fins are like the body in colour, except the pectoral and caudal ; the first is pale, the last is yellowish." The fin-rays in number are—

D. 77 : P. 11 : A. 44 : C. 18 or 20.

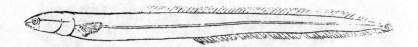

" This fish," Colonel Montagu observes, " does not appear to be very tenacious of life, like some of the Blennies, as it was placed in a tin box with the Crested and Smooth Blenny, covered with wet seaweed, and although these were lively, the Ophidium was dead before it could be got to his house. It died with its mouth shut, the pectoral fins thrown forward, and the body curved a little near the vent, throwing the head upwards."

" Little can be said of the natural habits of this fish ; but as it so rarely occurs, it is most probably an inhabitant of the rocky parts ; in such a situation, at low-water, the specimen here described was taken."

But little being known either of Montagu's or Pennant's Ophidium, the figure at the head of this subject is taken from Montagu's figure, and the outline at the foot of the preceding page is taken from Pennant's first figure, which Schneider appears to have adopted as the representative of the genus *Ophidium* in his Ichthyological work.

The *Ophidium barbatum*, or Bearded Ophidium, has also been included by Berkenhout in his Catalogue of British Fishes ; but whether on the personal authority of that author, or on what part of the British coast it was observed, no mention is made. The figure below is a representation of the Bearded Ophidium ; and the three woodcuts here given may assist investigators, should any species of Ophidium come to their hands.

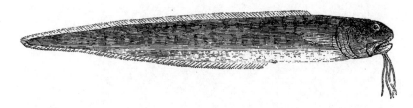

APODAL
MALACOPTERYGII.   *ANGUILLIDÆ.*

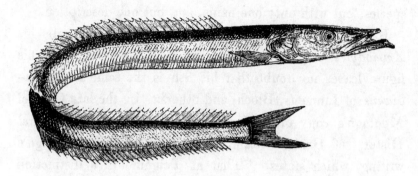

THE SAND-EEL.

HORNELS, (HORNEELS ?) *Edinburgh.*

Ammodytes Tobianus, Le Lançon, Cuvier, Règne An. t. ii. p. 360.
     ,,   ,, .   Linnæus. Bloch, pt. iii. pl. 75, fig. 2.
     ,,   *Anglorum verus,* True Sand-Eel, Jago. Ray, Syn. p. 165, pl. 2,
                                         fig. 12.
     ,,   *Tobianus,* Wide-mouthed Launce, Jenyns, Man. Brit. Vert. p. 482,
                                       sp. 170.
     ,,   ,,   *Sand-Eel,* Don. Brit. Fish. pl. 33 ?

Generic Characters.—Head and body elongated ; gill-openings large ; dorsal fin extending nearly the whole length of the back ; anal fin of considerable length ; dorsal and anal fins separated from the caudal fin.

WE are indebted, says Baron Cuvier, to M. Lesauvage, of Caen, for pointing out the true distinctions in the two species of Ammodytes belonging to the shores of the Channel, *A. Tobianus* and *A. Lancea*, the first of which is rare, but the second very common.

Our excellent naturalist and countryman Ray, has given us, in his Synopsis, from Jago's Catalogue of Cornish Fishes, a good figure of the true *Tobianus;* but it was not, I believe, till the recent publication of Mr. Jenyns' valuable Manual of British Vertebrate Animals, that any English zoologist had admitted two species among British Fishes. Shaw, in his General Zoology, vol. iv. p. 81, plate 9, has figured both species, but with only one name, and but one description.*

Ray, in his short notice from Jago's Catalogue, calls his *Ammodytes Anglorum verus* the true Sand-Eel; and his figure leaves no doubt that his fish is the same as the *Tobianus* of Linnæus, Bloch, and others. In the late Colonel Montagu's copy of Berkenhout's Synopsis of the Natural History of Great Britain, there is a note in Montagu's writing, which states, " That at Teignmouth a distinction is made between the Sand-Eel and Sand-launce, by the size and superior length of the head and gills in the one; it is also said to be much more rare." The rarity and greater length of the head are both on the side of the *Tobianus*, the Sand-Eel, which, as far as my own observation goes, is much more scarce than the smaller-sized species with the shorter head; I am therefore desirous of preserving the distinctive appellation of Sand-Eel to the longer fish, *A. Tobianus*, and continuing that of Sand-launce to the smaller species, bearing among naturalists the specific name of *Lancea*.

M. Lesauvage gave the name of *lanceolatus* to the species which had been previously called *Tobianus*, his trivial name will therefore only be used as a synonym.

Willughby's figure, G. 8, f. 1, appears to have been copied from Salvianus, and represents an *Ammodytes* with two small dorsal fins; I have not, therefore, referred it to either of our fishes.

The Sand-Eel is immediately to be distinguished from the Sandlaunce by its greater size, specimens now before me

* Both specimens are also figured by Klein.

measuring twelve inches in length ; Ray's fish was fifteen inches and a half long : it is further distinguished by the greater length of the head, and particularly of the lower jaw ; by the commencement of the dorsal fin being on a line with the end of the pectoral fin-rays : the dorsal fin of the Sandlaunce beginning in a line with the middle of the pectoral fin, and the head smaller and shorter, as shown in the two representations here given.

The habits of the two species are in many respects very similar, and will be more particularly referred to under Sandlaunce, which being exceedingly common on all our sandy shores, has afforded greater opportunities for observation.

Both species of *Ammodytes* are included by Professor Nilsson among the fishes of Scandinavia ; both species also occur in the Forth. Dr. Neill, in his account of the fishes of that locality, says, the Edinburgh fishermen call the large ones Hornels—probably an abbreviation of Horneels—in reference to the greater length of body and the horn-like elongation of the lower jaw, by means of which they are enabled to bury themselves in the wet sands of the sea-shore, from which they are scratched out with iron hooks for bait or sale.

Stephen Oliver the younger, in his agreeable Rambles in Northumberland and on the Scottish Border, when describing the fishing in the Tyne, says, Sand-Eels follow the young fry of the Coalfish into the harbour, and are frequently caught with the same bait as the Poodlers (young Coalfish), which is used in a manner similar to fly-fishing for Trout. The common length of the Sand-Eel in the Tyne is from twelve to fourteen inches ; and their jaws, by a peculiar conformation, admit of great expansion. They swim rapidly, and dash at a shoal of fry with the voracity and swiftness of a Pike. Mr. Couch says that a large specimen caught on a line by a Cornish fisherman had a small fish of its own species in its stomach.

From the extreme point of the lower jaw to the posterior end of the gill-cover is to the whole length of the fish as one to four and a half; the depth of the body rather less than one-third of the length of the head; the lower jaw very much elongated, with a strong, indurated projection at the extreme tip; the upper jaw much shorter than the lower, with a strong forked tooth of two points descending from the vomer: the nostrils double; both open on each side on a line, one before the other, about half-way between the eye and the point of the nose: the eyes rather small; the posterior margin exactly half-way between the point of the under jaw and the posterior angle of the gill-cover: the shape of the body very nearly round; covered with small scales: the pectoral fin arises under the posterior angle of the gill-cover, its length one-third that of the length of the head; the dorsal fin placed in a groove, with a prominent line extending along each side; the rays commence in a vertical line over the end of the pectoral fin-rays, and end near the tail; the lateral line indented and straight; the abdomen with three indented parallel lines extending to the anal aperture, which has another orifice behind it; along the whole line of the lower part of each side extends a narrow and slender membrane attached by one edge; the anal fin is about one-third of the whole length of the fish, ending short of the caudal fin, and nearly on the same plane as the dorsal fin; the tail forked.

The fin-rays in number are—

D. 55 : P. 15 : A. 29 : C. 17.

The irides, cheeks, gill-covers, lower part of the sides,

and the abdomen, bright silvery ; upper part of the head, back, and sides, light brown, reflecting tints of blue and green when held in different positions.

The vignette below represents the form of rake used to obtain Sand-Eels and Sand-Launce on some parts of the coast.

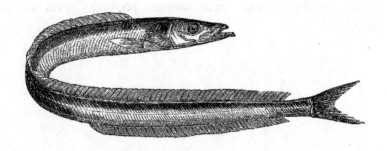

THE SAND-LAUNCE.

THE RIGGLE, *Sussex coast.*

Ammodytes Lancea,   *L'Equille,* Cuvier, Règne An. t. ii. p. 360.
    ,,   ,,   *Small-mouthed Launce,* Jenyns, Man. Brit. Vert. p. 483,
                                  sp. 171.
    ,,   *Tobianus, Sand-Launce,* Penn. Brit. Zool. vol. iii. p. 206, pl. 28.
    ,,   ,,   *Common Launce,* Flem. Brit. An. p. 201, sp. 113.

THE SAND-LAUNCE, as previously stated, is very abundant on many parts of the shore of the British Islands.   On account of its silvery brightness, it is in great estimation and constant use with fishermen as bait for the hooks of their sea and hand lines; and the habit peculiar to the species of burying themselves in the wet sand as the tide recedes affords easy means of capture.   The generic term *Ammodytes*, refers to this power of digging in sand.   With the projecting portion of the under jaw, aided by the muscular power of the fish, and its slender form, it is enabled to bury itself with rapidity five or six inches deep in the soft sand as the ebbing sea retires, and releases itself again on the approach of the ensuing flood-tide, apparently uninjured, though deprived of water for several hours : another instance of a low degree of respiration and great tenacity of life in a genus of fishes having very large gill-apertures.

In Orkney, Mr. Low says it is constantly used as a bait for other fish, and though of good flavour, is very seldom eaten. On the sands at Portobello, near Edinburgh, as well as at other localities in that vicinity, people of all ages may be seen, when the tide is out, diligently searching for the Sand-Launce, and raking them out with iron hooks. Some are used as bait ; but many are prepared for table, and considered delicate food.

Colonel Montagu mentions the Sand-Launce as being extremely plentiful at Slapton Sands, on the south coast of Devonshire, where the fishermen employ a small seine with a fine mesh, and are frequently so successful, that six or seven bushels are taken at one haul : these are usually sold to Dieppe fishermen for twenty-pence the bushel. Montagu adds, that on the part of the Devonshire coast here referred to, even the poorest people would not eat the Sand-Launce, while at Teignmouth it was in great request as food, and was counted out for sale by the score.

" It is only of late," says Mr. Couch, " that naturalists have learned to recognise two species, though it has been done long since by fishermen, who have been accustomed to observe that a small species, which keeps in larger bodies, and seldom goes far from land, is more followed by Mackerel than the others, and that its presence is a better sign of good fishing. On a calm evening it is an interesting sight to see the surface of the water broken by the repeated plunges of voracious fishes as they burst upon the little schull of Launces from beneath. Their only certain place of refuge from these pursuers is the sand.

I have obtained the fry of the Sand-Launce four inches

long in the month of April, and considered them to be the
young of the preceding year.   May, August, and December,
have each been named as the month in which the adult fish
deposit their spawn ; but the habits and economy of the two
species have been greatly confounded hitherto, under the
supposition that they were but examples of the same fish,
differing only in size.

The Sand-Launce has been noticed on the coasts of the
counties of Londonderry, Antrim, Dublin, and Cork ; I
learn also from F. C. Lukis, Esq. that both species are
found at Guernsey ; but that *Lancea* is the most common.
The search for them in the sand prevails there, but it is
usually made on moonlight nights.

The food of the Sand-Launce is marine worms and very
small fishes.

The usual length of this species is from five to seven
inches : the length of the head compared to the length of
the fish is less than as one to five : the lower jaw shorter in
proportion than in the other species ; the protractile portion
of the upper jaw much more free to move, and when the
lower jaw is pressed down, this moveable part comes forward
and downward : the posterior margin of the eye is less than
half-way between the point of the lower jaw and the posterior projecting angle of the gill-cover, being placed nearer the
nose than in *Tobianus ;* the dorsal fin commencing in a line
over the middle of the pectoral fin.

The fin-rays in number are—

D. 51 : P. 13 : A. 25 : C. 15.

In other respects, as to the lines along the body and the
colour of the various parts, the two species are very similar.

LOPHOBRANCHII. *SYNGNATHIDÆ.*

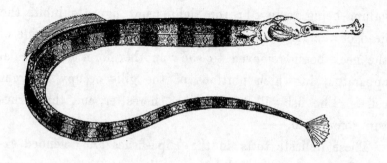

THE GREAT PIPE-FISH.

Syngnathus Acus, Linnæus. Bloch, pt. iii. pl. 91, fig. 1, young; fig. 2 adult.

,, ,, *Longer Pipe-fish,* Penn. Brit. Zool. vol. iii. p. 184, pl. 26. Two figures; upper, female; lower, male.

,, ,, *Pipe-fish,* Flem. Brit. An. p. 175, sp. 34.

,, ,, *Great Pipe-fish,* Jenyns, Man. Brit. Vert. p. 484, sp. 172.

Generic Characters.—Body elongated, slender, covered with a series of indurated plates arranged in parallel lines; head long; both jaws produced, united, tubular; no ventral fins. In the species of the first division, an elongated pouch under the tail in the males only, closed by two folding membranes.

In the species belonging to this family the jaws are united, forming a tube more or less cylindrical. The gills, instead of having the pectinated appearance so well-known to exist in the fishes previously described, are separated into small rounded tufts, which are arranged along the branchial arches, and the fishes of this family are therefore called *Lophobranchii.*

* The family of the Pipe-fishes.

The figure on the left hand of the vignette at the bottom of the page, represents one side of the pectinated gills of a Pike; that in the centre is drawn from the head of a specimen of *S. acus,* to show the gills in small tufts, the operculum being removed: the right hand figure exhibits the head of the river Lampern, part of the skin on the side of the neck being removed to show another form of branchial apparatus, in which portions of the gills occupy different cells.   The fish will be described hereafter, and this structure referred to.

These delicate tufts in the Pipe-fishes are defended externally by a large and hard operculum, having an aperture in the connecting membrane at its upper and posterior part. The fishes of this limited family are further remarkable for the extreme tenuity of their bodies, as well as for the number and arrangement of the indurated and sculptured plates by which their lengthened bodies are defended.   They are frequently called Needle-fish.

The five species of British *Syngnathi* require to be ar-

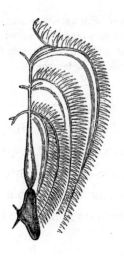  

ranged in two divisions; the first of which includes two species, having dorsal, pectoral, anal, and caudal fins; the three species of the second division possess the dorsal fin only: neither of the five species possess ventral fins. The mode in which the young are produced is very singular, but very different in the two divisions, which will be explained when describing the species.

The natural history of the *Syngnathi* appears not to have been so well understood, nor the species so clearly defined by the older authors as those of many other genera. By giving, in this work, figures taken from the specimens, and adding besides, as vignettes, enlarged representations of those parts which assist in determining specific distinction, five species, it is hoped, will be made out; and only those actually obtained on the British coast, and of which specimens are preserved, will be included. They are all marine.

Syngnathus acus, or the Great Pipe-fish, is one of the most common species, and is found on many parts of the coast, sometimes at low-water among seaweed, at other times in deep water. It is believed that the habit of proceeding to deep water at two different periods of the season has reference to important and interesting changes connected with the production of the young.

In a MS. History of British Fishes, written by the late John Walcott, Esq.* during his residence at Teignmouth in the years 1784 and 85, and which has been most kindly lent to me by his son, William Walcott, Esq. with liberty to make any use of it in the present work, I found a statement in reference to the sexes of *S. acus,* which has since been confirmed by four Continental naturalists, and which

* Author of various published works on Natural History.

I have verified by repeated examinations. Mr. Walcott's observation is as follows :—

" The male differs from the female in the belly from the vent to the tail fin being much broader, and in having for about two-thirds of its length two soft flaps, which fold together, and form a false belly (or pouch). They breed in the summer ; the females casting their roe into the false belly of the male. This I have asserted from having examined many, and having constantly found, early in the summer, roe in those without a false belly, but never any in those with ; and on opening them later in the summer, there has been no roe in those which I have termed the female, but only in the false belly of the male."

On dissecting males and females the proof of the correctness of this new view was obvious. The anal or sub-caudal pouch is peculiar to the males only, and is closed by two elongated lateral flaps. On separating these flaps, and exposing the inside, the ova, large and yellow, were seen lining the pouch in some specimens, while in others the hemispheric depressions from which the ova had been but recently removed were very visible. In each of these the opened abdomen exhibited true male organs. The females examined had no anal pouch, and the opened abdomen exposed two lobes of ova of large size. In a specimen of a male of *S. acus*, obtained at Dover on the 20th of July 1835, and for which I am indebted to W. Christy, Esq. the opened abdomen exhibited the preparatory organs of the male ; and the displayed sub-caudal pouch showed many eggs contained in it, the young of which were fully developed, and ready to escape from the capsules, while from others the young had actually escaped. They were rather more than one inch in length, and slightly barred with brown.

In the plate devoted to *Syngnathi*, in the last two octavo

editions of Pennant's British Zoology, the upper figure represents the female, and the second figure the male of *S. acus*. The enlargement on the under surface of the second figure, looking like an elongated fin, marks the situation of the distended pouch of a male. Pennant's third figure is the *S. ophidion*, and the fourth the *S. lumbriciformis* of this work. Neither *S. typhle* nor *S. æquoreus* are figured in the British Zoology.

At what time or in what manner the ova are transferred from the abdomen of the female to the sub-caudal pouch of the male is, I believe, unknown.

Mr. Walcott also adds, in his MS. that *S. acus* begins to breed when only four or five inches long. This I have also obtained proof of; and although examples of this species not uncommonly occur of eighteen inches long, and Bloch attributes to it a length of two to three feet, I have a specimen, four inches long only, a young fish apparently of the preceding year, in the opened abdomen of which the ova, in two small lobes, are full grown.

M. Risso notices the great attachment of the adult Pipe-fish to their young, and this pouch probably serves as a place of shelter to which the young ones retreat in case of danger. I have been assured by fishermen that if the young were shaken out of the pouch into the water over the side of the boat, they did not swim away, but when the parent fish was held in the water in a favourable position, the young would again enter the pouch.

The figures of *S. acus* and *typhle* are correctly represented by Rondeletius, and the characteristic difference in the form and size of the tubular mouth in each is well preserved. Below the figure, in that work, of the species now under consideration here, several of the young are represented as

VOL. II.   z

swimming near the abdomen of the parent fish. This figure
of Rondeletius is copied in Willughby, plate I. 25, fig. 6.

Mr. Couch says, " This species may be seen slowly moving about in a singular manner, horizontally or perpendi-
cularly, with the head downwards or upwards, and in every
attitude of contortion, in search of food, which chiefly seems
to be water insects."

From the great similarity in the form and size of the
mouth in all the species, it is probable that their food is
also similar. Worms, small mollusca, young and minute
thin-skinned crustacea, and the ova of other fishes, are
among the substances taken; and these *Syngnathi* are supposed to be able, by dilating their throat at pleasure, to draw
their food up their cylindrical beak-like mouth, as water is
drawn up the pipe of a syringe.

From the point of the tubular mouth to the posterior
edge of the indurated portion of the operculum, the length
is, when compared to the whole length of the fish, as one
to eight; if measured to the edge of the shoulder, it is
as one to seven and a half, and this proportion exists in
specimens of various ages or lengths, from six inches to
eighteen; from the mouth to a projecting point at the anterior edge of the eye, and thence to the origin of the pectoral fin, the distances are equal: the jaws united, tubular,
slightly compressed; in depth but one-third that of the
head at its deepest part, which is in a vertical line with
the centre of the operculum: the mouth small, placed at
the extremity of the tube, opening obliquely upwards; the
lower jaw the longest: eyes rather large, bony orbits prominent: operculum covered with radiating striæ: the head
between the eyes flattened; behind the eyes, rising into a
keel-like crest, which reaches to the neck: from the pectoral fin to the anal aperture the body is deepest and hept-

angular, with three ridges along each side, and one along the abdomen, which ends at the vent ; the surface defended by a series of nineteen plates ; throughout the short extent of the dorsal fin the body is hexangular, the ridge of the abdomen being discontinued ; thence to the end of the tail, tapering, slender, and quadrangular, with a series of forty-four plates ; the pectoral fins are small ; the dorsal fin commences at two-fifths of the whole length of the fish, and in a vertical line rather before the anal aperture ; the longest rays not equal in height to the depth of the body ; the anal fin very small ; the tail rounded and fan-shaped.

The fin-rays in number are—

D. 40 : P. 12 : A. 4 : C. 10.

The prevailing colour is pale brown, transversely barred with darker brown.

The vignette below represents the head and tail of the Great Pipe-fish from a larger specimen than that which is figured entire.

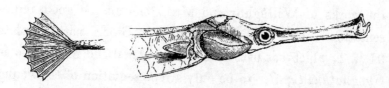

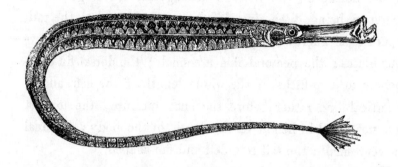

THE DEEP-NOSED PIPE-FISH.

Syngnathus Typhle,	Linnæus.
Acus Aristotelis,	*Typhle Antiquorum,* Willughby, p. 158, I. 25, fig. 1.
Syngnathus Typhle,	*Shorter Pipe-fish,* Don. Brit. Fish. pl. 56.
,, ,,	,, ,, Flem. Brit. An. p. 175, sp. 35.
,, ,,	*Lesser Pipe-fish,* Jenyns, Man. Brit. Vert. p. 485, sp. 173.

The Deep-nosed Pipe-fish is immediately distinguished from the preceding species by the more compressed form of the jaws, which are also so deep that the upper and lower edges are nearly parallel with the lines of the upper and under surface of the head. From the two large-sized Pipe-fish of the next division this species is easily known by the presence of pectoral, anal, and caudal fins. The figures in the works of Willughby and Mr. Donovan are good representations; but I believe the figure in Bloch, part iii. plate 91, f. 1, which has usually been considered and referred to as *Syngnathus typhle,* to be only a representation of the young of *S. acus.*

S. typhle has also been well figured by M. Laroche, in the Ann. de Mus. t. xiii. under the name of *S. Rondeletii.*

It is the *S. viridis* of M. Risso,* a term that seems liable to objection, even if a name were wanting, inasmuch as several other species are more or less green.

The Deep-nosed Pipe-fish does not differ materially in its habits, that I am aware of, from the species last described. The ova are transferred from the abdomen of the female to the sub-caudal pouch of the male, and there hatched in the same manner. When fishing in ten or twelve feet water over a soft surface covered with weeds, using the small net described and figured in vol. i. page 212, I have taken both sorts together, finding the deep-nosed species abundant on the Dorsetshire coast.

The whole length of the largest specimens I have seen was thirteen inches; from the point of the closed jaws to the posterior end of the indurated portion of the gill-cover, the distance is, compared to the whole length of the fish, as one to six; the head larger than in *S. acus*, and without the elevated ridge on the top of it; the distance from the point of the upper jaw to the projecting tubercle in front of the eye, and thence to the end of the pectoral fin, equal; the united jaws are very much compressed, and nearly as deep as the head, only slightly inclining to a slope before the eyes; the body hexangular; the middle lateral angle on each side becoming the upper angles of the quadrangular tail at the end of the dorsal fin. This fin commences farther back than in *S. acus*, the middle of the dorsal fin being very nearly the middle of the whole length of the fish; the series of indurated plates between the shoulder and the vent includes eighteen, thence to the end of the tail about thirty-seven; but both series are liable to a little variation in the number of these sculptured plates: the abdomen is almost rounded;

* Figured by M. Guerin, in illustration of the genera of the *Règne Animal*, *Poissons*, plate 65, fig. 1.

the anal fin minute ; the caudal fin pointed ; the two central
rays the longest ; the others graduated.

The fin-rays in number are—

D. 39 : P. 15 : A. 3 : C. 10.

The prevailing colour is olive green, mottled and spotted
with yellow brown and yellowish white.

As mentioned in the account of the Great Pipe-fish, last
described, the Deep-nosed Pipe-fish, *S. typhle*, is well
figured in the work of Rondeletius. The vignette below
represents the head and tail of this species of larger size than
the block of the whole fish would admit.

LOPHOBRANCHII. SYNGNATHIDÆ.

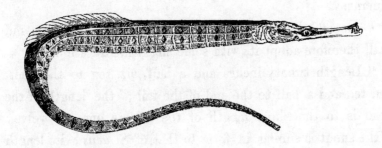

THE ÆQUOREAL PIPE-FISH.

Syngnathus æquoreus, LINNÆUS.
    ,,   ,,   Æquoreal Pipe-fish, MONTAGU, Wern. Mem. vol. i. p. 85.
                 pl. 4, fig. 1.
    ,,   ,,   ,,   ,,   PENN. Brit. Zool. vol. iii. p. 188.
    ,,   ,,   ,,   ,,   FLEM. Brit. An. p. 176, sp. 38.
    ,,   ,,   ,,   ,,   JENYNS, Man. Brit. Vert. p. 486,
                 sp. 174.

Characters. — The species belonging to the second division of the genus *Syngnathus* have a dorsal fin only; no pectoral, ventral, anal, or caudal fins; no sub-caudal pouch in either sex.

OF this division the British shores produce three species, the largest of which, the Æquoreal Pipe-fish, was described at length by Colonel Montagu from two specimens obtained on the Devonshire coast. It had been noticed as long ago as 1684 by Sir Robert Sibbald, in his *Scotia Illustrata*, part ii. book 3, page 24, who attributes to it a length of two feet.

Of this species I have not succeeded in taking any example; but I possess two, for one of which I am indebted to the kindness of Mr. Embleton, of the Berwickshire Naturalists' Club, who obtained it on the coast of that or the adjoining county; and for the other to Mr. Couch. By communication from F. C. Lukis, Esq. I learn also, that this species, which I believe to be rare in England, occurs at Guernsey.

I can add but little to the description of Montagu, and shall therefore adopt it, with some slight modifications.

" Length twenty inches and a half, viz. ten to the vent, and ten and a half to the end of the tail ;" the length of the head is to the whole length of the fish as one to twelve : " the snout is similar in form to that of *S. acus ;* its length to the eye three-quarters of an inch ; from thence to the end of the gill-cover, including the eye, one inch : the form of the body is rather compressed and angular, with an acute dorsal and abdominal ridge, which, together with three slight angles on each side, give it an octangular appearance : it is of equal size from the gills to the vent, which part contains about thirty plates ; from the vent to the extremity of the tail it is first quadrangular, and towards the end, round and taper, containing about thirty-six plates : immediately behind the vent, the body of this specimen suddenly decreases to one-third less in diameter ; but this may be a sexual distinction."

" The dorsal fin consists of forty rays, commencing in a vertical line considerably before the vent, and terminating behind it, so that three-fourths of the fin is before the ventral aperture. The end of the tail is extremely small and compressed, the rays of which are not visible to the naked eye. The colour is yellowish, with transverse pale lines and dark margins, one in each joint, and another down the middle of each plate, giving it the appearance of possessing double the number of joints it really has ; these markings, however, cease at the vent."

Mr. Couch, it appears, has not seen more than two or three specimens; but the Cornish fishermen say they find this species from ten to fifteen leagues from land, and in fine weather swimming at the surface over a depth of fifty fathoms or more.

My specimens being both females, the sexual peculiarities of this division of the *Syngnathi* will be explained when describing the next species.

The vignette below represents the head and tail of this species on a larger scale.

LOPHOBRANCHII.   *SYNGNATHIDÆ.*

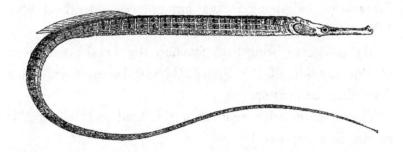

THE SNAKE PIPE-FISH.

Syngnathus ophidion, *Serpent de Mer,* BLOCH, pt. iii. pl. 91, fig. 3.
    ,,   ,,   *Snake Pipe-fish,* SHAW, Gen. Zool. vol. v. p. 453, pl. 179.
    ,,   ,,   *Longer Pipe-fish,* Low, Faun. Orcad. p. 179, sp. 1.
    ,,   ,,   *Snake Pipe-fish,* JENYNS, Man. Brit. Vert. p. 487, sp.
                        175.

No species of *Syngnathus* can better deserve the name of *ophidion,* snake-like, than the present. It is immediately distinguishable from the fish last described, with which alone it is likely to be confounded, by its much more slender as well as rounder body, which scarcely exceeds a goose-quill in size, and by the whole of the dorsal fin being, in a specimen of fourteen inches long, more than half an inch before the middle of the fish. Pennant has figured this fish, No. 61 of plate 26, but not described it.

In this species, as well as the two others belonging to this second division, neither male nor female possesses an anal pouch, but the ova after exclusion from the abdomen of the female are carried for a time by the male in separate hemispheric depressions on the external surface of the abdomen, anterior to the anus. The females have no such depressions. The sexes have been determined by examination of the internal structure. All the specimens examined having these external hemispheric cells proved to be males, the testes in the abdomen obvious; those without external depressions proved to be all females, internally provided with two lobes of enlarged ova. The males of this species when taken by me as late in the season as August, had one ovum of the size and colour of a mustard-seed lodged in each cup-shaped cell. These specimens were caught with a keer-drag net between Brownsey Island and South Haven, at the mouth of Poole Harbour. Many specimens of *S. acus* and *typhle* were obtained at the same time and place.

The length of the head in this species is, compared to the whole length of the fish, as one to eleven; the form of the body slightly octangular, but more slender and rounded than in that last described; the body uniform in size as far as the vent, then tapering gradually to the tail, which has a slightly flattened end; the divisions in the series of transverse plates, and the angles, of the body, almost obsolete; the dorsal fin, as before mentioned, entirely anterior to the middle of the fish; the number of rays thirty-eight: the vent in a line with the last fourth portion of the dorsal fin.

The colour of the body is a uniform olive green; the irides red, the pupils black. The specimens I possess vary in length from eight inches to fourteen inches.

LOPHOBRANCHII.   *SYNGNATHIDÆ.*

THE WORM PIPE-FISH.

Syngnathus lumbriciformis, Worm Pipe-fish, Jenyns, Man. Brit. Vert. p. 488,
sp. 176.

Acus lumbriciformis,   Willughby, p. 160.
Syngnathus ophidion,   *Little Pipe-fish,* Penn. Brit. Zool. vol. iii. p. 187,
pl. 26, No. 62.

,,   ,,   ,,   ,,   Flem. Brit. An. p. 176, sp. 39.

The Worm-like Pipe-fish is the smallest of the British
species, and is taken on various parts of the coast.   Mr. Low
describes it as found at Orkney under stones; and Mr. Couch
finds it in similar situations on the coast of Cornwall, where
it is considered common.

Pennant has figured this species with the ova attached to
the under and external surface of the abdomen, as in the
species last described.   There is little doubt that the young
are produced in the same mode as in the other species belonging to this division of the genus, and that the same
sexual peculiarities exist.   Pennant, not aware of the singular interchange which takes place, says, very naturally, " On
the belly of the female is a long hollow, to which adhere the
eggs disposed in three rows."

This species does not exceed five inches or five inches and
a half in length, and the wood-engraving at the head of the

preceding page represents this fish but little less than its natural size. It possesses no fin except that on the back, which in the specimen I examined contained thirty rays. The nose is very short, turned a little upwards; the eyes prominent; from the point of the jaws to the posterior edge of the orbit, and thence to the end of the operculum, the distances are equal; the length of these two portions together, compared to the whole length of the fish, is as one to twelve; the form of the body nearly cylindrical; the vent is situated at the end of the first third of the whole length, with a series of nineteen plates before it, and in a vertical line, with three-fourths of the dorsal fin behind it; from the vent the body tapers gradually all the way to the tail, which ends in a point; the number of plates forming the series between the vent and the tail-end, about fifty. The surface of the body is more smooth than in the two species previously described, and the colour is dark olive green.

LOPHOBRANCHII.   *SYNGNATHIDÆ.*

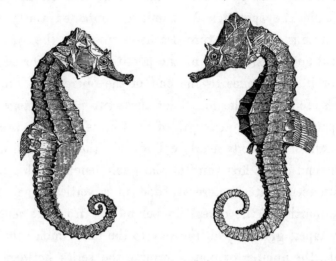

THE SHORT-NOSED HIPPOCAMPUS.

Hippocampus brevirostris, Cuvier, Règne An. t. ii. p. 363.
  ,,   *Rondeletii,* Willughby, p. 157, I. 25, fig. 3.
  ,,   *brevirostris, Sea-horse,* Jenyns, Man. Brit. Vert. p. 489, sp. 177.

Generic Characters.—The jaws united and tubular, like those of the *Syngnathi*; the mouth placed at the end; the body compressed, short, and deep; the whole length of the body and tail divided by longitudinal and transverse ridges, with tubercular points at the angles of intersection; both sexes have pectoral and dorsal fins; the females only have an anal fin; neither sex has ventral or caudal fins.

PENNANT, in the edition of his British Zoology, the three first volumes of which were published in 1776 and the fourth in 1777, states that he had been informed the *Syngnathus Hippocampus* of Linnæus, or what the English improperly call the Sea-horse, had been found on the southern shores of this kingdom." John Walcott, Esq. whose MS. History of British Fishes was written in the years 1784 and 1785, says, in reference to a drawing of a female specimen of what I believe to be the *Hippocampus brevirostris* of

Cuvier, " This was taken on the coast of Hampshire, and given me by the late Mr. Brander."

L. W. Dillwyn, Esq. obtained a specimen of *Hippocampus*, some years ago, in Swansea Bay ; and Messrs. C. and J. Paget, in their Sketch of the Natural History of Yarmouth, state that the *Hippocampus* is also occasionally met with there.

But the most valuable information I have received on this subject has been supplied by F. C. Lukis, Esq. of Guernsey, to whom I am indebted for the loan of the male and female specimens from which the figures at the head of the preceding page were taken.

By a comparison with M. Guerin's figure of the *Hippocampus brevirostris* of Cuvier's *Règne Animal*, I have little doubt that the two fishes here figured are examples of *H. brevirostris ;* and Mr. Lukis, in the autumn of 1835, obtained two specimens of *Hippocampus* on the Hampshire coast, one of which is stated to be identical with those here figured : there can therefore be no doubt, from these various authorities, that at least one species is found on our coast, and that this species is also obtained at Guernsey and the other Channel islands.

The circumstance of the same species occurring at Guernsey and on our southern coast, gives additional value to the following communication. At the time of writing, June 9, 1835, Mr. Lukis had two female specimens of *Hippocampus brevirostris*, then healthy and active, which had been living twelve days in a glass vessel, their actions equally novel and amusing. " An appearance of search for a resting-place induced me," says Mr. Lukis, " to consult their wishes by placing seaweed and straws in the vessel : the desired effect was obtained, and has afforded me much to reflect upon in their habits. They now exhibit many of their peculiarities, and few subjects of the deep have displayed, *in prison*, more sport or more intelligence."

" When swimming about, they maintain a vertical position ; but the tail is ready to grasp whatever meets it in the water, quickly entwines in any direction round the weeds, and, when fixed, the animal intently watches the surrounding objects, and darts at its prey with great dexterity."

" When both approach each other, they often twist their tails together, and struggle to separate or attach themselves to the weeds ; this is done by the under part of their cheeks or chin, which is also used for raising the body when a new spot is wanted for the tail to entwine afresh. The eyes move independently of each other, as in the chamelion ; this, with the brilliant changeable iridiscence about the head, and its blue bands, forcibly remind the observer of that animal."

The vignette in illustration of the habits here described was copied from a drawing by Mr. Lukis, most obligingly lent me for this purpose.

By the kindness of William Walcott, Esq. I learn that a gentleman of the Island of Jersey, an attentive observer of nature, remembers having more than once seen specimens of *Hippocampus* curled up in oyster-shells. About four years since, a specimen was shown at Southampton, which lived more than a fortnight in a glass globe. This was said to have been obtained on the French coast near Granville, and was brought to Southampton by one of the sailors of a steam-packet ; I have also heard of one that lived three weeks in confinement at Harwich, the undulating motion of which when swimming was performed with great ease, and was very interesting to observe.

The species of *Hippocampus* in their sexual peculiarities, as far as they have been investigated, appear to coincide with those of the *Syngnathi*. I had the pleasure of looking over,

with Mr. Owen, some specimens in the collection of the
Royal College of Surgeons, which had been examined and
the internal structure partly exposed to view by the dissections of John Hunter. The females with the abdomen
enlarged, as shown in the right-hand figure at the head of this
article, have a small anal fin of four rays, but no true pouch ;
the ova in the abdomen. Males have no anal fin, in any of
the specimens I have examined ; the pouch obvious; the
abdomen smaller than in the females, as shown in the left-
hand figure. The two specimens represented in the vignette
are both females.

Their food is unknown to me, but is probably very similar
to that taken by the *Syngnathi.*

The whole length from the point of the nose to the end of
the tail is about five inches : the connected jaws, forming a
tubular mouth, are considerably shorter than the rest of the
head : the eyes prominent, the irides straw yellow ; over
each eye a single prominent spinous tubercle : the operculum
covered with striæ, radiating from the front ; the pectoral fins,
placed immediately behind the operculum, are small, apparently containing about eight rays in each ; the form of the
body heptangular, three angles on each side, the seventh
longitudinal angular line being on the abdomen ; the back
flat ; the transverse segments of the body eleven, with tubercular projections at the points of intersection ; the rays of
the dorsal fin about sixteen : the anal fin is peculiar to the
female only, and probably performs some office at the time
of the transfer of the ova to the pouch of the male ; this
anal fin contains four rays : the abdomen as deep again as the
tail ; from the vent the form of the tail is quadrangular, ending in a point ; the number of segments about thirty.

The general colour is a pale ash brown, relieved by a

VOL. II.   2 A

changeable iridiscence, and variable tints of blue dispersed over different parts of the head, body, and tail.

I have not included any reference to Linnæus or Bloch in the synonymes, being doubtful that the species are identical with the one here described. Klein's figure, No. 10, very closely resembles the *H. brevirostris;* and by the description, the *H. antiquus* of Risso is also the *H. brevirostris* of Cuvier.

PLECTOGNATHI.   *GYMNODONTIDÆ.*

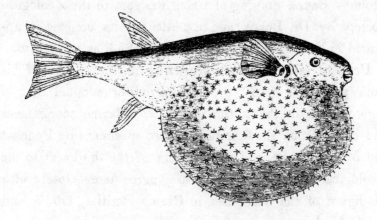

PENNANT'S GLOBE-FISH.

Tetrodon Pennantii,	*Pennant's Globe-fish,*	YARRELL.
,, *stellatus,*	*Stellated Globe-fish,*	DON. Brit. Fish. pl. 64.
Tetraodon lævigatus,	*Globe Diodon,*	PENN. Brit. Zool. ed. 1776, vol. iii. p. 132, pl. 20.
Tetrodon lagocephalus,	*Globe Tetrodon,*	PENN. Brit. Zool. ed. 1812, vol. iii. p. 174, pl. 23.
Tetraodon stellatus,	*Stellated Globe-fish,*	FLEM. Brit. An. p. 174, sp. 31. Proceedings Zool. Soc. 1833, p. 115.

Generic Characters.—Both jaws are divided in the middle by a suture, producing the appearance of four teeth in front, two above and two below. The skin, over part of the body, armed with numerous short spines. The branchial orifice small.

IN this order of fishes, the *Plectognathi* of Cuvier's arrangement, the principal distinctive character consists in the maxillary bones being firmly attached to the intermaxillaries, and both united to the palatine arch

Three examples of this singular-looking fish have been taken in this country, and all three in Cornwall. Pennant first described it as British from a specimen caught at Penzance. Mr. Donovan has recorded a second, taken on the

2 A 2

Cornish coast, and mentions another obtained in the European seas. Still more recently a specimen was taken in Mount's Bay, a drawing of which was sent to the Zoological Society by Dr. Boase, and a notice of its occurrence appeared in the Proceedings for October 1833, as referred to.

Pennant called his fish *lævigatus* in the edition of his work published in 1776, and his editor adopted that of *lagocephalus* in the edition of 1812, referring to Linnæus and Bloch ; but the figure of the two specimens by Pennant and Mr. Donovan, and the drawing of the third sent to the Zoological Society by Dr. Boase, agree more closely with the figure of the Globe-fish in Grew's Rarities, tab. 7, and the *Orbis lagocephalus* of Willughby, plate I. 2, which appear to be intended to represent the same fish, and being without spots or stripes, is, I think, distinct from the *lagocephalus* of Linnæus and Bloch, the spots of which are referred to in the description of the one, and both spots and stripes shown in the coloured figure of the other.

Mr. Donovan, when calling this fish *stellatus*, appears not to have been aware that this term had been previously appropriated to an Indian species with black spots ; and still considering this fish provisionally as a new species, I propose for it now the name of our highly-esteemed British zoologist, by whom, as far as I am aware, it was first made known.

" The species of this genus are remarkable for being provided with the means of suddenly assuming a globular form by swallowing air, which, passing into the crop or first stomach, blows up the whole animal like a balloon. The abdominal region being thus rendered the lightest, the body turns over, the stomach being the uppermost part, and the fish floats upon its back, without having the power of directing itself during this state of forced distension. But it is while thus bloated and passive, at the mercy of the waves, that this animal is really most secure ; for the numerous spines with which the surface of the body is universally beset are raised and erected by the stretching out of the skin, thus presenting an armed front to the enemy on whatever side he may venture to begin the attack."*

Pennant's fish measured one foot seven inches in length ; the belly when distended, one foot ; the whole circumference when in that state, two feet six inches. The form of the body is usually oblong ; but when alarmed, it assumes the shape which has been already referred to. The mouth is small ; the irides white, tinged with red ; the back from head to tail almost straight, or at least very slightly elevated ; there are no ventral fins ; the dorsal fin is placed low on the back ; the anal fin is opposite ; the tail almost even, divided by an angular projection in the middle.

The number of fin-rays according to Mr. Donovan—

D. 11 : P. 14 : A. 10 : C. 6.

The back is of a rich blue colour ; the belly and sides silvery white, studded over with straight spines arising from the centre of four rays ; the fins and tail brown. The spines in Pennant's representation of this fish are not so thickly set as in the figure of Mr. Donovan, or in the drawing by Dr. Boase ; but the space over which they are spread is alike in all three,—that is, bounded superiorly by the lower jaw and the base of the pectoral fin, and posteriorly by the anal aperture.

* Dr. Roget. Bridgewater Treatise, vol. i. p. 433.

PLECTOGNATHI.   *GYMNODONTIDÆ.*

THE SHORT SUN-FISH.

Orthagoriscus mola,	Schneider.	Cuvier, Règne An. t. ii. p. 369.
,,	*Rondeletii, Sun-fish,*	Willughby, p. 151, I. 26.
Tetrodon mola,	*Short Tetrodon,*	Penn. Brit. Zool. vol. iii. p. 172, pl. 22.
,, ,,	*Sun-fish,*	Don. Brit. Fish. pl. 25.
Orthagoriscus mola,	*Molebut,*	Flem. Brit. An. p. 175, sp. 32.
,, ,, ,,		Jenyns, Man. Brit. Vert. p. 490, sp. 179.

Generic Characters.—Jaws undivided, forming a cutting edge; body compressed, deep for its length, short, truncated, without spines; tail short, and very high vertically; rays of the dorsal and anal fins long and pointed, both united to the caudal fin at the base.

THE SUN-FISH, as this species has been called from the twofold circumstance of its almost circular form and shining surface, though occurring but occasionally, may be said to have been taken from John o'Groat's to the Land's End.

Sir Andrew Balfour and Sir Robert Sibbald have noticed this species in Scotland, and Dr. Neill mentions three examples that were taken in the Frith of Forth. I am indebted to Edward Jesse, Esq. for a memorandum of one caught on the coast of Northumberland in October 1834. Dawson Turner, Esq. and Mr. Paget have known it to be taken at Yarmouth. I have seen one that was brought to the London market. Colonel Montagu, in his MS. notes, mentions one that was caught at Salcombe in July 1799: this specimen was of large size, and weighed three hundred pounds. In the fifth volume of Mr. Loudon's Magazine of Natural History, page 315, there is a record of one that was taken at Plymouth; and Dr. Borlase, Willughby, and Mr. Couch have seen and described examples that were taken on the Cornish coast. Still farther to the westward and northward, the Sun-fish has been taken in the Bristol Channel, and one was caught during last summer at Tenby. On the Irish coast, it has been taken at Londonderry; and I am greatly indebted to the kindness of Dr. Arthur Jacob, Professor of Anatomy in the Royal College of Surgeons in Dublin, for his remarks on a specimen taken in the month of August 1826, between the south-west coast of England and Dublin Bay. This paper was published in the Dublin Philosophical Journal for November 1826, and is the best account of this fish that I am acquainted with.

When observed in our seas, they have generally appeared as though they were dead or dying, and floating along on one side, presenting the broad surface of the other side to view. Dr. Neill says, of one that was brought to him, " The fishermen informed him, that when they observed it, it was swimming along sideways, with its back-fin frequently above water. It seemed to be a stupid, dull fish : it made little or no attempt to escape, but allowed one of the sailors to put his hands under it, and lift it fairly into the boat. The Sun-fish has been generally mentioned as remarkable for its phosphorescence ; but this specimen did not exhibit that phenomenon so distinctly as a Haddock or a Herring." Pennant repeats Brunnich's account, that between Antibes and Genoa he saw one of this species lie asleep on the surface of the water : a sailor jumped overboard and caught it.

Mr. Couch says the Short Sun-fish is migratory, keeping probably at the bottom, and feeding on seaweeds in its ordinary habits ; but in calm weather it mounts to the surface, and lies, perhaps asleep, with its head and even its eyes above the water, floating with the tide. Mr. Couch has known the Sun-fish make powerful but awkward efforts to escape when attacked, bending and directing its motions in various ways.

The figure here given, and the description, are taken from a preserved specimen in the Museum of the Zoological Society. This is the smallest example I have seen. It measures but fourteen inches from the point of the nose to the end of the body ; the breadth of the caudal fin two inches ; the depth of the body eleven inches and a half : the length of the dorsal fin eight inches ; of the anal fin, seven inches and a half : the extension of skin connecting the fin-rays rather thick. The mouth small ; the branchial aperture just in advance of the pectoral fin, small and oval ; the vent just before the anal fin ; the caudal fin occupying the whole space between the anal and dorsal fins, and attached to the posterior vertical edge of the body as by a long hinge ; the surface of the body in this young specimen but slightly roughened, and somewhat wrinkled. The colour of the upper part of the body dusky bluish grey ; the lower part olive brown. The fin-rays in number are—

D. 15 : P. 11 : A. 15 : C. 13.

The figure at the head of this subject is an exact representation of the fish from which it was taken, differing only in size; and from some differences that appear in the descriptions of specimens of greater bulk, there is reason to believe this fish alters in appearance as it increases in age. In a much larger example the skin was of a uniform dirty pale brown; the texture hard, rough, coarse, and thick. According to Dr. Jacob, the irides in his specimen were dull greyish brown, with a silvery ring round the pupil.

I am indebted to Mr. Couch for the under jaw-bone of a Sun-fish of considerable size. The outer margin of this bone, for three inches round the front, in which there is no division, is covered to its edge by a narrow band of enamel: the inside, near the centre, contains various dull pearl-like teeth; some thin and flat, presenting an edge; behind them others, more cylindrical, short, and rather pointed.

Upon the external surface of the head of the example of the Sun-fish taken at Tenby, there were attached about twenty specimens of *Tristoma coccineum*. Two of these were given to me by H. E. Strickland, Esq. of Cracombe House, Gloucestershire, from one of which the representations in the vignette below of the upper and under surface were taken of the natural size. For an account of two species of these very rare parasitic animals, see the *Synopsis Entozoorum* of Rudolphi, page 427.

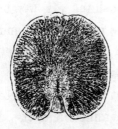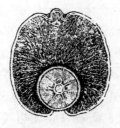

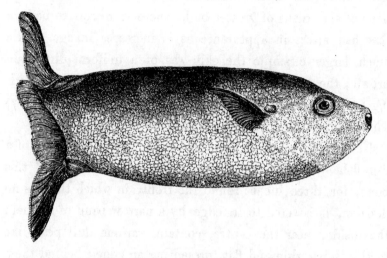

THE OBLONG SUN-FISH.

Orthagoriscus oblongus,	Schneider. Cuvier, Règne An. t. ii. p. 370.	
Tetrodon truncatus,	*Oblong Tetrodon,*	Penn. Brit. Zool. vol. iii. p. 170, pl. 22.
,, ,,	*Truncated Sun-fish,*	Don. Brit. Fish. pl. 41.
Orthagoriscus truncatus,	,, ,,	Flem. Brit. An. p. 175, sp. 33.
,, *oblongus,*	*Oblong Sun-fish,*	Jenyns, Man. Brit. Vert. p. 491, sp. 180.

It has been the opinion of some naturalists that this oblong Sun-fish is the same species as that last described, and that its greater length in proportion to its depth is but the consequence of greater age. The largest Short Sun-fish of which I have weight and measurement was that taken at Salcombe on the coast of Devonshire. It weighed three hundred pounds, was four feet five inches long, and six feet from the tip of the dorsal fin to the end of the anal fin. This was seen by Colonel Montagu, who was too keen an observer and too good a naturalist not to have detected the difference between the fish he examined and the described characters of the true *mola*, if any had existed; and this fish being probably very old, as well as very large, was the more likely to have assumed the elongated appearance of this second species.

The Oblong Sun-fish seems to be much more rare than that last figured. Dr. Borlase appears to be the first and almost the only English writer who has seen and described it. In his Natural History of Cornwall, he speaks of it under the title of the Sun-fish from Mount's Bay, after having described and figured the Short Sun-fish, and mentions that a specimen of this second species was taken at Plymouth in 1734, that weighed five hundred pounds.

Mr. Donovan, in his Natural History of British Fishes, says, " We have seen the dried skin of this species, the animal of which, when living, weighed between two and three hundred pounds. Our figure is taken from a small specimen, obtained in a recent state, in one of our fishing excursions on the Bristol Channel. This fish subsists on worms of the testaceous and other tribes, small crabs, &c. fragments of these being found on dissection in the stomach."

Mr. Couch, in his MS. says, " I have never met with this fish; but a fisherman informs me he once took a Sun-fish differing in figure from that with which he was familiar, and which from his description I judge to be this fish. His attention was particularly attracted by the curious and beautiful waved stripes which it acquired after death, but which he did not observe while it was alive."

Never having seen a specimen of this fish, the figure here given is from Mr. Donovan's work, and the description from that of Pennant.

" This fish grows to a great bulk; that which was examined by Salvianus was above a hundred pounds in weight. In form it resembles a Bream, or some deep fish cut off in the middle. The mouth is very small, and contains two broad teeth, with sharp edges."

" The eyes are little; before each is a small semilunar aperture: the pectoral fins very small, and placed behind them. The dorsal fin and the anal fin are high, and situated at the extremity of the body; the tail fin is narrow, and fills all the abrupt space between those two fins."

" The colour of the back is dusky, and dappled; the belly silvery; between the eyes and the pectoral fins are certain streaks pointing downwards. The skin is free from scales."

" When boiled it has been observed to turn into a glutinous jelly, resembling boiled starch when cold, and served the purposes of glue, on being tried on paper or leather. The flesh of this fish is uncommonly rank; it feeds on shell-fish."

" There seems to be no satisfactory reason for the old English name. Care must be taken not to confound it with the Sun-fish of the Irish,* which differs in all respects from this."

Dr. Turton describes the body of the Oblong Sun-fish to be nearly thrice as long as it is deep: the aperture of the gills semilunar. The fin-rays in number, according to Mr. Donovan, are—

D. 12 : P. 14 : A. 15 : C. 17.

---

* The Sun-fish of the Irish coast, particularly on the west coast of Ireland, is the Basking Shark, to be hereafter described, which sometimes attains a length of thirty feet, and is so called from its habit of basking and sunning itself at the surface of the water.

PLECTOGNATHI. *BALISTIDÆ.**

THE EUROPEAN FILE-FISH.

Balistes capriscus, Cuvier, Règne An. t. iii. p. 372.
Capriscus Rondeletii, Pesce Balestra, Willughby, p. 152, I. 19.
Balistes maculatus, File-fish, Bloch, pt. v. pl. 151 ?

Generic Characters.—Body compressed, covered with hard rhomboidal imbedded plates, which are not imbricated like scales; two dorsal fins, the first containing spines only, the second long; mouth with incisor-like cutting teeth in each jaw.

THE only example of this genus which has occurred in the English seas, that I am aware of, was taken off the Sussex coast in the month of August 1827; and the circumstance was made known by J. G. Children, Esq. who obtained the specimen, and who recorded this interesting capture in his address delivered at the anniversary meeting of the Zoological Club of the Linnean Society on the 29th of November of the same year. The specimen was exhibited.

This fish has since been deposited in the national collection at the British Museum; and by the kindness of the officers of the natural history department of that establish-

* The family of the File-fishes.

ment, I have been permitted to take a drawing and description from the specimen caught in our seas.

The *Balistes capriscus* is a species well known to the older authors as an inhabitant of the Mediterranean ; is figured by Salvianus ; by Grew, in his Rarities, tab. 7 ; and by Klein, tab. 3. It is, however, rather rare, though stated also to be an inhabitant of other seas. M. Risso says the flesh is tolerably good.

Baron Cuvier, in the *Règne Animal*, in part of the first note at the foot of page 372, says in reference to *Balistes capriscus*, " Je suis même tenté d'y rapporter le *B. buniva* de Lacépède." Possessing a dried specimen of *B. buniva* from the Mediterranean, which agrees exactly with the published descriptions of that species by Lacépède and M. Risso, I have compared it with the specimen of *B. capriscus* at the British Museum, and feel confident that the *B. buniva* of Lacépède is, as Cuvier suspected, identical with the *B. capriscus* of authors.

The first and strongest spine of the back in this fish is studded up the front with numerous small projections, which under the microscope have the appearance of so many points of enamel or pearl arising from the surface of the bone, giving a rough denticulated appearance ; and hence the name of File-fish. The second smaller spine has at the anterior part of the base a projection which, when the spines are elevated, locks into a corresponding depression in the posterior part of the base of the first spine, and fixes it like part of the work in a gun-lock ; and from this similarity this fish on the Italian shores of the Mediterranean is called *Pesce balestra*. The longest spine cannot be forced down till the shorter spine has been first depressed.

The length from the nose to the branchial orifice is to the whole length of the fish as one to four ; the depth of the body is rather less than half the whole length of the fish, the tail included in both measurements : the body compressed ; the surface hard ; the scales arranged in oblique lines over the whole breadth ; no lateral line observable, except along the middle of the fleshy portion of the tail : the mouth small and narrow ; the visible teeth four on each side the centre above and below, incisor-like or cutting ; the forehead wide between the eyes, which are small, enclosed in well-defined orbits ; the branchial orifice an elongated aperture commencing in the front at the base of the pectoral fin, and ascending obliquely backward ; pectoral fin of small size : first spine of the first dorsal fin in a vertical line over the branchial orifice, the second close behind and attached by a strong ligament ; the third spine removed to a distance, but connected by a membrane : the second dorsal fin is high anteriorly and long, commencing in a vertical line before the commencement of the anal fin, but both ending on the same plane, and far short of the base of the caudal rays ; in advance of the anal fin is a strong rough keel, which has some resemblance to ventral fins : the fleshy portion of the tail free, and rather long ; the rays nearly square at the end, large and strong.

The fin-rays in number are—

D. 3. 28 : P. 15 : A. 26 : C. 14.

The colour in the dried specimen is nearly a uniform pale brown ; rather darker on the back ; becoming lighter on the belly, and particularly on the under surface of the head : the naked gums smooth and dark brown. Living specimens are said to be tinged and even spotted with blue ; and it is probable that an individual in this state has furnished the material on which *B. maculatus* of Bloch is founded : the irides are described as green.

The whole length of the Museum specimen is nine inches and a half ; the depth four inches and three-eighths without the dorsal or anal fins.

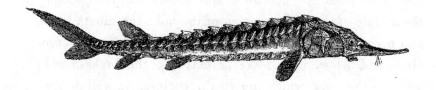

THE COMMON STURGEON.

Acipenser Sturio, *Common Sturgeon,* Linnæus. Bloch, pt. iii. pl. 88.
Sturio, *The Sturgeon,* Willughby, p. 239, P. 7, fig. 3.
Acipenser Sturio, *Common Sturgeon,* Penn. Brit. Zool. vol. iii. p. 164, pl. 22.
„ „ *L'Esturgeon,* Cuvier, Règne An. t. ii. p. 379.
„ „ *Common Sturgeon,* Don. Brit. Fish. pl. 65.
„ ,, *Sturgeon,* Flem. Brit. An. p. 173, sp. 30.
„ „ *Common Sturgeon,* Jenyns, Man. Brit. Vert. p. 493, sp. 182.

Generic Characters.—Body elongated and angular, defended by indurated plates and spines, arranged in longitudinal rows ; snout pointed, conical ; mouth placed on the under surface of the head, tubular, and without teeth.

All the remaining portion of the British Fishes to be yet described belong to Cuvier's division called *Chondroptery-giens,* or Cartilaginous Fishes, the skeletons of which are made up of cartilage, and not, as in the divisions of *Acan-thopterygiens* and *Malacopterygiens,* made up of true bone. The earthy matter in the hard parts of these fishes is smaller in quantity, is deposited in grains, and does not assume the form, as in other fishes, of distinct osseous fibres.

In the fishes of the families contained in this order, there are several interesting peculiarities. Some have their gills free, like those of ordinary fishes ; there are others in which the gills are fixed by having their outer edges attached to the

* The family of the Sturgeons.

skin. Several of them bring forth their young alive in a manner very different from any of the true bony fishes; while some, and these the last in the series, want that degree of organization in the bones of the upper jaw observable in those generally which have been hitherto described, but of which two or three examples of deficiency by malformation have been figured.

This order may be said to be further distinguished by including within its limits fishes exhibiting in certain points of their structure the highest as well as the lowest degrees of organization observable throughout the whole class. These different peculiarities will be pointed out on arriving at the different genera in succession.

The Sturgeon, the first of the cartilaginous fishes, allied to the Sharks in the elongated form of its body, resembles other fishes in having the gills free, and in being oviparous. It is caught occasionally on various parts of our coast, most frequently in the estuaries, or but a short distance up rivers; very seldom taken in the open sea, where it is believed to inhabit deep water, beyond the reach of nets, and is not, that I am aware, ever caught upon the fishermen's lines. Dr. Neill says that one or two are generally taken every summer about the mouth of the Almond or of the Esk, where they get entangled in the Salmon nets, and when of large size frequently doing the fishermen considerable damage by tearing their nets. They are otherwise harmless. One caught-in a stake net near Findhorn in Scotland in July 1833, measured eight feet six inches in length, and weighed two hundred and three pounds.

The Sturgeon is occasionally taken on the East coast, and frequently brought to the London market from various localities. When caught in the Thames, within the jurisdiction of the Lord Mayor, it is considered a Royal Fish;

VOL. II.   2 B

the term being intended to imply that it ought to be sent to the King, and it is said that the Sturgeon was exclusively reserved for the table of Henry the First of England.

On our Southern coast, Colonel Montagu mentions one taken in the estuary at Kingsbridge ; and Mr. Couch enumerates three instances at different periods of different years; one in the Tamar in June, one at Plymouth in August, and one near the Eddystone in January. In September 1802, a specimen, eight feet long, and weighing one hundred and ninety-two pounds, was caught in a weir below the castle at Shrewsbury. The largest specimen taken in this country is probably the fish recorded by Pennant, which was caught in the Esk, and weighed four hundred and sixty pounds.

In Ireland the Sturgeon has been taken on the south, the east, and the north coasts.

In the northern parts of Europe this fish is much more numerous than with us, and extensive fisheries are established for its destruction. Caviar is made of the roe of the female ; isinglass is obtained from the dense membrane forming the air-bladder ; and the flesh, besides being preserved by salting and pickling, is in request for the table while fresh, being generally stewed with rich gravy, and the flavour considered to be like that of veal. The flesh, like that of most of the cartilaginous fishes, is more firm and compact than is usual among those of the osseous families.

The Sturgeon, as has been before observed, is oviparous, spawning in winter. It has been frequently remarked that Sturgeons of very small size are seldom seen : by the kindness of Mr. George Daniell, however, I possess a small specimen, only twelve inches long, that is quite perfect, and exhibits all the characters of the mature fish. " It is presumed that the young, as soon as they escape from the eggs, which the female deposits in fresh water, descend immediately to the sea, and do not visit the places of their birth again till they come in their turn to deposit their spawn." The Sturgeon is said to subsist on small fishes; from the structure of the mouth it probably feeds also on any soft substance that it finds at the bottom.

The body is elongated; from the shoulders backward somewhat pentagonal in shape, with five longitudinal rows of flattened plates, with pointed central spines directed back-wards,—one row, larger than the others, along the ridge of the back, one row on each side, and another along the edge of the abdomen in a line from the pectoral fin to the ven-tral on each side; the flattened plates are marked with ra-diating striæ. The nose is long and pointed; the forehead with a longitudinal depression; the crown of the head ele-vated, the occiput rising into a sharp keel: the mouth placed on the under surface of the head, rather wider than long, with a projecting rim; no teeth within: about half-way between the mouth and the end of the nose, are four cirri ranged in a line across; the eyes small; the operculum hard and strong, covered with striæ radiating from a centre; dorsal fin placed very far back, but little in advance of the line of the anal fin: tail forked; upper lobe much the long-est, and pointed. The fin-rays in number—

D. 35 : P. 28 : V. 24 : A. 23 : C. 125.

The colours of the body are various shades of brown; the plates nearly white, the belly silvery.

The vignette represents the under surface of the head.

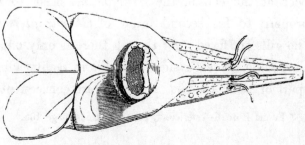

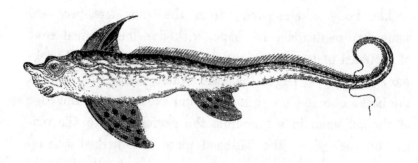

NORTHERN CHIMÆRA.

KING OF THE HERRINGS.   RABBIT-FISH, *Zetland.*

Chimæra monstrosa,	LINNÆUS.	BLOCH, pt. iv. pl. 124.	
,,	,,	*Northern Chimæra,*	PENN. Brit. Zool. vol. iii. p. 159.
,,	,,	*Sea Monster,*	DON. Brit. Fish. pl. 111.
,,	,,	*Rabbit-fish,*	FLEM. Brit. An. p. 172, sp. 29.

Generic Characters.—Body elongated, the tail ending in a lengthened filament; the first dorsal fin short at its base, but high; the second dorsal fin low,
commencing immediately behind the first, and extending to the tail.

THIS fish has considerable resemblance to the Sharks in
the form of the body, and the position as well as the shape
of the fins.

" The Chimæræ," says Dr. Richardson,* " though placed
by Cuvier at the end of the *Sturionidæ,* seem to belong
more properly to his second order of *Chondropterygii,* in
which the gills are fixed : for though there is only one apparent gill-opening on each side, the gills in reality adhere by
a large part of their borders, and there are consequently five

* Fauna Boreali-Americana, part iii. Fishes, page 285.

holes communicating with the external gill-opening. They have a rudimentary operculum concealed by the skin; and their jaws, still more reduced than those of the Sharks, are furnished with hard plates, four above and two below, in place of teeth. The males are distinguished by trifid bony appendages to the ventral fins, and produce very large leathery eggs, having flat velvety edges."

The Northern Chimæra is represented as a fish of singular appearance and beauty, a native of the northern seas only, where it seldom exceeds three feet in length, and is generally taken when in pursuit of shoals of Herrings, or other small roving fishes, upon which it principally subsists : Bloch says it feeds also on medusæ and crustacea. The flesh is described as hard and coarse. According to some authors, the Norwegians extract an oil from the liver which they consider of singular efficacy in disorders of the eyes.

Pennant received from a gentleman a drawing and particulars of one that had been taken among the Shetland Islands : this species was also known to Dr. Walker as an occasional visitor in that locality. Never having seen this fish, I avail myself of Dr. Fleming's description, taken from a specimen sent by L. Edmonston, Esq. from Unst, where it is termed the Rabbit-fish. A specimen taken from the same locality has lately been received by Mr. W. C. Hewitson of Newcastle, the author of a valuable work on the eggs of British Birds.

" Length nearly three feet. Body compressed. Head blunt; the snout sub-ascending, blunt. A narrow crenulated grinder on each side in the lower jaw, and a broad tubercular one corresponding above. Nostrils immediately above the upper lip contiguous, each with a cartilaginous complicated valve. Branchial openings in front of the pectorals. Eyes large, lateral. On the crown, in front of the eyes, a thin osseous plate, bent forwards, with a spinous disc at the extremity on the lower side. Lateral line connected with numerous waved anastomosing grooves on the cheeks and face. The first dorsal fin above the pectorals narrow, with a strong spine along the anterior edge. The second dorsal rises immediately behind the first, is narrow, and is continued to the caudal one, where it terminates suddenly. The pectorals are large, and subtriangular. Ventrals rounded ; in front of each a broad recurved osseous plate, with recurved spines on the ventral edge. Claspers pedunculated, divided into three linear segments ; the anteal one simple, the retral ones having the opposite edges covered with numerous small reflected spines. A small anal fin opposite the extremity of the second dorsal. Caudal fin above and below, broadest near the origin, gradually decreasing to a linear produced thread."

The representation here given was taken from the figure in Mr. Donovan's work ; and being that of a female fish, does not show the claspers described by Dr. Fleming as existing in his specimen, which was a male. These sexual and other peculiarities will be pointed out when describing other species of Sharks ; which being of much more frequent occurrence, have afforded opportunities for more detailed observations.

The appendage on the front of the head in this fish is peculiar to the males only, and has given rise to the name of King-fish, applied to it by the Norwegians ; who also call it Gold and Silver Fish, in reference to its beautiful colours : these are various shades of rich brown on a shining white ground. The eyes are large and brilliant ; the pupils green, the irides white.

This fish was first made known by Gesner.

CHONDROPTERYGII.   *SQUALIDÆ.* [*]

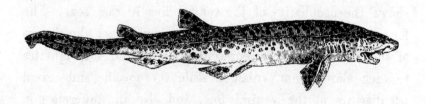

THE SMALL-SPOTTED DOG-FISH.

MORGAY, *Scotland.*—ROBIN HUSS, *Sussex coast.*

Scyllium canicula,	*La Grande Roussette,*	Cuvier, Règne An. t. ii. p. 386.
„ *catulus,*	*Morgay,*	Flem. Brit. An. p. 165, sp. 8.
„ *canicula,*	*Spotted Dog-fish,*	Jenyns, Man. Brit. Vert. p. 495, sp. 184.
Squalus canicula,	Linnæus.	Bloch, pt. iv. pl. 114.
„ „	*Spotted Shark,*	⎱ Penn. Brit. Zool. vol. iii. pl. 19.
„ *catulus,*	*Lesser Spotted Shark,*	⎰ Upper fig. male; lower fig. female.
;„ „	„ „ „	Don. Brit. Fish. pl. 55.
„ „	*Le Squale Roussette,*	Blainville, Faun. Franç. p. 69.

Generic Characters.—Head short and blunt; nostrils pierced near the mouth, and continued by a fissure to the edge of the upper lip, forming valves; teeth triangular, pointed, with a small lateral tubercle at the base on each side; branchial apertures partly over the pectoral fins: two dorsal fins; the first, about the middle of the whole length, placed, in a vertical line, behind the ventral fins; the second, behind the anal fin.

THE true Sharks, as previously stated, have their gills fixed, their margins being attached; the water escaping by five elongated branchial apertures, the form and position of which, in conjunction with modifications observed in the fins and other parts, furnish characters by which the different divisions forming this family are distinguished. Among the

* The family of the Sharks.

Sharks, the males differ from the females externally in having an elongated cylindrical appendage at the inner edge of each ventral fin, the uses of which are not understood. The third species of *Scyllium* here figured represents a male fish, and shows the peculiarity of the ventral fins in that sex. The females are not furnished with these appendages: the figures of the first two species represent females, and the vignettes to each show on an enlarged scale the specific and sexual peculiarities of the ventral fins, and also the difference in the form of the mouth in these two species.

Of the true Sharks, some produce their young alive, and are called viviparous; others, like those under present consideration, bring forth their young enclosed in horny cases, an example of which is here introduced, a portion of one side of the case being removed to show the young fish within.

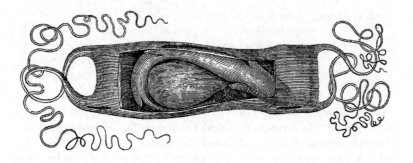

On examining adult females, the ova are observed in different stages of growth descending from the ovaries, usually in pairs, frequently one in each oviduct, becoming enclosed in the protecting covering when about to be excluded. These cases, which are frequently found on the sea-shore, and are called Mermaid's purses, sailor's purses, sea purses, &c. are oblong, of a pale yellowish horny colour, semitransparent, with an elongated tendril at each of the four corners: these are deposited by the parent Shark near the shore in the win-

ter months. The convoluted tendrils hanging to sea-weed
or other fixed bodies prevent the cases being washed away
into deep water. Two elongated fissures, one at each end,
allow the admission of sea-water ; and the young fish ultimately escapes by an opening at the end, near which the
head is situated. For a short time the young Shark continues to be nourished by the vitelline fluid contained in the
capsule attached to its body by the connecting pedicle, till,
having acquired the power of taking food by the mouth, the
remains of the ovum are taken up within the abdomen, as in
birds and some other animals.

A curious peculiarity has been observed in the young of
both Sharks and Skate during a very early stage of their existence. From each of the branchial apertures, branchial filaments project externally : each filament contains a single
minute reflected vessel, in which the blood is thus submitted
to the action of the surrounding medium. These appendages are only temporary, and the blood of the fish is afterwards aerated by the true gills. This very interesting discovery, which I believe is of recent date, forcibly reminding us
of the temporary external branchiæ in the young of Batra-
chian reptiles in the tadpole state, has been observed by Mr.
Richard Owen in the Blue Shark, *Carcharius glaucus*, by
Dr. John Davy in the Torpedo, and by Dr. Allen Thompson of Edinburgh in the Thornback. Cuvier had previously
noticed it, and in the *Règne Animal* has referred to a figure
published by Schneider of a very young Shark in this condition, for which, regarding it as the normal state of this fish,
that industrious pupil of Bloch had proposed the name of
Squalus ciliaris.

Among the Sharks, as among the truly predacious birds,
the females are larger than the males ; and almost all the species have received some name resembling Beagle, Hound,

Rough Hound, Smooth Hound, Dog-fish, Spotted Dog, Penny Dog, &c. probably from their habit of following their prey or hunting in company or packs. All the Sharks are exceedingly tenacious of life. Their skins, of very variable degrees of roughness, according to the species, are used for different purposes ; in some instances by cabinet-makers, for bringing up and smoothing the surfaces of hard wood.

The two British species of Spotted Sharks appear to have been frequently confounded with each other. The terms Greater and Lesser seem sometimes to have been considered as referring to the size of the spots, and at others to the size of the fish. A slight alteration in the names, which is here suggested, will assist in defining the two species, and other decided specific distinctions will be pointed out. Both species are called *Roussette* by the French, on account of their prevailing reddish brown colour.

The Small-spotted Dog-fish, the subject of the present notice, is one of the most common species on our shores, particularly along the Southern coast. Its station in the water is near the bottom ; its food, small fish and crustacea. It takes a bait freely, and is often caught on the fishermen's lines, but is a useless capture to them. It is troublesome and annoying from its numbers, and injurious to the fisheries from its voracity.

The teeth of the Sharks are very formidable weapons, generally constructed decidedly either for cutting or holding. The representation here introduced shows on the left hand

two teeth formed for cutting; these are flattened, thin, sharp, and serrated on both edges; the teeth represented belong to the left of the centre when viewed in front. The two teeth on the right hand are formed for holding, being generally curved inwards, and provided with a small lateral tubercle at the base on each side. The teeth of the Sharks belonging to the genus *Scyllium*, now under consideration, are of this last form, but smaller, depending on the size of the fish. The outside tooth in the front row of each jaw in the Sharks is supported on the inside by various other teeth, which supply deficiencies as necessity may require.

The specimen from which the description was taken measured eighteen inches in length; the body, from the base of the pectoral fins, where it is thickest, tapering all the way to the end of the tail. The head is flattened on the top; the eyes large; the orbits elongated, with a distinct aperture behind each; the form of the under surface of the nose, the nostrils, and upper lip, as shown in the left-hand figure of the vignette at the end; the mouth in the form of a horse-shoe, the extreme angles only being directed outwards; the teeth numerous, small, pointed, and sharp, like those on the right hand in the representations of Shark's teeth, but very minute; the pectoral fins large : the branchial apertures on the sides of the neck elongated vertically, five in number, the first rather the largest, the last the smallest; the fourth aperture over the anterior edge of the pectoral fin : the ventral fins united almost to the posterior extremity in the males, less completely united in females; the elongated anal aperture in the middle between them : the outer posterior margins, in both sexes, are as oblique as those of the front : the right-hand figure of the vignette at the end shows the lozenge-shape of the fins when seen from below. The first dorsal fin is over the space between the ventral and anal fins, and occupies nearly the middle of the whole length of the fish ; the anal fin is under the space between the first and second dorsal fins ; the posterior edge of the second dorsal fin half-way between the commencement of the first dorsal fin and the end of the tail ; the vertebral portion of the tail nearly in a line with the body, with a narrow elongated membranous expansion above it, and one long and one short triangular expansion below it. All the upper part of the body marked with numerous small, dark, reddish brown spots, on a pale reddish ground ; the spots on the fins rather larger and less numerous than those on the body ; the lower part of the sides and the under surface yellowish white. The skin, to the finger passed from the head towards the tail, is smooth ; in the opposite direction it is rough. The appearance of the skin under a lens is that of being covered with minute spiculæ, all the points of which are directed backwards.

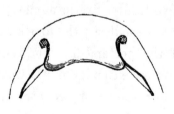

CHONDROPTERYGII. *SQUALIDÆ.*

THE LARGE-SPOTTED DOG-FISH.

ROCK DOG-FISH. BOUNCE, *Scotland*.

Scyllium catulus,	*La Petite Roussette,*	Cuvier, Règne An. t. ii. p. 386.
,, *stellaris,*	*Le Rochier,*	,, ,, ,, ,,
,, *stellare,*	*Bounce,*	Flem. Brit. An. p. 165, sp. 7.
,, *stellaris,*	*Rock Dog-fish,*	Jenyns, Man. Brit. Vert. p. 496, sp. 185.
Squalus catulus et stellaris,		Linnæus.
,, *canicula,*	*Le Squale Panthère,*	Blainville, Faun. Franç. p. 71.
,, *stellaris,* ,,	,, *Rochier,*	,, ,, ,,

THIS Shark is at once distinguished from the species last described by its larger but less numerous spots, by the greater bulk of its body for the same length, and by the ventral fins, which are truncated or nearly square at the end. Like the Small-spotted Dog-fish, its haunts are near the bottom, and its food similar; but it also frequents rocky ground, and has accordingly been distinguished on the Continent by the term *Rochier*, as shown in the list of synonymes.

Mr. Jenyns, in his valuable Manual of British Vertebrate Animals, has so clearly pointed out the specific distinctions of this fish, from examples obtained at Weymouth, that, having no specimen, this Shark being by much the more rare of the two, I avail myself, by permission, of the comparative description therein given.

" Length from two to three feet. According to M. Blainville, this species attains to a larger size than the last. Differs essentially from *S. canicula* in the structure of the lobes of the nostrils, and in the form of the ventrals :* the former are not united as in that species, and of a smaller size, leaving the whole of the mouth and the upper lip visible : the ventrals, instead of being cut obliquely, are cut nearly square, their posterior margins meeting at a very obtuse angle ; they are united or separate according to the sex, in a similar manner : the snout is rather more elongated ; and, according to some authors, the tail rather shorter, giving the dorsal a more backward position ; but this last character I have not noticed myself. Upper parts brownish grey, with very little of the red tinge observable in the last species : back, flanks, and tail, sparingly marked with large spots of a deep brown or black colour : under parts whitish."

* See the vignettes of the nostrils and the ventral fins of both species.

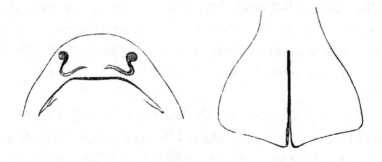

CHONDROPTERYGII.   *SQUALIDÆ.*

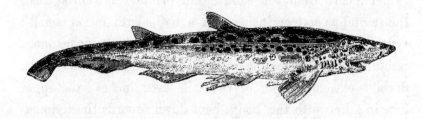

THE BLACK-MOUTHED DOG-FISH.

EYED DOG-FISH, *Cornwall.*

Scyllium melanostomum,   Fauna Italica of PRINCE MUSIGNANO.
    ,,   ,,   *Eyed Dog-fish,*   COUCH'S MS.
Squalus melastomus,   *Le Squale Melastome,* BLAINVILLE, Faun Franç. p. 75.

THE following account of this species, new to the British catalogue, is from the MS. of Mr. Couch, who is probably the only English naturalist into whose hands it has fallen.

" The specimen from which my description has been taken was caught on a line by a fisherman of Polperro on the 8th February 1834. Its length was twenty-five inches and three-quarters, and seven inches round where stoutest. The head flat on the top, rather wide posteriorly; snout thin, protruded one inch and three-quarters from the anterior angle of the eye; nostrils one inch and a quarter from the snout, double, one beneath linear, the other on the margin, the hinder edge prominent, a depression in the head immediately above it; eye rather large, oval, close behind it a moderately sized temporal orifice; mouth one inch and three-quarters wide; teeth numerous, small, sharp, at each side of the base of each tooth a small sharp process; spiracles five, open ; the back somewhat elevated close behind the head ; the skin rough against the grain ; pectoral fins wide, much like those of the Picked Dog : the first dorsal begins at twelve inches from the snout, and behind the ventral fins ; the second at sixteen inches and a half, both rather small : ventrals ten inches from the snout ; anal fin four inches long, rather narrow, terminating just opposite the end of the second dorsal : extreme length of the tail seven inches ; the upper lobe in a line with the body, bent down towards the termination, rounded, incised, or jagged ; under lobe rather narrow in its course, expanded beneath ; the upper ridge of the superior lobe has a double row of prickles pointing outward and downward on each side ; lateral line suddenly bent opposite the origin of the caudal fin. Colour, a light brown on the head and along the back : on each side two rows of ocellated spots ; one row beginning at the side of the neck, and continued along the side of the back ; the second row commencing behind the eye and passing along the upper side of the belly, becoming obsolete near the ventral fins ; these rows are separated by numerous irregular spots, which, however, assume somewhat of a straight direction ; the fins and hinder part of the back are finely barred and clouded with various tints of brown and yellow ; the mouth dark-coloured within.

This species is well known in the Mediterranean. Mr. Couch's specimen was that of a male, and the figure is taken from a drawing lent for that purpose.

CHONDROPTERYGII.　　　　　　　　　　*SQUALIDÆ.*

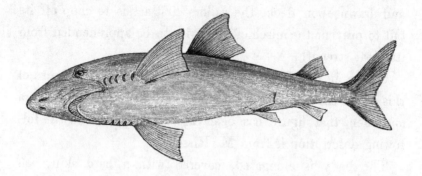

THE WHITE SHARK.

Carcharias vulgaris, *Le Requin,*   Cuvier, Règne An. t. ii. p. 387.
　　,,　　　　,,　　　　*White Shark,* Flem. Brit. An. p. 167, sp. 12.
Canis Carcharias,   Willughby, p. 47, B. 7.
Squalus   ,,　　　　*White Shark,* Penn. Brit. Zool. vol. iii. p. 139.
　　,,　　　,,　　　　　,,　　,,　　Jenyns, Man. Brit. Vert. p. 497, sp. 186.
　　,,　　　,,　　　*Le Squale Requin,* Blainville, Faun. Franç. p. 89.

Generic Characters.—Jaws and head depressed ; nostrils pierced in front ; teeth pointed and cutting, often serrated at the edges : first dorsal fin large and placed behind the pectorals and before the ventrals ; pectoral fins large.

THIS species has been noticed by several English authors, but apparently not from specimens which had been examined by them personally. Low says that, according to information given him, it is sometimes met with among the Orkney Islands. Grew, in his Rarities of Gresham College, page 90, states that they are sometimes found upon our own coast near Cornwall. As it appears to be well known in the Mediterranean, and to be a great wanderer, the Cornish coast is a very probable locality, and it, or the fish figured in the vignette, may prove to be the Rashleigh Shark of Mr. Couch.

This fish acquires a large size, and with another species, not very dissimilar in shape and equally powerful, are the terror of mariners in most of the warm countries of the globe.

VOL. II.　　　　　　　　　　　　　　　　2 c

It swims with great ease and swiftness from the large size of its pectoral fins ; and when caught with a baited hook at sea, and drawn upon deck, the sailors' first act is to chop off its tail to prevent the mischief otherwise to be apprehended from its great strength.

Cuvier, in the *Règne Animal,* says the only good figure of this fish is that in Belon, page 60 ; and having no access to a specimen, that figure has been carefully copied, and the following description is from M. Risso.

The body is elongated, covered with a hard skin, ash brown above and whitish below. The head is large ; the muzzle depressed, short, and pierced with numerous pores : the mouth is large and wide ; the tongue short and rough : the upper jaw furnished with six rows of triangular teeth, thin, nearly straight at the edges, and serrated ; in the under jaw four rows, sharper than those above, but less compressed : the irides are pearl white ; pectoral fins very large ; the first dorsal fin elevated ; the ventral fins small ; the anal fin is opposed to the second dorsal ; the tail is divided, forming two lobes, of which the upper lobe is the longest.

It is most frequently seen in the Mediterranean during spring and autumn.

The vignette represents another species, which has also been called White Shark, and may assist observers on the coast.

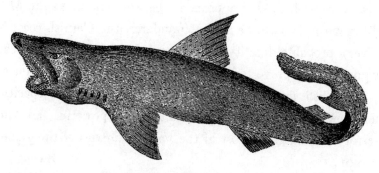

CHONDROPTERYGII. *SQUALIDÆ.*

THE FOX SHARK.

SEA-FOX. THRESHER. SEA-APE.

Carcharias vulpes,	*La Faux, ou Renard,*	Cuvier, Règne An. t. ii. p. 388.
,, ,,	*Thresher,*	Flem. Brit. An. p. 167, sp. 14.
Vulpes marina,	Willughby, p. 54, B. 6, fig. 2.	
Squalus vulpes,	*Long-tailed Shark,*	Penn. Brit. Zool. vol. iii. p. 145, pl. 17.
,, ,,	*Le Squale Renard,*	Blainv. Faun. Franç. p. 94.

This species is occasionally met with on the British coast: Pennant examined one that measured thirteen feet in length; and specimens have been seen of fifteen feet long. It is called the Sea-Fox from the length and size of its tail; and, according to Dr. Borlase, has received the name of Thresher from its habit of attacking other animals, or defending itself, by blows of the tail.* It is an inhabitant of the Mediterranean, as well as other seas; and a specimen has been taken near Belfast.

The extreme length of a specimen examined by Mr. Couch, " was in a straight line ten feet ten inches, and along the curve eleven feet eight inches; three feet four inches round where thickest; solid at the chest; conical from the

* See vol. i. page 144.

2 c 2

snout to the pectoral fins, and thick even to the tail, which organ from the root was five feet and a half long, and consequently more than half the length of the body ; eye prominent, round, hard, four inches from the snout ; iris blue, pupil green : the nostrils small, and not lobed ; mouth five inches wide, shaped like an horse-shoe ; teeth flat, triangular, in two or three rows, not numerous ; spiracles five ; pectoral fins wide at the base, pointed, eighteen inches and a half long.　Measured along the curve, from the snout to the first dorsal fin, was two feet five inches, the fin triangular ; from the first dorsal to the second, fourteen inches and a half; this and the anal fin small ; ventral fins also rather small, triangular ; above and below at the base of the tail a deep depression ; skin smooth ; lateral line central and straight ; breadth of the tail, including both lobes, thirteen inches ; the upper lobe narrow throughout its great length, and on the lower margin, at four inches from the extremity, is a triangular process.　Colour of the body and fins dark blue, mottled with white over the belly."

Mr. Couch says it is not uncommon for a Thresher to approach an herd of Dolphins *(Delphini)* that may be sporting in unsuspicious security, and by one splash of its tail on the water put them all to flight like so many hares before a hound.

" The specimen here described was taken at the entrance of the harbour of Looe in Cornwall, in October 1826, having become entangled in a net set for Salmon.　The mouth seemed more feeble than in most of its genus, which is rendered more probable by the circumstances of its capture ; for the Blue Shark (next to be described) would in an instant have cut its way through an obstruction that proved fatal to the Thresher.　The stomach was filled with young Herrings."

CHONDROPTERYGII. *SQUALIDÆ.*

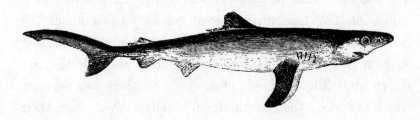

THE BLUE SHARK.

Carcharias glaucus, *Le Bleu,* Cuvier, Règne An. t. ii. p. 388.
Galeus glaucus, *Blew Shark,* Willughby, p. 49, B. 8.
Squalus ,, *Le Cagnot bleu,* Bloch, pt. iii. pl. 86.
,, ,, *Blue Shark,* Penn. Brit. Zool. vol. iii. p. 143.
Carcharias ,, ,, ,, Flem. Brit. An. p. 167, sp. 13.

The affection of the Blue Shark for its young was the theme of several of the older writers, ichthyologists as well as poets; and mariners of the present day believe that, when danger appears, the young brood enter the mouth of the parent fish, and take shelter in its belly. Living young have doubtless been found in the stomachs of large Sharks: their extraordinary tenacity of life is proverbial, and will account for this; but the safety to be expected from incarceration in such a prison is somewhat problematical.

The Blue Shark is an inhabitant of the Mediterranean, and appears to occur much more frequently on the Devonshire and Cornish coasts than on any other part of the British Islands; it has also been taken off the south coast of Ireland, and has been known to wander even as far north as Zetland.

Mr. Couch, who has had frequent opportunities of seeing this species, makes the following observations :—" The Blue

Shark is migratory, and I have never known it arrive on the coast of Cornwall before the middle of June ; but afterwards it becomes abundant, so that I have known eleven taken in one boat, and nine in another, in one day. The injury they inflict on the fishermen is great, as they hover about the boats, watch the lines, (which they sometimes cut asunder without any obvious motive,) and pursue the fish that are drawn up. This, indeed, often leads to their own destruction : but when their teeth do not deliver them from their difficulty, they have a singular method of proceeding, which is by rolling the body round so as to twine the line about them throughout its whole length ; and sometimes this is done in such a complicated manner, that I have known a fisherman give up any attempt to unroll it as a hopeless task. To the Pilchard drift-net this Shark is a still more dangerous enemy, and it is common for it to pass in succession along the whole length of the net, cutting out, as with shears, the fish and the net that holds them, and swallowing both together."

The specimen described measured fourteen inches ; the head depressed, broadest between the eyes, which are lateral ; half-way between the eyes and the point of the nose are the nostrils, linear, directed obliquely downwards and backwards, the most inferior portion covered with a valvular fold of skin ; the eyes round and rather large ; the mouth forming half a circle, the teeth in this specimen very minute,—the cutting teeth on the left hand in the representation at page 370 belong to this species, and were from a specimen about six feet in length, in each jaw of which there are three rows, those immediately in the centre, to the number of four, being calculated more for holding than cutting ; the number of rows of teeth in the Sharks are said, and I believe correctly, to increase with age, and vary in this species from one to six. The branchial apertures are five, the fourth placed over the line of the anterior edge of the pectoral fin ; the pectoral fins large and falciform ; the body of the fish deepest in the line of their origin, but becoming more compressed and tapering from thence to the tail ; the first dorsal fin situated over the space between the pectoral and anal fins, rather small, low and rounded above, with a horizontal projecting elongation at the base behind : the ventral fins small, obliquely truncated, and placed under the space between the first and second dorsal fins ; the anal fin placed half-way between the ventral fins and the lower lobe of the tail, opposed to or under the second dorsal fin, and each ending in a prolongation directed backwards ; the tail divided, the upper lobe two-thirds longer than the lower, the vertebral column continued along it ; the inferior lobe somewhat triangular in shape ; the upper lobe falciform, and with an extension of the membrane towards the extreme end.

The whole of the upper surface of the head, back, both dorsal fins, and most of the tail, are of a fine slate blue ; the irides, upper surface of the pectoral and ventral fins, are also blue ; the lower part of the sides, under surface of the head, neck, pectoral fins, belly, ventral fins, and the anal fin to the base of the tail, white. The skin of this Shark has a granulated appearance on the surface, and is only slightly rough to the touch on passing the finger in the direction from the tail towards the head.

For a reference to habits see volume i. page 150.

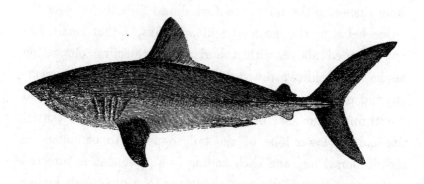

THE PORBEAGLE.

Lamna Cornubica, Le Squale Nez, Cuvier, Règne An. t. ii. p. 389.
    ,,   ,,   *Porbeagle,*   Flem. Brit. An. p. 168, sp. 15.
Squalus Cornubicus,   ,,   Penn. Brit. Zool. vol. iii. p. 152.
  ,,   ,,   ,, *Shark,* Don. Brit. Fish. pl. 108.
  ,,   ,,   ,,   Goodenough, Linn. Trans. vol. iii. p. 80,
                       tab. 15.
  ,,   ,,   ,,   Jenyns, Man. Brit. Vert. p. 500, sp. 189.

Generic Characters.—Point of the nose conical, nostrils pierced on its under surface ; all the five branchial apertures in advance of the origin of the pectorals ; the first dorsal fin placed much nearer the line of the pectoral than the anal fins ; lobes of the tail nearly equal.

The Porbeagle occurs more frequently on the northern than on the southern coasts of this country, and is mentioned as having been taken at Belfast. The specimen described and figured by Dr. Goodenough in the Transactions of the Linnean Society, as quoted, was taken at Hastings ; Mr. Couch has seen it occasionally in Cornwall, and it was figured by Dr. Borlase in his history of that coast. Mr. Couch states of this species, " That it associates in small companies in pursuit of prey, from which circumstance, and a distant resemblance to the Porpus, they derive their name. I have found the remains of cartilaginous fishes and Cuttles in their stomachs, and in one instance three full-grown Hakes. This species attains a large size at an early age, so that I have found it cutting its second row of teeth when nearly full-grown."

On the northern and north-east coast it occurs most frequently during autumn, and, not to multiply descriptions already in print, I shall here insert one furnished by Dr. George Johnston of Berwick, who examined two specimens in the autumn of 1834, both of which were taken in Berwick Bay, and who also very kindly sent me, with his description, a portion of a jaw, from which the teeth on the right hand at page 370 were drawn. Of these teeth there were three rows, the third or inner row being much smaller than the teeth of the two preceding rows, and perhaps only recently exposed.

Body fusiform, very narrow at the tail, and strongly keeled there on each side; skin smooth when stroked backwards, of a uniform greyish black colour, the belly white : snout obtusely pointed, with a band of punctures on each side of the forehead terminating above the eyes, a few similar punctures behind the eyes, and a triangular patch of them before the nostrils ; they are the apertures of canals filled with a transparent jelly : eyes round, dark blue ; branchial slits five, cut across the neck, the posterior oblique and close to the pectoral fin ; back rounded ; dorsal fin triangular, with a free pointed pale-coloured process behind ; posterior dorsal fin also pointed posteriorly ; pectorals somewhat triangular, obliquely sinuate on the posterior edge, black ; ventral fins rhomboidal, meeting at the mesial line, on which are the anal and generative apertures ; anal fin small, pointed behind : tail lunate, with unequal lobes, the superior and largest with a projecting outline near the tip ; above the tail there is a flat space bounded by a short transverse ridge, and a similar one opposite on the ventral side : lateral line straight ; the keel on the body runs forward on the tail, and there is a small keel beneath this confined to the tail itself. The length along the lateral line, five feet eight inches and a half ; circumference in front of the dorsal fin, two feet eight inches and a half ; from the snout to the eye, four inches and three-quarters ; diameter of the eye, one inch and one-tenth : breadth between the eyes, five inches and one-quarter ; from the snout to the margin of the upper lip, four inches and a half, thence to the angle of the mouth also four inches and a half ; breadth of the mouth from angle to angle, eight inches and one-quarter ; from the snout to the first gill-aperture, one foot three inches ; snout to pectoral fin, one foot six inches and a half ; length of pectoral fin, one foot one inch ; breadth of pectoral fin, six inches and a half ; snout to dorsal fin, two feet one inch and three-quarters ; height of dorsal fin, nine inches and three-quarters ; length of dorsal fin, ten inches and one-quarter ; length of the free portion of it, three inches ; space between the first and second dorsal fins, one foot eight inches ; length from the snout to the anal aperture, three feet eight inches ; extreme breadth of the tail, one foot eight inches ; length of the tail in the mesial line, six inches and one quarter.

CHONDROPTERYGII. *SQUALIDÆ.*

THE BEAUMARIS SHARK.

Lamna Monensis, Cuvier, Règne An. t. ii. p. 389, *note* 2.
Squalus ,, *Beaumaris Shark,* Shaw, Gen. Zool. vol. v. pt. 2, p. 350.
    ,,   ,,   ,,   ,, Penn. Brit. Zool. vol. iii. p. 254, pl. 20.
    ,,   ,,   ,,   ,, Jenyns, Man. Brit. Vert. p. 501, sp. 190.

Baron Cuvier, in his *Règne Animal,* as above quoted, considers the *Monensis* of Shaw and Pennant distinct from the *Cornubicus* last described, on account of its shorter muzzle and sharper teeth; to this may be added that the eye is much larger, the pectoral and dorsal fins are placed farther back on the body, and the bulk of the fish is greater in proportion to its length. The editor of the last edition of Pennant's British Zoology, published in 1812, possessing at that time the original drawing of the Rev. Hugh Davies of Beaumaris, has borne testimony to the correctness of Pennant's figure of this fish, which had been questioned; and the Beaumaris Shark is therefore considered, provisionally, as a distinct species.

Two specimens, the only examples known, having both occurred on the Anglesey side of the Menai, confirm the propriety of the terms Beaumaris Shark and *Monensis*, by which this fish is known. Some particulars of both examples are here added from Pennant.

" The first specimen obtained was seven feet long; the snout and body of a cylindrical form; the greatest circumference four feet eight inches; the nose blunt; the nostrils small; the mouth armed with three rows of slender teeth, flatted on each side, very sharp, and furnished at the base with two sharp processes; the teeth are fixed to the jaws by certain muscles, and are liable to be raised or depressed at pleasure. The first dorsal fin was two feet eight inches distant from the snout, of a triangular form; the second very small, and placed near the tail; the pectoral fins strong and large; the ventral and anal small; the space between the second dorsal fin and the tail much depressed, the sides forming an acute angle; above and below was a transverse fossule or dent. The tail was in the form of a crescent, but the horns of unequal length; the upper, one foot ten inches; the lower, one foot one inch. The whole fish was of a lead colour. The skin comparatively smooth, being far less rough than that of the lesser species of this genus."

" The second example was nine feet six inches in length, that is, two feet and a half longer than the first, but each part of this bore an exact proportion to the corresponding parts of the other; except that the nose of this, although above one-third a larger animal than the former, was smaller in every respect, being more abruptly tapering, but blunt and shorter, as it measured but four inches and eight-tenths from the eye to the end, whereas the snout of the smaller fish was six inches in length from the end to the eye. This was a vast animal; its general circumference seemed greater in proportion to its length, than that of the former, but it was particularly so at the region of the abdomen. This is readily accounted for, when we say that it was a female, and had in its belly four young ones, each about eight-and-twenty or thirty inches long. Seventeen quarts of oil were obtained from the liver. As it is supposed, with reason, that in this tribe of ferocious animals, the female is invariably the largest, I am induced to conclude, that the specimen which I observed near forty years ago, might have been a full-grown male, and that the difference between the two sexes is inferiority of size with regard to the male, but with a front in every respect larger than that of a female."

" In the third volume of the late edition of Mr. Pennant's Tour in Wales, the Rev. Hugh Davies has furnished some further observations on the Beaumaris Shark, and a comparative outline is given of that species and of the Porbeagle Shark."

The latter species appears to be by much the more common fish of the two.

The vignette represents the sort of hand-line used at sea on the Hampshire coast for Mackerel and Whiting fishing, and is usually called the Portsmouth pattern.

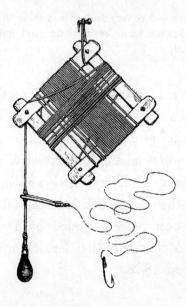

CHONDROPTERYGII.   *SQUALIDÆ.*

THE COMMON TOPE.

PENNY DOG, *Hastings.* — MILLER'S DOG, *Cornwall.*

Galeus vulgaris, Le Milandre, CUVIER, Règne An. t. ii. p. 389.
Canis galeus Rondeletii,   WILLUGHBY, p. 51, B. 6, fig. 1.
Squalus ,,   LINNÆUS. BLOCH, pt. iv. pl. 118.
     ,,   ,,   *Tope Shark,* PENN. Brit. Zool. vol. iii. p. 146, pl. 18.
Galeus vulgaris, Common Tope, FLEM. Brit. An. p. 165, sp. 6.
Squalus galeus,   ,,   ,,   JENYNS, Man. Brit. Vert. p. 501, sp. 191.

    Generic Characters.—Not very dissimilar to those of the genus *Carcharias* : disposition of the fins the same, but have temporal orifices ; the teeth serrated on the outer edge only.

    THE TOPE is a common species along the southern coast, where it is known by the names of Penny Dog and Miller's Dog ; it has also been noticed by Pennant in Flintshire ; and by others on the south coast of Ireland, about the shore of the county of Kerry. It is not, however, considered so plentiful in the north, but has been taken about Berwick Bay, and its occurrence recorded by Dr. Johnston in his address to the Members of the Berwickshire Natural History Club for the year 1832.

On the Cornish coast this is a common and rapacious species ; but it is not so destructive as the Blue Shark. The larger specimens, which are about six feet long, abound chiefly in summer ; and the young, to the number of thirty or more, according to Mr. Couch, are excluded all at once from the female in May and June. They do not reach the full size until the second year, and continue with us through the first winter, while those of larger size retire into deep water. No use is made of this fish beyond melting the liver for oil. When caught on a fisherman's line, this fish sometimes has recourse to the same attempt at deliverance as the Blue Shark, by twisting the line throughout the whole length round its body.

Body fusiform ; the skin almost smooth ; lateral line straight ; the first and second dorsal fins rather small, triangular, very slightly convex on their posterior edges, both ending in points directed backwards ; the first dorsal fin placed over the interval between the pectoral and ventral fins ; the second immediately over the anal fin, and a little larger in size : the head is rather large ; the muzzle elongated and depressed ; nostrils pierced very near the mouth, in part closed by a membrane ; the eyes moderate, and over the mouth ; temporal orifices small ; the jaws semicircular ; teeth small, in several rows, and very nearly alike both above and below, triangular and denticulated on the outer side ; the branchial apertures are small, placed near together, the four first nearly equal in size, the fifth the smallest, and placed over the anterior edge of the pectoral fins ; the pectoral fins are of moderate size, and triangular in shape ; the ventral fins small, near the middle of the whole length, and under the space between the first and second dorsal ; the tail rather less than half the length of the body, with a bi-lobed fin ; the upper lobe terminal, oblique, and truncated ;

the inferior lobe with one deep triangular elongation, and a smaller one near the end.

All the upper part of the body and sides are of a uniform slate grey, the under surface lighter in colour, inclining to greyish white.

The vignette represents a boat of the Lake of Geneva.

CHONDROPTERYGII.   *SQUALIDÆ.*

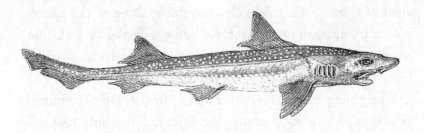

THE SMOOTH HOUND.

SKATE-TOOTHED SHARK.   RAY-MOUTHED DOG, *Cornwall.*

Mustelus lævis,	*L'Emissole,*	CUVIER, Règne An. t. ii. p. 389.
Squalus mustelus,	LINNÆUS.	WILLUGHBY, p. 60, B. 5, f. 2.
,, ,,	*Smooth Shark,*	PENN. Brit. Zool. vol. iii. p. 151.
Mustelus lævis,	*Smooth Hound,*	FLEM. Brit. An. p. 166, sp. 4.
Squalus mustelus,	,, ,,	JENYNS, Man. Brit. Vert. p. 502, sp. 192.

Generic Characters.—The same as in the last genus, *Galeus,* except the pointed teeth, which in this are flat, like those of the Skate.

THIS Shark is rather a common species round our coast. It is occasionally taken in the Frith of Forth ; and Dr. Fleming says the flesh of it is used as food in the Hebrides, and is esteemed a delicate dish. I have received this Shark from Dr. Johnston of Berwick, and have seen it at various places on the coasts of Kent and Sussex. It is called Smooth Hound, from the comparative softness of its skin in reference to British Sharks in general ; and it is also called Ray-mouthed Dog in Cornwall, from the form of its teeth, which are flat and without prominent points, like those of the female or young male of the Thornback. The vignette represents an inside and an outside view of one half of the mouth and teeth of this Shark, which are so different from those of

VOL. II.   2 D

any other British Shark as to serve the purpose of a distinguishing character. The peculiarity in the form and arrangement of these teeth, so closely resembling those of the Skate, is seen by comparing the vignette before referred to with that representing the teeth of the Thornback, which is given hereafter, when describing the first species of true Skate.

The young of the Smooth Hound frequently have numerous small white spots above the lateral line; but the teeth and other characters agree so closely with the spotless grey examples of larger size, that I am induced to consider these spots only as marks of youth, which may also be observed in other species, particularly in the Picked Dog-fish, *Spinax Acanthias*, Cuvier: and in this view I am further confirmed by the opinions of Mr. Couch and Dr. Johnston.

Mr. Couch says of this species, in reference to its habits, that it is common, but not abundant, and keeps close to the bottom on clean ground, where it feeds on crustaceous animals, which it crushes previous to swallowing, and for which its flat pavement-like teeth are well adapted: it also takes a bait, but is less rapacious than most of the tribe. The young are produced alive in November, the whole coming to perfection at once; but they are few in number, not perhaps exceeding a dozen, and soon after birth they all go into deep water, from which they do not emerge until the following May.

This species has been taken on the coasts of the counties of Antrim and Londonderry.

The specimen described measured eighteen inches in length; the top of the head flat and rather broad; the beginning of the back elevated and rather rounded; the eye large, lateral, elongated horizontally; temporal orifices rather small, and placed immediately behind the posterior angle: first dorsal fin considerably larger than the second; both of the same shape, with an elongated free point at the base projecting backward, the centre of the first dorsal at the distance of six inches, and that of the second at twelve inches, from the point of the nose. Under surface of the head flat; nostrils semilunar in shape, with a central free cutaneous valve; the mouth half the width of the whole under surface, rather angular in shape than semicircular; upper lip on each outside ending in a free elongation of the membrane; the teeth small, flat, like those of a young Skate; pectoral fins large, commencing at three inches and a half from the point of the nose: ventral fins under the space between the two dorsals; the anal fin begins in a line under the middle of the second dorsal fin, but being only half its size, ends but a little behind it: the upper part of the caudal fin is a long narrow horizontal slip; the free part of the under portion is made up of two triangular portions, the first of which is long, the second and last short. The surface of the body smoother than that of Sharks in general: the colour of the upper part of the head, body, and fins, pearl grey; under parts greyish yellow white: lateral line prominent; above it the body along its whole length is marked with numerous small circular white spots, which, as before stated, are most conspicuous while this fish is young.

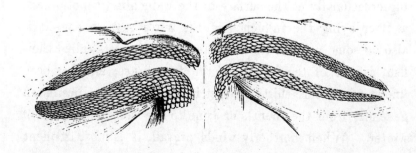

THE BASKING SHARK.

THE SUN-FISH, *and* SAIL-FISH.

Selachus maximus,	*Le Pelerin,*	Cuvier, Règne An. t. ii. p. 390.
Squalus ,,	Linnæus?	
,, ,,	*Basking Shark,*	Penn. Brit. Zool. vol. iii. p. 134, pl. 16.
,, ,,	*Common Sail-fish,*	Flem. Brit. An. p. 164, sp. 5.
,, ,,	*Basking Shark,*	Jenyns, Man. Brit. Vert. p. 503, sp. 193.

Generic Characters.—Branchial apertures elongated, nearly surrounding the neck, all placed in advance of the pectoral fins; temporal orifices present; teeth very numerous, small, conical, without serrated edges; anal fin present.

THE BASKING SHARK, so called from its habit of remaining occasionally at the surface of the water almost motionless, as if enjoying the influence of the sun's rays, whence it is also on some parts of the Irish and Welsh coasts called Sun-fish, is one of the largest of the true fishes, and has been known to measure thirty-six feet in length. It has been seen generally from the month of June to the commencement of winter. When northerly winds prevail, it is most frequent on the west coast of Scotland. It has also been seen on the north and on the west coasts of Ireland. If westerly winds prevail, it is not unusual to see them along the whole line of the southern coast. It has been taken on the coasts of Waterford, Wales, Cornwall, Devonshire, Dorsetshire, and several times at different places on the coast of Sussex. The specimen described and figured by Sir E. Home, in the Philosophical Transactions for 1809, was taken off Hastings ; and the largest specimen I have seen, which measured thirty-six feet in length, was caught some years since off Brighton. From our southern coast it frequently wanders as far to the eastward and south as the coast of France ; and the fish described and figured by M. de Blainville in the eighteenth volume of the *Annales du Muséum,* I have very little doubt was of the same species as that described by Sir E. Home, which has been already referred to.

The difficulty of obtaining a perfect view of this unwieldy fish, either when floating in water, or when from its great weight it lies partly imbedded in the soft soil of the sea-shore, has led to the differences which appear in the representations of it which have been published by different naturalists.

The Basking Shark is said to exhibit but little of the ferocious character of the Sharks in general, and is so indifferent to the approach of a boat as to suffer it even to touch its body when listlessly sunning itself at the surface. From its habit of swimming slowly along with its dorsal fin, and sometimes part of its back, out of water, it has obtained in the North the name of Sail-fish. In Orkney it is called Hoe-mother, and by contraction Homer,—that is, the mother of the Picked Dog-fish, which is there called the Hoe. If deeply struck with a harpoon, the Basking Shark plunges suddenly down, and swims away with such rapidity and violence as to become a difficult as well as a dangerous capture.

This species has the smallest teeth in proportion to its size of any of the Sharks. No remains of fish have been found in its stomach. One examined by Mr. Low contained a red pulpy mass, like bruised crabs, or the roe of *Echini*. Mr. Low adds, that this Shark's appearance, manners, and weapons do not indicate it to be a ravenous fish. Linnæus says that its food is *Medusæ*, and Pennant considered that it subsisted on marine plants.

The body is thickest about the middle, and diminishes towards both extremities; when afloat the form is nearly cylindrical; the skin thick and rough, of a brownish black colour, with tints of blue. The head conical, the muzzle short, rather blunt, smooth, and pierced with numerous circular pores; eyes near the snout, small, oval, the elongation horizontal, the irides brown; half-way between the eye and the first branchial opening is the temporal orifice, oblique and small; branchial openings five on each side, of great vertical length, each set including the whole side of the neck, and leaving only a small space above and below; nostrils oval, small, placed rather laterally, and opening on the edge of the upper lip; pectoral fins of moderate size for so large a fish,—perhaps, as before stated, the largest of the true fishes, —the form somewhat triangular, placed close to the last branchial orifice, convex anteriorly and thick, slightly concave and much thinner behind; the ventral fins also of moderate size, rather elongated at the base, placed behind the middle of the whole length of the fish, convex in front, concave behind, the inner and posterior half free, exhibiting in the figure chosen the cylindrical appendages peculiar to the male. The first dorsal fin, placed before the middle of the whole length of the fish, is much the larger of the two, forming an elevated triangle; anterior edge but slightly convex, posterior edge concave, with an elongated point at the base directed

backwards : the second dorsal fin much smaller than the first,
rounded above, attached throughout half its base only, and
placed at two-thirds of the distance from the first dorsal to
the caudal fin ; the anal fin is still smaller than the second
dorsal, but of the same shape. From the line of the anal
fin to the base of the tail there is a strong and prominent
keel-like edge on each side ; and just in advance of the base
of the caudal fin, both above and below, is a groove,—that
underneath rather smaller than that above. The caudal fin
divided into two lobes, the upper one larger than the lower ;
the posterior edge of the caudal fin appears to become notched and abraded by age and use, and is frequently found unequal at its margin, and variable in shape.

The vignette below represents the *Argulus foliaceus* of
Jurine ; another species of parasitic animal occasionally found
attached to fresh-water fishes. I have specimens that were
taken from the Pike and the Trout. The figure on the left-
hand represents the upper surface of a male : by the powers
of the microscope some of the vessels of the body are rendered visible through the external tunic. The figure on the
right-hand represents the under surface of a female : the ova
are very conspicuous. The small figure between the two is
of the natural size.

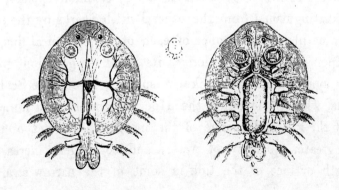

THE PICKED DOG-FISH.

BONE-DOG, *Sussex.*—HOE, *Orkney.*

Spinax acanthias, L'aiguillat, Cuvier, Règne An. t. ii. p. 391.
Galeus ,, sive spinax, Willughby, p. 56, B. 5, f. 1.
Squalus ,, Linnæus. Bloch, pt. iii. pl. 85 ; the young, pt. iii. pl. 75, fig. 1.
,, spinax, Picked Shark, Penn. Brit. Zool. vol. iii. p. 133.
,, acanthias, ,, ,, Don. Brit. Fish. pl. 82.
Spinax ,, Common Dog-fish, Flem. Brit. An. p. 166, sp. 10.
Squalus ,, Picked Dog-fish, Jenyns, Man. Brit. Vert. p. 505, sp. 194.

Generic Characters.—Two dorsal fins, with a strong spine at the anterior edge of each ; no anal fin ; temporal orifices present ; teeth in several rows, small and cutting.

THE PICKED DOG-FISH is a very common species, at once distinguished from the other British Sharks by the single spine placed in advance of each of its two dorsal fins,—a weapon from which it derives its specific appellation, pick being synonymous with pike or spike. Among the Scotch islands, where it is called the Hoe, it appears most numerous at the full and change of the moon, on account of the then greater quantity of water, and consequent increased strength or race of the tide in some of the narrow straits.

Being gregarious, they frequently make their appearance in such shoals that the fishermen load their boats to the water's edge with them ; and, according to Mr. Low, they prove a valuable capture. The flesh is dried and eaten : the livers yield a large quantity of oil, while their intestines and other refuse parts are strewed over the land as manure. Dr. Neill says this species is very common in the Forth during the Herring season, where numbers are caught ; but their flesh is not eaten in that neighbourhood.

The Picked Dog-fish is found in numbers at most of the fishing stations along the east coast, round to Kent and Sussex on the south coast, where it is almost universally called the Bone Dog. According to Montagu's MS. it is very numerous in Devonshire and in Cornwall : Mr. Couch says, " It is the most abundant of the Sharks, and is sometimes found in incalculable numbers, to the no small annoyance of the fishermen, whose hooks they cut from the lines in rapid succession. I have heard of twenty thousand taken in a sean at one time ; and such is the strength of instinct, that little creatures not exceeding six inches in length may be found, in company with the larger and stronger, following schulls of fish, on which at that time it is impossible they should be able to prey. The Picked Dog bends itself into the form of a bow for the purpose of using its spines, and by a sudden motion causes them to spring asunder in opposite directions; and so accurately is this intention effected, that if a finger be placed on its head, it will strike it without piercing its own skin. This fish is subject, like many others, to occasional monstrosity. A friend of mine was in possession of a Picked Dog-fish with two heads, the separation continuing so far back as behind the pectoral fins. The fisherman who found it informed me that there was only one egg attached to it, and that it must have been dropped from the mother after she was taken.   The young are produced at various periods from June to November."

The whole length of the specimen described was eighteen inches ; the top of the head flat ; the temporal orifices large, and seen from above : first dorsal fin commencing at one-third of the whole length ; rather small in size ; front edge convex, concave behind ; the point of the spine preceding the fin half as high as the fin : the second dorsal fin half-way between the first and the end of the tail ; small in size, with a spine as high as the fin : the nose rather pointed ; the eyes lateral, elongated horizontally ; temporal orifices behind, large, but above the line of the eye ; nostrils small, with a minute valve ; mouth semicircular, when quite open nearly round ; the teeth from the centre of both jaws with points projecting outward on each side, the edges sharp ; pectoral fins large, commencing half-way between the snout and the first dorsal ; anal fins small, placed intermediate, in a vertical line, between the first and second dorsal ; no anal fin ; tail powerful, upper membrane broad, the lower anterior part triangular, ending in a slip prolonged backward.   The upper part of the head, body, and fins, slate grey ; under parts yellowish white ; young specimens generally exhibit a few white spots. Skin moderately rough on passing the finger upwards towards the head ; in the contrary direction quite smooth.

CHONDROPTERYGII. *SQUALIDÆ.*

THE GREENLAND SHARK.

Scymnus borealis, *Greenland Shark,* Flem. Brit. An. p. 166, sp. 11.
Squalus ,, ,, ,, Scoresby, Arctic Regions, vol. i. p. 538,
        pl. 15, figs. 3, 4.
  ,, *glacialis,* *Faber,* Nilsson, Prod. Icht. Scand. p. 116,
        sp. 7.
  ,, *borealis,* *Greenland Shark,* Jenyns, Man. Brit. Vert. p. 506, sp. 195.

Generic Characters.—Two rows of cutting teeth in the lower jaw ; several rows of slender, pointed teeth in the upper : temporal orifices present ; two dorsal fins, the first near the middle of the body, the second but little behind the line of the ventrals ; no anal fin ; tail short.

This species of Shark, which is a native of the Northern Seas, has been twice noticed in Scotland, and is therefore entitled to a place in this work. According to Dr. Fleming, one was caught in the Pentland Firth in 1803 ; and another, measuring thirteen feet and a half long, found dead at Burra Firth, Unst, was seen by Mr. Edmonston.

This Shark appears to be well known to several Northern zoologists ; and the following account, as well as the figure, of this fish, is derived from the valuable work on the Arctic Regions by Captain W. Scoresby.

" The *Squalus borealis* is twelve or fourteen feet in length, sometimes more, and six or eight feet in circumference. The opening of the mouth, which extends nearly across the lower part of the head, is from twenty-one to twenty-four inches in width. The teeth are serrated in one jaw, and lancet-shaped and denticulated in the other. It is without the anal fin, but has the temporal opening; the spiracles on the neck are five in number on each side. The colour is cinereous grey. The irides are blue, the pupil emerald green."

" This Shark is one of the foes of the Whale. It bites it and annoys it while living, and feeds on it when dead. It scoops hemispherical pieces out of its body, nearly as big as a person's head; and continues scooping and gorging lump after lump, until the whole cavity of its belly is filled. It is so insensible of pain, that though it has been run through the body with a knife and escaped, yet, after a while, I have seen it return to banquet again on the Whale, at the very spot where it received its wounds. The heart is very small: it performs six or eight pulsations in a minute, and continues its beating for some hours after taken out of the body. The body, also, though separated into any number of parts, gives evidence of life for a similar length of time. It is therefore extremely difficult to kill. It is actually unsafe to trust the hand in its mouth, though the head be separated from the body. Though the Whale-fishers frequently slip into the water where Sharks abound, there has been no instance, that I have heard of, of their ever having been attacked by the Shark."

" Besides dead Whales, the Sharks feed on small fishes and crabs. A fish, in size and form resembling a Whiting, was found in the stomach of one that I killed; but the process of digestion had gone so far, that its species could not be satisfactorily discovered. In swimming, the tail only is used : the rest of its fins being spread out to balance it, are never observed in motion but when some change of direction is required."

" To the posterior edge of the pupil of the eye is attached a white vermiform substance, one or two inches in length. Each extremity of it consists of two filaments, but the central part is single. The sailors imagine this Shark is blind, because it pays not the least attention to the presence of a man ; and is, indeed, so apparently stupid, that it never draws back when a blow is aimed at it with a knife or lance."

The eyes of this Greenland Shark, with the appendages, were brought home by Captain W. Scoresby, preserved in spirits, and submitted to Sir David Brewster, who gave one specimen to Dr. Grant. The appendage proved to be a new species of parasitic animal, which Dr. Grant named *Lernæa elongata*, and described it, adding a figure of it, in the seventh volume of the Edinburgh Journal of Science. The imperfection of the vision of the fish was probably produced by the various perforations made in the cornea by the tentacula of this new species of *Lernæa ;* as it is by those organs that these parasitic animals retain their hold and live upon the fluids extracted from the animal to which they adhere. This species of *Lernæa* is perhaps the largest known : it measured three inches in length.

The genus of Sharks next in order, according to Cuvier's arrangement in the *Règne Animal,* is that of *Zygæna,* or Hammer-headed Sharks, of which a single specimen is recorded by Messrs. C. and J. Paget, in their Sketch of the Natural History of Yarmouth, page 17, to have been taken there in October 1829, the head of which is now preserved in the Norwich Museum.

The specific name of the example taken, and here referred

to, has not, I believe, been determined.   A reference to a paper by M. Valenciennes in the ninth volume of the *Mémoires du Muséum*, which supplies detailed descriptions of four species of this genus, would probably settle this point.

A representation of the most common species, *Zygæna malleus*, Val. is here given as a vignette to draw the attention of observers to the subject.

CHONDROPTERYGII. *SQUALIDÆ.*

THE ANGEL-FISH.

MONK-FISH, SHARK-RAY, *and* KINGSTON.

Squatina angelus, DUMERIL. CUVIER, Règne An. t. ii. p. 394.
     ,,   *Monk,* or *Angel-fish,* WILLUGHBY, p. 79, D. 3.
Squalus squatina, LINNÆUS. BLOCH, pt. iv. pl. 116.
     ,,   ,,   *Angel Shark,* PENN. Brit. Zool. vol. iii. p. 130, pl. 15,
                       male.
     ,,   ,,   ,,   ,,   DON. Brit. Fish. pl. 17.
Squatina vulgaris, *Monk-fish,* FLEM. Brit. An. p. 169, sp. 16.
     ,,   *Angelus,* *Angel-fish,* JENYNS, Man. Brit. Vert. p. 507, sp. 197.

Generic Characters.—Body very much depressed ; head flat, rounded anteriorly ; both eyes on the upper surface ; temporal orifices present ; mouth at the
end of the snout ; pectoral fins large ; two dorsal fins, both behind the ventrals ;
no anal fin.

THIS fish, certainly more remarkable for the singularity of
its form than for its beauty, is called Angel-fish in England,
France, and Italy, and is said to have acquired that name
from the extended pectoral fins having the appearance of
wings : it is also called Monk-fish, because its rounded head
looks as if enveloped in a monk's hood. Mr. Donovan says
the form of its body has obtained for it in some places the

name of Fiddle-fish ; and it is also called Shark-Ray, from its partaking of the characters of both Shark and Ray, though in some respects distinct from either.   It is, however, by no means so truly osculant between those families as the exotic genus *Rhinobatus*.

It is most numerous on the southern coast of our island ; but is occasionally taken in the Forth, and some other parts of the east coast, particularly about Cromer and Yarmouth. It is common on the coasts of Kent and Sussex, where it is called a Kingston,—a name for it that occurs in Merrett's Pinax.   It is also taken in Cornwall ; and is recorded as occurring in Ireland on the coasts of Kerry, Waterford, Dublin, and Belfast.

This fish is very voracious, and feeds on the smaller flat-fishes, which, like itself, swim close to the bottom ; occasionally, like them also, hiding itself in the loose, soft soil that floats over it.   The Angel-fish sometimes attains a large size.   Cuvier, Pennant, and others, mention having seen specimens that would have weighed one hundred pounds. The flesh is now considered indifferent and seldom eaten, but is said to have been formerly held in high estimation.   The skin is rather rough, and is used for polishing, and other works in the arts : Mr. Donovan also says that the Turks at the present time make shagreen of it.

A second species of this genus has been supposed to occur on our coast ; but the Angel-fish is probably liable to some variation in colour, depending on the nature of the ground in the locality in which it is found : the sexes also exhibit some differences.   The females produce their young alive in June.

This species is said to attain the length of seven or eight feet ; the specimen described measured but fourteen inches ; the breadth of the head in the line of the temporal orifices three inches, whole breadth across the pectoral fins from angle to angle seven inches and a half, breadth across the ventral fins four inches and one quarter; head depressed, rounded at the anterior margin; eyes on the upper surface, distance between them one inch and one quarter; temporal orifices very large, one inch and a half apart, elongated transversely, about as far behind the eyes as these are from the anterior margin of the head; pectoral fins large, lateral, pointed in front, triangular on the outer edge, and rounded posteriorly; sides of the body of the fish parallel behind their free edges; ventral fins elongated, slightly rounded, contracted in breadth behind; commencement of first dorsal fin even with the posterior edge of the ventrals; the second dorsal fin begins at the half of the distance between the commencement of the first dorsal and the caudal fin; tail with an equal-sized triangular lobe above and below. The mouth is very wide, opening on the anterior margin of the head; the angles of the mouth under the external angles of each temporal orifice; teeth long and pointed; branchial apertures elongated; the parallelism of the sides of the fish most conspicuous from below; anal orifice rather before the middle of the whole length; the colour of all the under parts dirty white; the surface smooth; all the upper surface granulated, rough, of a dark mottled chocolate brown; a row of short spines, directed backwards, are ranged along the central line of the back between the ventral fins.

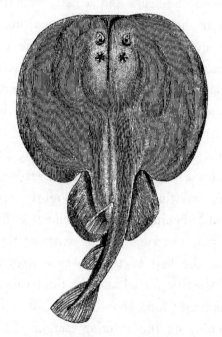

THE ELECTRIC RAY.

COMMON CRAMP-FISH. NUMB-FISH, *Weymouth.*

CRAMP RAY, *Cornwall.*

Torpedo ——— Cuvier, Règne An. t. ii. p. 369.
,, ——— *Cramp-fish,* Willughby, p. 81, D. 4.
Raia Torpedo, Linnæus. Bloch, pt. iv. pl. 122.
,, ,, *Electric Ray,* Penn. Brit. Zool. vol. iii. p. 118, pl. 12.
,, ,, ,, ,, Don. Brit. Fish. pl. 53.
Torpedo vulgaris, Common Cramp-fish, Flem. Brit. An. p. 169, sp. 17.

Generic Characters.—The disk of the body nearly circular ; pectoral fins large ; two dorsal fins placed so far back as to be on the tail ; surface of the body smooth ; tail short, and rather thick ; teeth small and sharp.

THE earliest notices of this fish on our coasts by English writers were made by Smith in his History of Waterford, and afterwards by Pennant and Walsh ; but as, according to

* The family of the Rays or Skate.

Baron Cuvier, several species have been included under the name of the *Raia Torpedo* of Linnæus, the true name of the British species is still doubtful, and it remains therefore for some naturalist who is fortunate enough to obtain a British specimen to determine the particular species of our coast.

Colonel Montagu, in his MS. notes, mentions having met with two examples of the Torpedo; but no description of either of them is given. The first was of small size, and was taken at Torcross, where it was so rare as to be unknown to the oldest of the fishermen of that place. Of the second, the notice is as follows:—" I observed a very large specimen that was taken on a Turbot-hook off the coast of Tenby, in Wales. It was dead when disengaged from the hook, or the fisherman would certainly have had a shock that would have made him remember the species again. It appeared, however, so rare an occurrence here, that no one knew the fish, which was exhibited as an extraordinary creature. Its weight was about one hundred pounds."

The figure at the head of this subject was taken from a small specimen which appears to be of the same species as that figured by Pennant in the British Zoology; but Pennant's plate, which exhibits in the two outside figures the under and upper surface of a female, the third and middle figure being that of a male of smaller size, appears to have been copied from a larger print representing specimens taken on the sea-shore in the neighbourhood of La Rochelle. Mr. Donovan's figure differs from that of Pennant in exhibiting a marbled appearance on its upper surface, with five distinct dark spots: it differs also in its form and proportions.

The electrical powers of the Torpedo are so well understood by the different names that have been applied to it, as well as by the various and voluminous accounts that have been published, that it is unnecessary to repeat here what

2 E 2

has already appeared so often in print elsewhere. The situation of the apparatus or structure from which these species derive their extraordinary power is indicated by the two elevations, one of which is placed on each outside of the eyes and temporal orifices, and extending to the lateral external rounded edges. The apparatus occupies the whole of the space between the upper and under surface of the body, and is composed, as shown by the figures of Walsh and Pennant, of a great number of tubes arranged perpendicular to the plane of the upper and under surfaces, which when exposed by a transverse section have very much the appearance of a portion of honeycomb. The tubes contain a mucous secretion, and the structure is largely provided with nerves derived from the eighth pair. It is said that when the shock is given, the convex part of the upper surface is gradually depressed, the sensation is then felt, and the convexity suddenly returns.

The whole use of the electrical apparatus and power to the fish can only be conjectured. That it serves as a means of defence, is very probable ; that it also enables a slow, inactive fish to arrest and obtain as food some of the more active inhabitants of the deep, is also probable. Mr. Couch thinks other powers may be derived from it, and his opinion is thus expressed :—" One well-known effect of the electric shock is to deprive animals killed by it of their organic irritability,* and consequently to render them more readily disposed to pass into a state of decomposition, in which condition the digestive powers more speedily and effectually act upon them. If any creature more than others might seem to require such a preparation of its food, it is the Cramp-Ray, the whole canal of whose intestine is not more than half as long as the stomach."

* The bodies of animals killed by lightning do not become stiff.

" So long ago as the time of Dioscorides, the physician of Anthony and Cleopatra, the shock of this fish was recommended for medical purposes, and especially for pains of the head ; and this may be considered as the earliest record of the application of electricity to medicine. In later times, it was applied to the cure of gout ; the patient being directed to keep his foot on the fish until the numbness extended to the knees. Baron Humboldt remarks, that the will of the fish directs the effect to whatever part it feels most strongly irritated, but only under the influence of the brain and heart. When a fish was cut through the middle, the fore part of the body alone gave shocks."

But little of its habits are known : it is said to prefer soft and muddy ground, where its actions are slow and inert. It is rare on the British coast ; but two or three species inhabit the Mediterranean, and others have been found in various parts of the world. Not being certain of the British species, no description is ventured upon. The observer on our coast who obtains a specimen will find specific characters in the *Règne Animal*, and in the *Faune Française*.

The vignette below represents a view of Erith Church.

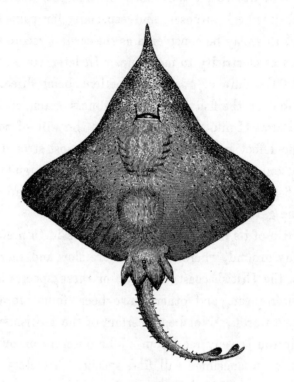

LONG-NOSED SKATE.

SHAGREEN RAY.

Raia chagrinea,	*Shagreen Ray,*	MONTAGU, Wern. Mem. vol. ii. p. 420, pl. 21.		
,,	,,	,,	,,	PENN. Brit. Zool. vol. iii. p. 117.
,,	*aspera,*	,,	,,	FLEM. Brit. An. p. 172, sp. 25.
,,	*chagrinea,*	,,	,,	JENYNS, Man. Brit. Vert. p. 513, sp. 202.
,,	*mucronata,*	*Long-nosed Skate,*	COUCH's, MS.	

Generic Characters.—Form of the body rhomboidal, very much depressed; tail long and slender, generally armed on the upper surface with one or more rows of sharp spines; two small fins near the end of the tail, and sometimes a small terminal or caudal fin; the eyes and temporal orifices on the upper surface of the head; nostrils, mouth, and branchial apertures, beneath; teeth flattened, lozenge-shaped, the inner angle elongated in old males.

THE RAYS, or Skate, as they are popularly called, are remarkable for the rhomboidal form and consequent breadth of their bodies, contrasted with their long narrow tails, frequently furnished with two and sometimes three small fins, and mostly armed with one or more rows of sharp spines along the whole length. The whole body is very much depressed; the great breadth of it is produced by the expansion of what are considered as the pectoral fins, the base of each of which is equal to the whole length of the side of the fish. The Skate may almost be considered as having no true head or neck, the sides of both being included and thus protected by the expanded anterior margin of each pectoral fin. The nostrils, mouth, branchial and anal apertures, are on the under surface; the eyes and temporal orifices on the upper surface. The texture of the skin of the body varies considerably, and will be referred to when describing the different species. From the peculiar form of the body, admirably adapted to exist at the bottom of the water, the Skate may with more propriety be called a Flatfish than any species of the *Pleuronectidæ*. Their mode of progression is not very easily described: it is, when they are not alarmed, performed with a slight motion of the pectoral fins, something between a slide and a swim. I once heard a North-country fisherman call it sluddering. When a Skate makes the best of its way either to gain a prize in the matter of food, or to escape an enemy, great muscular exertion is evident. The mode of defending itself, as described by Mr. Couch, is very effectual: the point of the nose and the base of the tail are bent upwards toward each other; the upper surface of the body being then concave, the tail is lashed about in all directions over it, and the rows of sharp spines frequently inflict severe wounds.

Some sexual peculiarities require particular notice. The woodcut introduced overleaf represents in the left-hand portion an inside view of one-half of the mouth of an adult male;

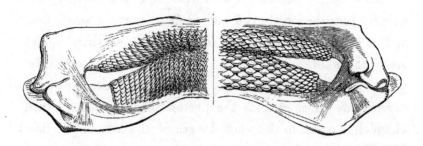

that on the right, an inside view of one-half of the mouth of an equally adult female of the Thornback Ray. While both are young, the teeth in both sexes are alike broad and flat ; but as the male acquires age and sexual power, the teeth that are nearest the centre begin to alter in form and become pointed, as will be seen on close examination, by an elongation of the internal angle ; all the points being directed backwards or towards the throat. Some exceptions to this apparent rule will be pointed out.

Another sexual peculiarity in which the Skate resemble the Sharks is the cylindrical appendage to each ventral fin in the males. The figure at the head of this subject is taken from the under surface of a female, in which no appendages exist : the second figure of the true Skate, the next in succession, is from the under surface of a young male, and small appendages lying on each side close to the tail may be seen ; and even in very young specimens, not more than three inches in breadth, the sexes may be determined by the constant existence of these appendages in the males. The figure of the third species is taken from the upper surface of an adult male, and exhibits these appendages of much greater length : their use may be inferred from the name they bear—they are usually called claspers. This third figure of the Sharp-nosed Ray exhibits also other peculiarities common to males : these are the clusters of spines outside the

eye and temporal orifice on each side, and the regular rows
of spines towards the upper outer surface of the pectoral fins.
The elongation of the central teeth, the development of the
cylindrical appendages, and the appearance and growth of the
clusters and rows of spines on the upper surface at the parts
pointed out, may be considered analogous to those sexual
distinctions which exist in many species of birds and mam-
mals, and which have been called by John Hunter and
others, secondary sexual characters. These spines on the
upper surface of the males occur in the different species of
Skate with smooth skins, as well as in the others, and are
entirely independent of those spinous productions of the
cuticle which distinguish two British species, and will be
more particularly noticed hereafter. It may here be stated
generally, that the Skate are very voracious : their food consists of any sort of fish that they can catch, with mollusca,
testaceous or naked, and crustacea. So powerful are their
muscles and jaws, that they are able to crush the strong shell
of a crab with ease. As in the Sharks, the females are
larger than the males.

The under surface of the Skate at the head of this subject
presents two central circular cavities. The upper one just
below the transverse mouth is bounded laterally by the five
branchial apertures on each outside ; within this cavity the
gills are placed. The circular cavity below is the abdomen,
and contains the stomach, intestines, and other viscera. The
heart is placed immediately in the centre between the two
cavities, and is protected by a broad and strong transverse
cartilaginous arch, the situation of which is indicated.

The young are produced towards the latter part of spring,
or during summer. They are deposited by the parent
fish in thin horny cases, like those of some of the Sharks
already described ; but they are more square in form, as

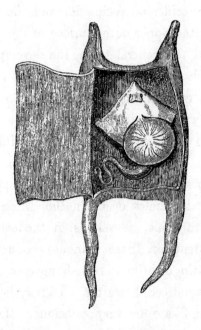

the representation here inserted will evince. These horny cases of the Rays, like those of the Sharks, are also called purses; and on the coast of Cumberland bear the name of Skate-barrows, from the resemblance in shape to a four-handed machine by which two men carry goods. As the young Skate increases in size, the angular parts of the body curve over for a time, till the fish ultimately escapes to provide for itself in a much wider but more dangerous region.

The eight species of true Rays which are found on the coasts of this country will be arranged here in two divisions; the first of which contains four species, having the snout more or less elongated and sharp: the second division contains also four species with blunt noses, the two last of which are furnished with numerous sharp spines on various parts of the surface of the body, and thus lead to the more powerfully armed short-nosed species which will follow.

The Skate, as food, are held in very different degrees of

estimation in different places. In London, particularly, large quantities are consumed, and the flesh is considered delicate and well-flavoured ; but on some parts of the coast, though caught in considerable numbers, both by lines and nets, the flesh is seldom devoted to any purpose beyond that of baiting pots for catching crabs and lobsters.

Skate are in the best condition for the table during autumn and winter. In spring, and in the early part of summer, they are usually maturing eggs or young, and their flesh is then soft and woolly.

The Long-nosed Skate is immediately distinguished from any other Skate found on the British coast, not only by the great length of the nose, but also by the distance between its most extreme point and the transverse line of the mouth ; characters particularly observable in comparison with the species next in order, with which it most assimilates in colour. The snout is very much produced, narrow and sharp, slender as far as the eyes, from whence the body dilates gradually to its greatest breadth, which is behind the centre ; the whole length of the body and tail one-third longer than the width. On the upper surface the body is slightly roughened, and of a light lead colour ; the tail rather more rough, with a row of large crooked spines on each side of the central line, and this species is observed never to have more or less than these two lateral rows ; the small fins on the tail not far removed from each other, the second about its own length from the end. The under surface is a dirty greyish white, marked with dusky specks like the true Skate next to be described ; but the body is thinner in substance than either that or the Sharp-nosed Skate : the nostrils are lobed ; the mouth narrow ; the teeth in old males sharp, with frequently nine or ten spines above the eyes ; on the snout two rows of minute tubercular spines ; behind the head seven or eight spines ranged in a line along the dorsal ridge; towards the outer upper edge of the pectoral fins on each side are the usual rows of sharp hooked spines, and close to the tail the long pendent claspers. Montagu having figured in the Wernerian Memoirs, as referred to, the upper surface of a male, the figure here given represents the under surface of a female.

By some of the West-country fishermen this species is called the Dun Cow: it attains considerable size, and is said to feed on Sand-eels and Sand-launce. According to Mr. Couch, it frequents deep water, and is not caught through the winter: fishermen state that it is exceedingly violent when hooked. I may here state generally, that the greater part of the Skate brought to market are taken in the trawl-nets.

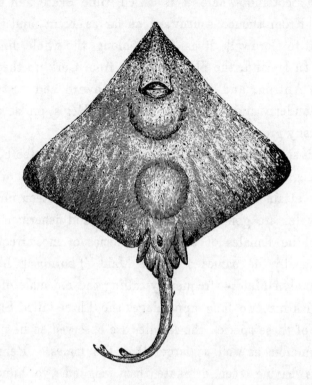

THE SKATE.

BLUE SKATE, *and* GREY SKATE, *Scotland.*

TINKER, *Lyme Regis.*

Raia batis, LINNÆUS. BLOCH, pt. iii. pl. 79, female.
   ,,   ,,   *La Raie cendrée,* CUVIER, Règne An. t. ii. p. 398.
   ,, *lævis seu cinerea,*   WILLUGHBY, p. 69, C. 5, male.
   ,, *batis,* *The Skate,*   PENN. Brit. Zool. vol. iii. p. 111.
   ,,   ,,   ,,   FLEM. Brit. An. p. 171, sp. 24.
   ,,   ,,   ,,   JENYNS, Man. Brit. Vert. p. 510, sp. 199.
   ,,   ,,   ,,   COUCH'S MS.

THIS species, which is frequently called the True Skate, to distinguish it from the Thornback and Homelyn, which are also popularly called Skate, is not so commonly taken as either, but is still better than either as an article of food.

It appears to be found among the Orkneys, and on the
coast of Scotland, where it is called Blue Skate and Grey
Skate. From thence southward as far as Kent, and again
westward to Cornwall, it is found along the whole line of
coast. In Ireland, the Skate is taken from Cork up the east
coast to Antrim, and from thence northward and westward
to Londonderry and Donegal. At Lyme Regis, on account
of its dusky grey colour, it is called the Tinker.

In this species both sexes when adult have sharp teeth, the
points beginning to elongate by the time the body of the
fish has attained the breadth of twelve or fourteen inches.
The females are generally called Maids; and fishermen distinguish the females of the three species of most frequent
occurrence by the names of Skate Maid, Thornback Maid,
and Homelyn Maid,—frequently calling the old male of the
Skate with his two long appendages the Three-tailed Skate.
In each of these species the females are observed to be much
more numerous as well as larger than the males. Pennant
mentions having seen a Skate that weighed two hundred
pounds: it is very voracious, and Mr. Couch has known five
different species of fish, besides crustacea, taken from the
stomach of a single individual. There is reason to believe
that the true Skate produces its young later in the season
than either the Thornback or the Homelyn.

The breadth of the body is to its length nearly as four to
three; the form of the nose conical: the lines from the extreme lateral angle of each pectoral fin being nearly straight,
similar lines taken in a direction backward to a point on the
tail two inches below the end of the ventral fins, would form
a true rhomb: the eyes are slightly elevated above the line
of the upper surface of the body, with a short, hard tubercle
in the front of each, and a second on the inner side of each;
the irides yellow; the temporal orifices valvular, and placed

close behind : the dorsal ridge of the body without spines till near the origin of the ventral fins ; then commence a single row on the centre, reaching along the tail as far as the first of the two small fins, all the points of the spines directed backwards ; one spine between the two small dorsal fins. On the sides of the tail of a female of small size there were no lateral spines ; but in a young male of the same size, there were several lateral spines on each side, the points of which were directed forwards, and are in that respect characteristic of this species. The colour of the upper surface of the body and tail greyish brown : the margins anterior to the angles of the pectoral fins tinged with reddish brown ; those behind the angles brownish black, darker than the body : the colour on the under surface is sooty white, with dark lines in various directions, and numerous blue specks with small sharp points disposed among them over the surface. The nostrils are valvular, half the width of the mouth in advance of each of its angles ; the mouth rather wide ; the teeth in this species are sharp in both sexes when adult, the inner angles of the central teeth beginning to elongate in specimens when they are about twelve inches in breadth across the body.

I may here add that the true Skate, the subject of the present article, the Long-nosed Skate which precedes it, and the Sharp-nosed Skate which is next to be described, are, in some localities, included under the general term of Skate, from their similarity in colour.

THE SHARP-NOSED RAY.

WHITE SKATE, *Scotland.*—BURTON SKATE, *Cornwall.*

Raia oxyrhynchus,	*Sharp-nosed Ray,*		MONTAGU, Wern. Mem. vol. ii. p. 423
,,	,,	,, ,,	PENN. Brit. Zool. vol. iii. p. 113.
,,	,,	,, . ,,	FLEM. Brit. An. p. 171, sp. 21.
,,	,,	,, ,,	JENYNS, Man. Brit. Vert. p. 511, sp. 20 .
,,	,,	*Burton Skate,*	COUCH's MS.

THIS species, says Mr. Couch, from whose drawing the
figure is taken, " may be easily recognised by its sharp
snout, by the waved line of the margin of the body from
the snout to the extremity of the expansion, and by its pure
white colour on the lower surface. It is the largest of the
British Rays; for though in length and breadth it may not
exceed the common Skate, its superior thickness renders it
heavier."

Colonel Montagu, in the Wernerian Memoirs already quoted, says, by way of further distinction, the snout in this species is slender, the lateral margins in a moderately-sized fish running nearly parallel to each other for three or four inches at the extremity. The skin is smooth, with the exception of the spines on the upper surface, peculiar to the males, as shown in the figure; the colour a plain brown without spots or lines, and never so dark as the Skate last described, with which it is sometimes confounded. The teeth of the males, according to a specimen of the mouth very kindly sent to me by Mr. Couch, are longer, more pointed, and sharper than those of any other species I have had an opportunity of examining. The tail is armed with three rows of spines.

Mr. Couch states that the smaller-sized specimens are taken throughout the year; but those which are larger keep in deep water, and are only taken in summer and autumn.

The French are great consumers of Skate, and this species is their favourite fish : their boats come to Plymouth during Lent to purchase Skate, which they preserve fresh and moist during the run back to their own coast by keeping them covered with wet sand.

This species is the White Skate of the Orkneys, and of Scotland generally; and is said to have been taken on the south-east coast of Ireland.

It is doubtful whether this species be the *oxyrhynchus* of the older authors: it is certainly not the *oxyrhynchus* of Bloch, part iii. plate 80, which is the *R. chardon* of other authors; a species with a short nose, and further remarkable for the intense black colour of its upper surface, and which there is reason to believe has been taken once in Cornwall; but the specimen had been too much mutilated before Mr Couch saw it to enable him to determine correctly.

VOL. II. 2 F

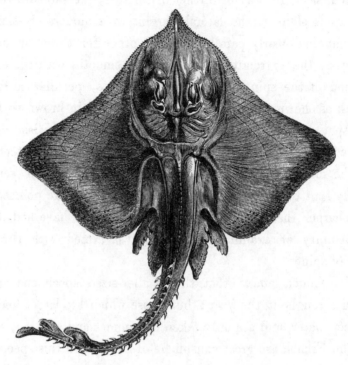

THE BORDERED RAY.

Raia marginata, Bordered Ray, FLEM. Brit. An. p. 172, sp. 27.
,, ,, ,, ,, JENYNS, Man. Brit. Vert. p. 512, sp. 201.

THE BORDERED RAY, as it is called from the broad
dark marginal edge of its pectoral fins, has been taken at
Liverpool, Brighton, and Weymouth; it has also been
taken at Dieppe, and noticed by M. Noel and Lacépède.
It is a well-known species in the Mediterranean, described
by M. de Blainville, by the Prince of Musignano, and M.
Risso. But little is known of its habits, and it does not
attain a large size. M. Risso states that the flesh is considered pretty good.

I avail myself, by permission, of Mr. Jenyns' description

of this species, taken from a specimen obtained at Weymouth by Professor Henslow.

" Total length fifteen inches six lines : length of the head from the end of the snout to the spiracles behind the eyes, three inches six lines ; of the tail from the vent to its extremity, seven inches nine lines : greatest breadth across the pectorals, eleven inches three lines. The total length of M. de Blainville's specimen was two feet. The form rhomboidal ; the transverse diameter rather more than one-third greater than the length from the end of the snout to the vent : snout elongated, projecting considerably from between the pectorals, terminating in a sharp point, with the lateral margins nearly parallel for the last quarter of their length : mouth moderately wide ; jaws transverse ; teeth numerous, closely set, in several rows, roundish or somewhat quadrilateral at the base, each terminating in a sharp point : nostrils in a line with the corners of the mouth, and rather more than half-way between them and the upper margins of the pectorals ; a channel from the nostrils to the mouth, covered by a membranous flap : eyes and spiracles both large : skin perfectly smooth above ; and beneath also, excepting along the anterior margins of the pectorals and the surface of the snout, which are set with very minute spines and denticles : one large spine above each eye, inclining backwards, and another smaller one behind each eye : no spines on any part of the back, but three rows on the tail, one occupying the middle ridge, the two others the sides ; the spines on these rows strong and sharp, and mostly inclining backwards : tail scarcely longer than the body, depressed, rather stout, with two moderately-sized finlets of equal form, nearly contiguous ; scarcely the rudiment of a caudal : pectorals broad, with the anterior margins hollowed out, and not prolonged beyond the basal half of the

2 F 2

snout; ventrals moderate, deeply notched or bilobated. General colour of the upper part reddish brown, somewhat paler on the pectorals, with a faint indication of round whitish spots; beneath white, with a broad border all round, especially beneath the angles of the pectorals, of dark reddish brown, approaching to dusky: tail entirely black."

Since the preceding portion of this article was printed, I have received a specimen of the Bordered Ray from Lyme Regis, for which I am indebted to the kindness of Lord Cole.

The vignette below represents a view taken near Hungerford market.

CHONDROPTERYGII. *RAIIDÆ.*

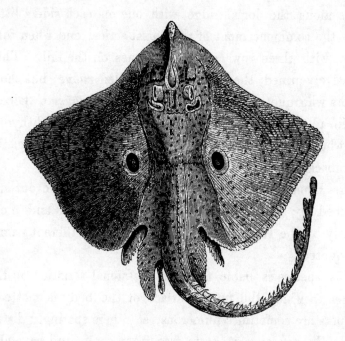

THE HOMELYN RAY.

THE HOME, SAND RAY, AND SPOTTED RAY.

Raia maculata,   *Sand Ray,*   MONTAGU, Wern. Mem. vol. ii. p. 426.
   ,,   *miraletus,*   *Mirror Ray,* DON. Brit. Fish. pl. 103.
   ,,   *rubus,*   *Rough Ray,*   ,,   ,,   ,,   pl. 20.
   ,,   *oculata,*   *Mirror Ray,* FLEM. Brit. An. p. 172, sp. 26.
   ,,   *maculata,*   *Spotted Ray,* JENYNS, Man. Brit. Vert. p. 514, sp. 203.

THIS smooth-skinned spotted Ray, called *Raia lævis* and Homelyn so long ago as the time of Merrett,* and one of our most common species along the line of our southern coast, has not been so well distinguished or so clearly defined by some authors as its obvious characters admit and require. The males, though they have, like the females, a perfectly

* Pinax Rerum Naturalium Britannicarum. London, 1667, p. 185.

smooth skin, have also spines about the eyes, rows of small
hooks on the upper surface of the pectoral fins, one row of
spines along the dorsal ridge, with one on each side a little
below the commencement of the dorsal series, and when full-
grown, with three rows of strong spines on the tail. Thus
extensively armed, the male has been called *rubus:* but those
authors who quote as a synonyme the *R. rubus* of Bloch,
part iii. pl. 84, have been misled by the German ichthyologist, whose figure proves his fish to have been a male of the
Thornback, of which his plate 83 is the female.

The Homelyn of our coast has been best made out and
described by Mr. Donovan, Colonel Montagu, and more
recently by the Rev. Mr. Jenyns, under the different names
here quoted.

This species is liable to some occasional variation in the
manner in which the upper surface of the body is spotted;
the spots are sometimes numerous, at others sparingly distributed: I have seen it quite free from spots, and have also
seen it with only one eye-like spot on each side, not far removed from the line of the back. I have mentioned that
the skin, independent of the accessory organs, is quite
smooth. These variations have given rise to the different
trivial names *miraletus, oculata, lævis,* and *maculata,* which
have been applied to it by different authors, from the appearance of the particular specimens examined.

Colonel Montagu, referring to the *miraletus* and *rubus* of
Mr. Donovan, had no doubt that they were both identical
with his own *maculata,* since, being a common species on the
Devonshire coast, he had ample opportunities of seeing it
under its different appearances.

Mr. Donovan has given correct figures both of the Mirror
Ray and of the Homelyn, as quoted, the latter under the
trivial term *rubus;* but he believed with Montagu, that they

were not distinct species. Mr. Donovan had noticed two similar eye-like spots on several small examples of the true Skate (*batis*); and I possess young specimens of the Thornback (*clavata*) with the same sort of ocellated lateral spots, and have seen many others of the three most common species. Mr. Donovan's remark accords so closely with my own view, that I insert it here in his own words :—

" Although we present this as the *Raia miraletus* of Linnæus with perfect confidence, it is not without some hesitation at least that we can offer it as a distinct species. In every respect, except the ocellar spot on the wings, it perfectly agrees with the Homerling Ray, and may possibly prove, on further examination of other specimens, to be only a *lusus*, or remarkable variety of that fish."

The figure given at the head of this subject exhibits the eye-like lateral spots, from the possession of which it has been called *miraletus* and *oculata* : the smoothness of the surface of the skin, and its numerous smaller spots, sufficiently warrant the terms *lævis* and *maculata*. The words *oculata* and *lævis* were combined by some of the older authors, and probably referred to this species.

The Homelyn and the Thornback, which are not very dissimilar in shape, though otherwise perfectly distinct, are the two species most common in the London market : a large proportion of both are taken in the trawl-nets.

The form is rhomboidal; the diameter of the body about one-fourth greater than the length : the nose short and blunt, its projection beyond the outline of the pectorals but small : in a young male specimen of twelve inches in breadth the secondary sexual characters begin to appear ; there are numerous small spines about the nose, and some extending along part of the anterior edge of the pectoral fins ; two or three prominent spines before and behind each eye, with rough granulations on the surface of the skin before and between them : the more conspicuous characters that distinguish the males have been already noticed. The eyes and temporal orifices are large : the central row of dorsal spines commence above the middle of the body, with one strong spine on each side of it about the middle of the body and in the line of its greatest diameter : the series of spines on the dorsal ridge extend along the centre of the tail, with a row along each side of it in adult specimens ; in young examples the series on each side is not complete. On the tail are two small fins, with two spines between ; the points of all the spines on the central line and on the tail directed backwards.

The colour of the upper surface is a pale yellowish or reddish brown, with spots of darker brown, subject to the variations that have been already pointed out ; the colour of the under surface plain white ; the skin smooth ; nostrils and mouth near the end of the nose ; the mouth transverse, rather small. Montagu says, both sexes of the *maculata* have sharp teeth ; but this refers to examples that are perfectly adult : young males of small size, and females when larger, have the teeth blunt ; in old males, and very probably also in old females, from the operation of those laws which influence the secondary sexual characters, the teeth become pointed.

The term Sand Ray is in some localities exclusively applied to the males of this species, from their greater roughness.

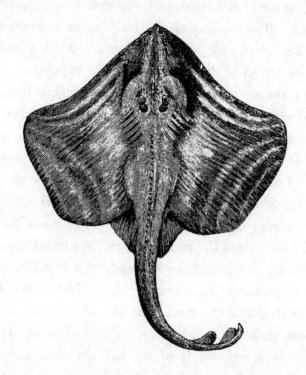

THE SMALL-EYED RAY,

OR PAINTED RAY.

Raia microcellata, *Small-eyed Ray,* Montagu, Wern. Mem. vol. ii. p. 430.
,,   ,,   ,,   ,,   Flem. Brit. An. p. 171, sp. 23.
,,   ,,   ,,   ,,   Jenyns, Man. Brit. Vert. p. 515, sp. 204.
,,   ,,   *Painted Ray,*   Couch's MS.

Colonel Montagu and Mr. Couch appear to be the only British naturalists who have obtained this species; and it must be considered a rare one, since the first of these gentlemen saw but two examples, and the latter has only seen one. The very small size of the eye is stated by both to be a remarkable and striking distinction.

The length of the specimen obtained by Mr. Couch was thirty-three inches and a half, of which the tail measured thirteen inches ; breadth across the fins twenty-four inches ; the eyes three inches apart, and five inches and a half from the snout. The outline of the body much resembles that of the Thornback, *R. clavata ;* snout a little prominent, the margin waved to the extremity of the expansion, behind rounded ; the eyes very small ; temporal orifices large : the body covered with rough granulations, but altogether without spines, either on its surface or about the eyes, except a row that runs along two-thirds of the back, and down the middle of the tail to the fins ; an irregular row of similar hooked spines extends along each side of the tail ; along the tail is a border on each side, like a membranous fin ; two rounded fins towards the end of the tail, somewhat separated, the hindmost one inch from the end, with which it is continuous by means of an elevated ridge. In the distribution of its colours this is the most beautiful of the British Rays. The upper surface is a light grey, with a lighter line running along the back and middle of the tail, enclosing the central row of spines. The disk is beautifully and regularly quartered, first by three white lines enclosing each other, and passing from near the eye circularly to near the extremity of the expansion, the convexity of the arch inwards, and consequently the shorter line nearer the margin ; on the hinder edge of the disk, formed by the pectorals, are two other lines passing from behind the expansion circularly to the neighbourhood of the abdominal fins, the convexity of the arch inwards ; on the more central part of the disk are a few whitish spots, those of both sides answering to each other ; the extreme edge of the disk posterior to its greatest expansion, and also the abdominals, as well as the fin-like margin of the tail, are edged with white. The nostrils have a prominent expanded membrane ; width of the mouth three inches ; teeth flat, like those of the Thornback ; mucous orifices on the under surface numerous, and as if punctured with a pin ; the colour of the skin a pure white.

Such is the description Mr. Couch gives of his specimen, which was a female, and which was taken by a line on the 28th of January 1835. In it numerous eggs were found, some of which had attained their full growth ;—a circumstance which fixes the period for the production of the young in this species.

Montagu says both his examples were females, resembling his *R. maculata* in form ; Mr. Couch refers to the Thornback for shape : the figure here given is taken from Mr. Couch's drawing, and it will be observed that all three have considerable similarity of outline. A few extracts from Montagu's description will exhibit further resemblance. The proportions by measurement are very nearly alike ; the upper surface pale brown, with a few scattered spots and lines of a lighter colour on the margins of the wings ; the skin covered with minute spines, which make it feel rough : the eyes remarkably small, at once pointing out a material distinction ; those of the specimen described did not exceed half an inch in diameter from the opposite angles of the eyelids ; whereas the *R. maculata*, and most others of similar size, have eyes nearly double that diameter : one row of small hooked spines on the tail, continuing along the dorsal ridge to the head. Colonel Montagu's specimens being younger than that obtained and described by Mr. Couch, had not acquired the lateral marginal rows of spines on the tail ; the under part smooth and white ; the teeth obtusely cuneiform, with a broad edge that felt rough to the finger as it was withdrawn from the mouth.

THE THORNBACK.

Raia clavata, *Thornback,* Willughby, p. 74.

    ,,   ,,   Linnæus. Cuvier, Règne An. t. ii. p. 398.

    ,,   ,,   *Thornback,* Bloch, pt. iii. pl. 83, female.

    ,,   *rubus,*   *Rough Ray,*   ,,   ,,   pl. 84, male.

    ,,   *clavata,* *Thornback,* Montagu, Wern. Mem. vol. ii. p. 416.

    ,,   ,,   ,,   Penn. Brit. Zool. vol. iii. p. 122, pl. 14, female.

    ,,   ,,   ,,   Don. Brit. Fish. pl. 26, female.

    ,,   ,,   ,,   Flem. Brit. An. p. 170, sp. 19.

    ,,   ,,   ,,   Jenyns, Man. Brit. Vert. p. 516, sp. 205.

    ,,   ,,   ,,   Couch's MS.

The Thornback exhibits very marked distinguishing characters, and being also a very common fish, is one of the best known of the species of Rays,—a term which Mr. Couch considers to be derived from the Anglo-Saxon ' Reoh,' which means ' rough,' and is particularly appropriate to the

Thornback, which, on the Cornish coast, is preeminently distinguished as the Ray. The Thornback is also taken commonly both on the coast of Scotland and Ireland. From the good quality of the flesh of this fish, and the immense quantity taken every year, the Thornback, and its female, the Maid, is one of the most valuable of the species. Mr. Couch says that the flesh takes salt well, and in this preserved state affords the poor fishermen and their families many wholesome meals when stormy weather prevents them obtaining fresh supplies. The Thornback is taken in the greatest abundance during spring and summer, because the fish then frequent sandy bottoms in shallower water and nearer the shore than usual, for the purpose of depositing their eggs ; but the flesh of the Thornback at this season is not, as before noticed, so firm as in autumn and winter. It is in the best condition for table about November. Their food is various other fish, particularly Flatfish, testaceous mollusca, and crustacea.

Bloch's figure, plate 83, represents the female of this species, under the name of *R. clavata;* and the fish next in succession in that work, plate 84, is the male of the same species, but is called *R. rubus,* although most of the synonymes quoted are those of *clavata.*

The figure here given was taken from a young male measuring fourteen inches in breadth. The point of the nose is but little produced : the anterior margins of the pectorals are undulated ; the outline behind each lateral angle of the pectorals nearly straight, or slightly rounded : the eyes and temporal orifices rather large, with two or three strong hooked spines both before and behind them. The whole upper surface of the body rough with small points, which when examined with a lens have stellated bases. Besides these, there are distributed over this upper surface numerous nail-like tubercular

spines, each of which has an oval osseous base ; the margin of
the base entire, with a central projecting crooked shank or
spine directed backward. Two of these broadly-based spines
occupy the central ridge of the nose ; others, to the number
of thirteen or fourteen, are distributed over each side with
some regularity, and similarly disposed on the two sides.
The dorsal ridge of unequally-sized spines begins a short distance between and behind the temporal orifices, one or two
small spines occurring between each of the larger ones : this
single line of spines extends to the origin of the tail, where
three rows of spines begin and are continued along it, forming
a series of powerful weapons. The tail is furnished with two
membranous fins on the upper central ridge, and ends with a
small dilatation. The prevailing colour of the upper part is
brown, with numerous lighter-coloured spots, and sometimes,
as has been noticed already, with one larger rounded spot on
each pectoral. Young males and females have fewer spines
on the body than old males, and both sexes attain some size
before they put forth any ; they have frequently also but one
row of spines along the tail. The colour of the under side
is pure white, with a few spines only on each side. The teeth
of the adult male in this species are decidedly different from
those of the female, as shown in the woodcut at the top of
page 416 ; those represented on the left hand being from a
male fish, and those on the right from a female fish of the
same size, and representing one-half of the inside of the
mouth of each as seen from behind.

THE STARRY RAY.

Raia radiata,	*Starry Ray,*	Don. Brit. Fish. pl. 114, female.		
,,	,,	,,	,,	Flem. Brit. An. p. 170, sp. 20.
,,	,,	,,	,,	Jenyns, Man. Brit. Vert. p. 517, sp. 206.

THIS very beautiful and well-marked species was made known by Mr. Donovan in his History of British Fishes, and a very good coloured representation of it is added in that work, which will prevent its being confounded with any other.

Whether this species was really known to other authors their descriptions do not afford unequivocal proof, and I have therefore only quoted those synonymes which I know to refer to this fish. Mr. Donovan's specimen, which was not more than four inches across and seven inches in length, was caught on the north coast of Britain, and was communicated to him by Mr. Stuchbury.

Dr. Fleming, in his History of British Animals, quotes as a synonyme to *radiata* the *R. Fullonica* of the Fauna of Greenland, by Fabricius; and it is probably a Northern species, the only three examples of it I have seen having been received, one from Berwick Bay, and two from the Frith of Forth. The first was a female, for which I am indebted to the kindness of Dr. George Johnston, and from this example the figure here given was derived. In 1835, Dr. Parnell sent me from Edinburgh two examples, a male and a female, which had been obtained in the Forth, and obligingly permitted me to retain the male for my own collection, which came marked accordingly. On comparing these three examples with Mr. Donovan's figure, no doubt remained that they were of the same species.

The habits of this fish are but little known, and the figure here given being that of a female, I shall closely describe the male, which was nineteen inches long from the point of the nose to the end of the tail, and fourteen inches in breadth; the snout but little produced, almost falling in with the line of the anterior margin; the lateral expansion of the pectorals and their posterior margins rounded; the pelvic fins rather large: the central ridge of the nose, and a great portion of the pectoral fins or wings, are covered with asperities of different sizes, but the form of which are all alike, being a single spine bent backwards, arising from a stellated base of many radii; these appear to be nearly symmetrical, and about equal in number on the two sides: the eyes are blue and rather large, placed about half-way between the central transverse cartilaginous arch of the body and the end of the snout; before each eye one large spine, and two large spines behind, with several smaller ones along

the inner edge of each eye ; temporal orifices rather large ;
one large spine above the line of the transverse cartilaginous
arch, one upon the centre of it in the line of the dorsal ridge,
and two spines at each lateral extremity of the transverse
arch : below this cross-bar commences a series of equally
large spines on the dorsal ridge, which extends to the first fin
on the tail; between these large spines are a few smaller ones,
and on each side the central row of large spines is another
row of spines about half the size of the large ones, but more
numerous, forming three distinct rows down the back and
tail ; but all of them, though differing in size, have the same
character in respect to the beautifully radiated form of the
base from which the ascending spine arises: the upper surface
of the body independently of this arming is perfectly smooth;
the colour pale brown, with a tinge of orange brown.

On the under surface the colour is uniformly white ; the
skin soft and smooth ; the nostrils large, defended by a cutaneous valve ; the mouth rather small ; the teeth in the male
with the internal angle elongated and sharp, and in a second
specimen, a female of ten inches only in length, the teeth are
becoming pointed.

The sexual appendages in the male here described are half
as long as the tail ; and as these, as well as the other sexual
distinctions, are well marked in this fish, which only measures
fourteen inches in breadth, I am induced to believe, from
the early acquisition of these characters, that this species
does not attain a very large size. This species is probably
the *Raia asteria aspera Rondeletii* of Willughby, p. 73,
pl. D. 5, f. 4, and the *Raia aspera* of M. de Blainville, in
the *Faun. Franc.;* but I have not included these names
among the synonymes at the head of the subject, for the
want of that additional certainty which good figures would
have supplied.

VOL. II.   2 G

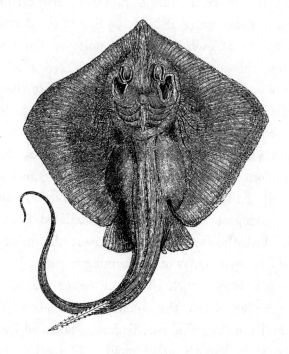

THE STING RAY.

COMMON TRYGON. FIRE FLAIRE.

Trygon pastinaca,	*La Pastinaque,*	Cuvier, Règne An. t. ii. p. 399.
„ „	*Common Trygon,*	Flem. Brit. An. p. 170, sp. 18.
Pastinaca marina,	*Rondeletii,*	Willughby, p. 67, C. 3.
Raia pastinaca,	Linnæus.	Bloch, pt. iii. pl. 82.
„ „	*Sting Ray,*	Penn. Brit. Zool. vol. iii. p. 125.
„ „	„ „	Don. Brit. Fish. pl. 99.
„ „	„ „	Jenyns, Man. Brit. Vert. p. 518, sp. 207.

Generic Characters.—Head enclosed laterally by the pectorals; posterior portion of the disk of the body somewhat rounded; tail armed near its origin with a long and sharp flattened spine, serrated on both edges; the rest of the tail slender, without fins, and ending in a point; teeth small.

From the Rays whose bodies are more or less covered and protected with sharp spines supported on broad bases, and which spines, continued along the upper surface of the tail, are defensively or offensively used, the transition to those species of Rays which are still more powerfully armed is easy and natural.

The Sting Ray was well known to the ancients, who entertained many curious notions of the power and venom of its spine ; and this fish is also noticed as an inhabitant of the shores of this country so long ago as the days of Merrett and Sibbald. At present it is more frequently taken on the southern coast than elsewhere, from Sussex even as far west as the county of Cork in Ireland. It appears, however, otherwise to occupy an extensive range, being found in the Mediterranean, and from thence to a high degree of north latitude on the coast of Norway.

Colonel Montagu, in his notes, mentions obtaining a specimen, taken at Hastings, which was presented to him by the Rev. Mr. Whitear. " At the base of the bony process in the tail of this fish, was a smaller one ready to replace the original if by accident it should be lost ; or possibly this weapon may be deciduous and occasionally discharged."

Mr. Couch in his MS. says, " This species keeps on sandy ground at no great distance from land, and in summer wanders into shallow water, where it is often entangled in the fishermen's nets,—the only way in which it is usually caught, for it rarely swallows a bait. The manner in which this fish defends itself, shows its consciousness of the formidable weapon it carries on its tail. When seized or terrified, its habit is to twist its long, slender, and flexible tail round the object of attack, and with the serrated spine tear the surface, lacerating it in a manner calculated to produce violent inflammation." Other authors state that it is capable of striking its weapon with the swiftness of an arrow into its prey or its enemy, when with its winding tail it secures its capture. These spines, as may be supposed, possess no really

2 G 2

venomous quality : when lacerated wounds happen to men of a bad habit of body, the symptoms are frequently very severe. In some countries, serrated fish spines, admitting of easy application by tying, are used to point arrows and spears, which when thus mounted become very formidable weapons.

A specimen examined and described by Pennant was two feet nine inches long from the tip of the nose to the end of the tail ; to the origin of the tail, one foot three inches : the breadth, one foot eight inches. The body is quite smooth, except, according to M. de Blainville, a few small tubercles along the central line of the back and tail, as well as on the upper and posterior part of the pectoral fins—probably a male fish ; the shape almost round, and of a much greater thickness and more elevated form in the middle than any other of the Rays, but grows very thin towards the edges ; the nose is very sharp-pointed, but short ; the irides are of a gold colour ; behind each eye the temporal orifice is very large : the colour of the upper surface of the body is a dirty yellow ; the middle part, of an obscure blue: Mr. Donovan says the young are spotted with brown. The tail and spine are dusky ; the former very thick at the beginning : the spine, placed at about one-third of the length of the tail from the body, is about five inches long, flat on the top and bottom, very hard, sharp-pointed, the two side edges thin, and closely and sharply serrated the whole way ; the tail extends four inches beyond the end of this spine, and becomes very slender at the extremity. The under surface is white ; the nasal lobes very large ; mouth and teeth small. The flesh is said to be rank and disagreeable, and when laid bare by skinning or cutting into, is very red,—a circumstance which may account for the old name of Fire Flaire.

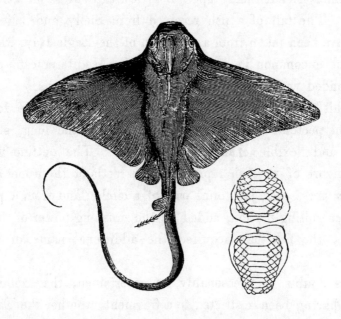

THE EAGLE RAY.

THE WHIP RAY.   MILLER.

Myliobatis aquila,	*Aigle de mer,* Cuvier, Règne An. t. ii. p. 401.
Aquila Bellonii,	Willughby, p. 64, C. 2.
Raia aquila,	Linnæus. Bloch, pt. iii. pl. 81.
,, ,,	*Whip Ray,* Penn. Brit. Zool. vol. iii. p. 128.
,, ,,	,, ,, Jenyns, Man. Brit. Vert. p. 519.

Generic Characters.—Head partly disengaged from the pectoral fins ; teeth flat ; the central plates much larger than those which are lateral ; pectoral fins wing-like ; the tail armed with a serrated spine, as in the last genus *Trygon.*

Pennant, in his British Zoology, states that Mr. Travis, surgeon at Scarborough, had the tail of a Ray brought to him by a fisherman of that town : he had taken it in the sea off that coast, but threw away the body.　It was above three feet long, entirely covered with hard obtuse tubercles, extremely slender and taper, and destitute of a fin at the end.　The tail of a fish received from Sicily, and believed to have been taken from a specimen of the Eagle Ray, which is not uncommon in some parts of the Mediterranean, corresponded with the description given by Mr. Travis.

This fish is called Eagle Ray from the wing-like form of the pectoral fins; and Whip Ray, from the long, slender, and flexible character of its tail.　The outline near the figure of the fish represents the teeth of the upper and under jaw; each jaw forms part of a circle; and from a particular rolling motion, added to the crushing power of these teeth, the fish has acquired the additional name of the Miller.

As doubts may reasonably be entertained, the examination having been restricted to a fragment, whether the Eagle Ray has really been taken on our coast, the figure of this fish, as well as that which forms the subject of the vignette, are here inserted to invite the attention of observers on the coast.

To William Thompson, Esq. of Belfast, one of the Vice Presidents of the Natural History Society of that town, I am indebted for many valuable and interesting notices of the fishes of the Irish lakes and coast which are distributed in various parts of this work.　In 1835, Mr. Thompson made the following communication to the Zoological Society of London, which is published in the Proceedings for that year, at page 78.

" *Cephaloptera*, Dumeril.—A fish of this singular genus, taken about five years ago on the southern coast of Ireland, and thence sent to the Royal Society of Dublin, is at present preserved in their museum. In breadth it is about forty-five inches. The specimen being imperfect, and the characters of some of the species being ill defined, I hesitate applying to it a specific name. It somewhat resembles the *Cephaloptera giorna* as figured by M. Risso.

A representation of this fish is given below.

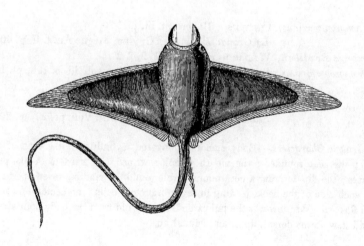

CHONDROPTERYGII.   *PETROMYZIDÆ.**

THE LAMPREY.

Petromyzon marinus, Linnæus.   Bloch, pt. iii. pl. 77.
      ,,   ,,   *La Grande Lamproye,* Cuvier, Règne An. t. ii. p. 404.
Lampetra Rondeletii, Willughby, p. 105, G. 2, f. 2.
Petromyzon marinus, Sea Lamprey, Penn. Brit. Zool. vol. iii. p. 102, pl. 10.
      ,,   ,,   Spotted Lamprey, Don. Brit. Fish. pl. 81.
      ,,   ,,   Lamprey,   Flem. Brit. An. p. 163, sp. 1.
      ,,   ,,   Sea Lamprey, Jenyns, Man. Brit. Vert. p. 520, sp. 209.

Generic Characters.—Body smooth, elongated, cylindrical, like that of an Eel; the head rounded; the mouth circular, armed with hard tooth-like processes; the lip forming a continuous circle round the mouth; seven apertures on each side of the neck, leading to seven branchial cells; no pectoral or ventral fins; the skin towards the tail extending in a fold from the body both above and below, forms dorsal, anal, and caudal fins.

The last family of the cartilaginous or chondropterygian fishes contains the Lampreys, and some cylindrical fishes very closely allied to them. These fishes are, in reference to their skeleton, and in some other respects, the lowest in the scale of organization among vertebrated animals. The form and peculiarities of the mouth will be best understood

* The family of the Lampreys.

by a reference to the vignette; the figure on the left hand of which shows the flexible lip concealing the mouth; the figure on the right hand represents the rounded mouth, the small and numerous tubercular teeth, and the central aperture leading by the throat to the stomach.

The situation of the branchial cells, and the gills or branchiæ within these cells as they exist in the common River Lamprey, or Lampern, as it is also called, are shown in the right-hand figure at the bottom of page 326: and the water obtains access and egress by seven small apertures on each side of the neck, by the mouth of the fish, or by an aperture through the upper part of the head which communicates with the pharynx, and which communication is distinctly seen in a divided head.

The Lampreys, like the Sharks and Rays, have no swimming-bladder; and being also without pectoral fins, are usually seen near the bottom. To save themselves from the constant muscular exertion which is necessary to prevent them being carried along by the current of the water, they attach themselves by the mouth to stones or rocks, and were in consequence called *Petromyzon*, or Stone-sucker; while the circular form of the mouth induced the name of *Cyclostomes*, or Round-mouthed Fishes, which was bestowed upon them by M. Dumeril.

In reference to the respiratory apparatus in the species of this genus, Mr. Owen has remarked,* that " when the Lamprey is firmly attached, as is commonly the case, to foreign bodies by means of its suctorial mouth, it is obvious that no water can pass by that aperture from the pharynx to the gills; it is therefore alternately received and expelled by

* Descriptive and Illustrated Catalogue of the Physiological Series of Comparative Anatomy contained in the Museum of the Royal College of Surgeons in London, vol. ii. page 80.

the external apertures. If a Lamprey, while so attached to the side of a vessel, be held with one series of apertures out of the water, the respiratory currents are seen to enter by the submerged orifices, and, after traversing the corresponding sacs and the pharynx, to pass through the opposite branchiæ, and to be forcibly ejected therefrom by the exposed orifices. The same mode of respiration must take place in the Mixine," (a species of this family to be described hereafter,) " while its head is buried in the flesh of its prey. The cyclostomous fishes thus present an obvious affinity to the *Cephalopoda*, inasmuch as the branchial currents are independent of the actions of the parts concerned in deglutition."

The intestinal canal is small, and extends in a straight line along the abdomen to the anal aperture without any convolution. The Lampreys are oviparous, spawning late in the spring ; the roe escaping, in both sexes, by a small membranous sheath, which has internally at its base five apertures, one leading upward to the intestine, one to each kidney, and one to each lateral cavity of the abdomen.

The Marine Lamprey, which from its mottled appearance was called *P. maculosus* by Artedi, has a very extensive geographical range. It is found in the Mediterranean, and from thence northwards in most of the rivers of Europe as far as Scandinavia, during the spring. Professor Reinhardt includes it among the fishes of Iceland, and our countryman Pennant gives it a place in his Arctic Zoology. From a description and figure in the Natural History of the Fishes of Massachusets, by Dr. Smith of Boston, this fish appears to be common in the rivers of North America, attaining a large size in those of the more southern states, but not exceeding seventeen or twenty inches in length in a high northern latitude. Dr. Mitchell also includes this species among his fishes of New York. It is rather common during spring and summer in some of the rivers on the southern coast of England, particularly the Severn, and is found in smaller numbers in several of the rivers of Scotland and Ireland about the same period of the year.

I have received specimens of large size from the Severn in April and May, during which months it ascends that river to a great distance from the sea for the purpose of depositing its spawn. At this time it is considered in perfection as food, and considerable quantities are prepared in various ways for the table : the potted Lampreys and Lamperns of Worcester are in high estimation. A few are caught in the Thames almost every year, up which river it travels notwithstanding all the numerous and various obstacles which the port of London presents. I am indebted to my friend Mr. Broderip for a note of one taken in June 1834, and another in the same month of 1835, as high up the Thames as Sunbury Weir. A fisherman saw the Lamprey, and struck at it with his punt pole, and supposed he hit it, as the fish rose to the surface and was taken as it was swimming along. The haunt of this Marine Lamprey at Sunbury is a little above the church, and nearly opposite the vicarage, in a place called the Church Deep.

In Scotland, the appearance of the Lamprey in the fresh water is rather later in the year than in the rivers of the south. Sir William Jardine says, " They ascend our rivers to breed about the end of June, and remain until the beginning of August. They are not furnished with any elongation of the jaw, afforded to most of our fresh-water fish, to form the receiving furrows at this important season ; but the want is supplied by their sucker-like mouth, by which they individually remove each stone. Their power is immense. Stones of a very large size are transported, and a large furrow is soon formed. The *P. marinus* remain in pairs, two on each spawning place ; and while there employed, retain themselves affixed by the mouth to a large stone."

After the spawning season is over, the flesh of the Lamprey, like that of other fish, loses for a time its firmness and other good qualities, and the weakened fish makes its way back to the sea, to recruit its wasted condition.

The food of the Lamprey consists generally of any soft animal matter ; and in the sea it is known to attack other fishes even of large size, by fastening upon them, and with its numerous small rasp-like teeth eating away the soft parts down to the bone. It is not very often caught while it remains at sea.

This species usually measures from twenty to twenty-eight inches in length ; the head is rounded ; the form of the body long and cylindrical, slightly compressed towards the tail ; on the top of the head, rather before and between the eyes, is an external aperture, which if examined with a blunt probe is found to pass downward and backward, opening into a tube on a line with the internal orifice of the first branchial sac : along each outside of the neck are seven rounded apertures, leading to as many branchial cells lined with a membrane constructed like that of the gills in fishes ; each of these cells has an internal opening into a tube which is closed by a car-tilaginous pericardium at the bottom, but communicates up-wards with the mouth : the lips surrounding the mouth, and the numerous small teeth within, have been already referred to : on the lower third portion of the body are two distinct membranous dorsal fins, the second of which is the most elevated, the edges of both convex ; a continuation of this membrane round the extreme fleshy portion of the tail forms a caudal fin, and a narrow slip passing upwards on the under side forms an anal fin.

The skin is perfectly smooth ; the colour of the body olive

brown, mottled and spotted on the back and sides with darker green and dark brown; the margins of the fins inclining to reddish brown; the irides golden yellow.

In slowly-running water, the Lamprey swims with a lateral undulating motion of the body, assisted by its fins: where the current is rapid, it makes successive plunges forward, attaching itself quickly to any fixed substance that offers to secure the advantage gained.

The figure of the fish at the head of this subject was taken from an excellent drawing made by Mrs. Ley.

Pennant states that it has been an old custom for the city of Gloucester annually to present his Majesty with a Lamprey pie, covered with a large raised crust.

 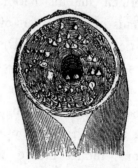

THE LAMPERN, OR RIVER LAMPREY.

Petromyzon fluviatilis, Linnæus.　Bloch, pt. iii. pl. 78, fig. 1.
　　　,,　　　　　,,　　　Cuvier, Règne An. t. ii. p. 404.
　　　,,　　　　　,,　　　*Lesser Lamprey,* Penn. Brit. Zool. vol. iii. p. 106, pl. 10.
　　　,,　　　　　,,　　　*Lampern,*　　Don. Brit. Fish. pl. 80.
　　　,,　　　　　,,　　　*River Lamprey,* Flem. Brit. An. p. 163, sp. 2.
　　　,,　　　　　,,　　　　,,　　　,,　　Jenyns, Man. Brit. Vert. p. 521,
　　　　　　　　　　　　　　　　sp. 210.

The River Lamprey, or Lampern, as it is called by
fishermen for distinction, is a well-known species which
abounds in many rivers of England, particularly the Thames,
the Severn, and the Dee : it is also abundant in several
rivers of Scotland and Ireland.

Some authors state that this species, like that last described,
visits our rivers in spring, and returns to the sea after spawning ; but the recorded opinions of others, and my own observations, induce me to believe that it generally remains all
the year in the fresh water.　In the Thames I am certain it
is to be obtained every month in the year ; but is considered
in the best condition for the table from October to March,
during which time it is permitted to be caught, according
to the rules adopted for the conservation of the fishery.

Formerly the Lampern was considered a fish of considerable importance. It was taken in great quantities in the Thames from Battersea Reach to Taplow Mills, and was sold to the Dutch as bait for the Turbot, Cod, and other fisheries. Four hundred thousand have been sold in one season for this purpose, at the rate of forty shillings per thousand. From five pounds to eight pounds the thousand have been given; but a comparative scarcity of late years, and consequent increase in price, has obliged the line fishermen to adopt other substances for bait. Formerly the Thames alone supplied from one million to twelve hundred thousand Lamperns annually. They are very tenacious of life, and the Dutch fishermen managed to keep them alive at sea for many weeks.

If this species, which is very easily obtained, be examined in the months of March or April, the distinction of the sexes will be immediately evident on opening them. The female may generally be known externally by the larger size of the abdomen, and the male by his lips being more tumid and the mouth larger than that of the female.—The season of spawning is May, and the process has been described by several observers. This sometimes takes place in pairs only, and at others by many of both sexes occupying one general spawning bed.

The food of this species, according to Bloch, is insects, worms, small fish, and the flesh of dead fish.

The adult fish is usually from twelve to fifteen inches in length; the body rather slender, cylindrical for two-thirds of its length, then compressed to the end of the tail; the head rounded, with a single aperture on the crown, leading to the tube between the cells, as in the other species: the eye rather large; the seven lateral openings ranged in a line behind, but a little obliquely and below it, on each side: the lip surrounding the mouth has a continuous row of small points on its margin; the mouth and teeth as represented near the figure of the fish: the back furnished with two rather elongated dorsal fins, with a separation between them; the tail furnished with an extension of the membrane above and below.

The skin is quite smooth, of a blue colour on the back and sides, passing into silvery white underneath.

In "The Treatyse of Fysshynge wyth an Angle," attributed to Dame Juliana Berners, and first printed by Wynkyn de Worde, in his edition of the Book of St. Albans, in 1496, the learned lady, after recommending a minnow and a worm as proper baits for the Trout in the month of March, adds, " In Aprill take the same baytes: and also Juneba, other wyse named VII. eyes."

Seven eyes and nine eyes, in reference to the apertures about the head, are common names for the Lamprey in this and some other countries; but a derivation for the term Juneba is a desideratum.

The vignette represents a fisherman of South Wales bearing his coracle; see vol. ii. page 27.

CHONDROPTERYGII. *PETROMYZIDÆ.*

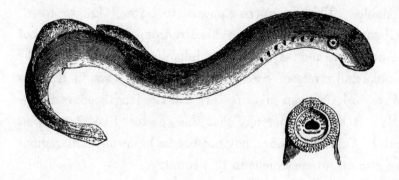

THE FRINGED-LIPPED LAMPERN.

PLANER'S LAMPREY.

Petromyzon Planeri, *Planer's Lamprey,* Bloch, pt. iii. pl. 78, fig. 3.
  „ „ *La Petite Lamproye,* Cuvier, Règne An. t. ii. p. 404.
  „ „ *Planer's Lamprey,* Jenyns, Man. Brit. Vert. p. 522, sp. 211.

This species, when adult, is easily distinguished from the Lampern last-described, by its being much shorter in length, and yet equally thick in substance : it may also be recognised at all ages, on comparison with it, by its having the whole broad edge of the circular lip furnished with numerous papillæ forming a thickly-set fringe, and by the depth and close connexion of the two dorsal fins.

I am indebted to the kindness of Sir William Jardine for two specimens of the young of this species, which were sent from the Tweed. I subsequently obtained some from a brook in Surrey, which were rather larger than those sent from the North ; and have also received some specimens from Lancashire, the males of which measure near eight inches in length, and the females nine inches.

VOL. II. 2 H

This species was named by Bloch after his friend Planer, a professor at Erfort, who sent him specimens; but if Bloch's species be the same as our British fish, his figure is exceptionable. This Lampern appears to be well known to M. Nilsson, who includes it in his Prodromus of the Fishes of Scandinavia, and says it is an inhabitant of almost all the brooks and rivers of Sweden, and that it spawns in April or May. M. Nilsson gives to this fish the length of six inches only: it appears therefore that this species, like *P. marinus* and *P. fluviatilis*, does not acquire in high northern regions the size of our specimens in this country.

When compared with *P. fluviatilis*, Planer's Lampern has the orifice on the forehead, the eye, and the first of the branchial apertures, much nearer the anterior edge of the lip than in the other species; the lip broad and fringed, and the disposition of the teeth as shown in the additional figure of the mouth only: the first dorsal fin begins about the middle of the whole length of the fish, and is in close contact with the second dorsal fin, which in its base is as long again as the first: the tail is furnished with an extension of membrane above and below, forming a caudal fin; and a narrow slip passing forwards towards the anal sheath, forms a rudimentary anal fin.

In its colours this species agrees with the common Lampern, being dusky blue on the back and sides, passing into silvery white on the belly, the fins having a brown tint.

In its habits, Planer's Lampern so closely resembles the common Lampern, as frequently, no doubt, to have been mistaken for it. Both may go to the salt or brackish water from that part of a river within the influence of the tide.

CHONDROPTERYGII. *PETROMYZIDÆ.*

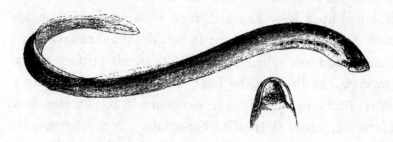

THE PRIDE, AND SANDPRIDE.

SANDPREY, AND MUD LAMPREY.

Ammocætes branchialis,	*Lamprillon,*	Cuvier, Règne An. t. ii. p. 406.
,,	,,	Pride,	Flem. Brit. An. p. 164, sp. 3.
,,	,,	,,	Jenyns, Man. Brit. Vert. p. 522, sp. 212.
Petromyzon	,,	Linnæus.	Bloch, pt. iii. pl. 78, fig. 2.
,,	,,	Pride,	Penn. Brit. Zool. vol. iii. p. 107, pl. 10.
,,	*cæcus,*	*Mud Lamprey,*	Couch, Loudon's Mag. Nat. Hist. vol. v. p. 23, figs. 9 & 10.

Generic Characters.—Form of the body, the branchial apertures and fins, like those of the Lampreys ; upper lip semicircular, with a straight, transverse under lip ; mouth without teeth, but furnished with numerous short membranous cirri.

This small fish is very similar in its general appearance to the young of the Lampreys found in fresh water ; but its prominent lip is in the form of a horse-shoe, and the circle not being complete, it has not the power of adhering to stones and other substances like the true Lampreys, but generally hides itself in the mud or loose sandy bottoms of rivers and brooks in this country, in most of which it will be found, but requires close search to detect it. It is of little value, seldom exceeding six or seven inches in length, and is about as thick as a large quill.

2 H 2

It was formerly considered to be a Lamprey, and was called *Petromyzon cæcus* by Ray, on account of its very small eyes : it afterwards had the trivial name of *branchialis* bestowed upon it by Linnæus, from a notion that it attached itself to the gills of fishes. It is said to be common about Oxford, and was called by Dr. Plot, in his History of Oxfordshire, the Pride of the Isis ; Prid being an ancient diminutive for Lamprey. It is very common in the Thames about Hampton, where it is called Sandpride. Mr. Jesse says the Eel is one of its greatest enemies, and feeds greedily upon it. I have received it from Hertfordshire, and some other inland counties. It spawns at the end of April or the beginning of May, and feeds upon worms, insects, and dead or even putrid animal matter.

Mr. Couch says, " I find this species frequents our smaller streams in Cornwall, living in the muddy bottom, from which it rarely, if ever, willingly emerges. I have kept it for months in stagnant water, with mud at the bottom, without injury to its health or activity. The only apparent use of its fins is to enable it to regain its station, when forced from it by violent torrents. When kept in clear water it seems to sleep much. I have never found this species to attach itself to any object by the mouth ; but the lips are capable of extensive and complicated motions. Our fishermen collect them to use as bait for their hooks when whiffing for Pollacks."

The upper lip and the mouth in this species, as shown in the enlarged representation of the lower surface of the head under the figure of the whole fish, is in the form of a horseshoe ; the inner part furnished with numerous short and slender membranous cirri ; " the lingual and palatine plates somewhat harder than the other portion, but no true teeth :" on the top of the forehead is a small orifice and canal, which leads to the internal tube between and connected with each lateral set of branchial cells, as in the Lampreys; the eye is very small, so much so as to have been occasionally overlooked, and it is situated at the bottom of a small and deep depression: the branchial apertures are seven on each side, arranged along a kind of lateral groove: the body of the fish at this part is rather tumid; behind this the form of the body is nearly round, the portion beyond the anal opening compressed: there are two dorsal fins, the first short and low, the second longer and higher, with a distinct diminution between it and the first dorsal fin, and also with the dilated membrane forming the caudal fin, which is somewhat rounded, the fleshy portion of the tail being pointed; a narrow slip of membrane forms an anal fin.

Some variations occur in specimens of this fish from different localities, and there is even reason to suppose that two species may exist. The most frequent colour is yellowish brown, approaching to blackish brown on the top of the head and upper part of the back, much lighter underneath and on the fins.

The vignette was copied from a sketch of a fishing party made by T. Stothard, R.A. about the year 1780.

THE MYXINE.

GLUTINOUS HAG, AND BORER.

Gastrobranchus cæcus, *Blindfish,* Bloch, pt. xii. pl. 413.
    ,,   ,,   Cuvier, Règne An. t. ii. p. 406.
    ,,   ,,   *Glutinous Hag,* Penn. Brit. Zool. vol. iii. p. 109.
Mixine glutinosa,   Linnæus.
    ,,   ,,   *Glutinous Hag,* Flem. Brit. An. p. 164, sp. 4.
    ,,   ,,   ,,   ,, Jenyns, Man. Brit. Vert. p. 523,
                 sp. 213.

Generic Characters.—Body elongated, cylindrical, smooth ; dorsal fin very low, continued round the tail to the vent ; a single spiracle on the head communicating with the interior ; lips surrounded with eight barbules or cirri ; mouth with one hook-like tooth ; tongue with two rows of teeth on each side ; branchial apertures two, placed under the commencement of the belly.

The worm-like form of the fish figured above induced several systematic authors, including Linnæus, to class it with the Worms ; and it was not till after dissections and published descriptions that its true relations with the Lampreys were acknowledged.   Of these memoirs, that furnished to the French Institute in 1797 by Bloch, the ichthyologist of Berlin, will be read with interest ; and the substance of it will be found in the twelfth part of his valuable work on Fishes, in which the internal structure is rendered obvious by various coloured illustrations.

In the family of fishes now under consideration, the last of vertebrate animals, the spinal column is in a rudimentary condition. In the Lampreys it is but indistinctly divided into rounded portions. In the Myxine, in place of a series of bones composing the vertebral column, there is merely a soft and flexible cartilaginous tube ;* while in the diminutive fish next to be described, which is the last of the British species, this support is reduced to a small and slender semi-transparent column, extending throughout and connecting the whole length of the body like the flexible horny pen in some species of Cephalopods, and to which class other relations of structure both in the Myxine and in this small fish will be pointed out.

As a British fish, the Myxine occurs most frequently on the eastern coast. It enters, says Pennant, " the mouths of other fish when on the hooks attached to the lines which remain a tide under water, and totally devours the whole except the skin and bones. The Scarborough fishermen often take it in the *robbed fish* on drawing up their lines." On this part of our coast it is called Hag, and also Borer, because, as others say, it first pierces a small aperture in the skin, and afterwards buries its head in the abdomen or body. It is most usually found in the body of the Cod, or some other equally rapacious fish.

For the only specimens of the Myxine I ever possessed, I am indebted to the unremitting kindness of Dr. George Johnston of Berwick, who has assisted me most materially

* If a section be made, a probe passes readily in either direction.

by sending me examples of many interesting species which have been referred to throughout this work. The Myxine is not uncommon at Berwick ; but it is only to be obtained at a particular season of the year in one or two particular localities, when during fine weather, at the end of spring or the beginning of summer, the fishermen lay their long lines on a bank with a soft mud bottom near that coast when fishing for Cod and Haddock. It is considered by some that the Myxine, which is without eyes, obtains access to the interior of the body of the fish by passing in at the anal aperture ; others endeavour to account for its being found in the belly of a voracious fish by supposing it had been swallowed ; while many experienced fishermen still repeat their belief that the Myxine enters the mouth of the Codfish while it is hanging on the line. It is conjectured that the Myxine does not fasten upon any fish unless it be either dead or entangled on a hook : but how a fish that is blind is able to find its way to a particular aperture, is a matter not easily explained. The eight barbules or cirri about the mouth of the Myxine are, there is no doubt, delicate organs of touch, by which it obtains cognizance of the nature and quality of the substances with which they are in contact ; and its single hooked tooth upon the palate enables it to retain its hold till the double rows of lingual teeth are brought into action to aid the desire to obtain food.

The high northern geographical range of this singular fish is shown by M. Nilsson including it among the fishes of the shores of Scandinavia, where, he adds, four and even six examples have been found within the body of one Haddock, the flesh of which was entirely consumed. The Myxine is oviparous, and the ova are of the same colour, size, and form as those of the Lampern,—that is, small, round, and yellow.

Along the whole length of the under surface of the body, from head to tail, there are two rows of mucous pores, from which a large quantity of a gelatinous secretion is expressed occasionally at the will of the animal, and by which, either in reference to its quality or quantity, or both, this fish is said to escape its enemies. So copious and so thick in its consistence is this jelly-like secretion, that some of the older naturalists believed this fish had the power of converting water into glue, and it obtained in consequence the name of the Glutinous Hag.

The body is elongated, eel-like in form, cylindrical throughout the greater part of its length, tapering and compressed towards the tail; the whole length from twelve to fifteen inches; the skin perfectly smooth and unctuous; the head obtusely pointed, with a single spiracle connected with the interior of the mouth and branchiæ; eyes wanting; eight tentacula, cirri, or feelers, as they are called, are placed about the lips, four near the front, and two on each side; lips soft, extensible, inclining to a circle in their figure; one single hooked tooth on the palate; the tongue furnished with four rows of small pointed teeth, two rows on each side: at the division between the thoracic and abdominal cavities are two external apertures, each of which is connected by a membranous tube with the six branchiæ on its own side; hence Bloch's name of *Gastrobranchus*. The anal aperture is an elongated fissure situated about two inches before the end of the tail; along the whole under surface of the body are ranged two rows of pores, which afford egress to the secretion of the numerous glands within; the dorsal fin is low and rudimentary, except towards the tail, where the membrane is dilated, and being continued round the end of the tail, and thence upwards to the anal aperture, forms in addition a caudal and an anal fin, which no. doubt materially assist this fish in swimming. In colour the Myxine is dark brown along the back, lighter chestnut brown on the sides, and yellowish white underneath.

The vignette here added is from a drawing by Mr. Clift, engraved for the Philosophical Transactions for the year 1815, where it illustrates a paper by Sir E. Home on the organs of respiration in the Lamprey and Myxine.

The upper angle of the figure exhibits the single spiracle, about it the eight tentacula, on the centre of the palate the single hooked tooth ; to the right and left are the double rows of lingual teeth : an inch below, on each side, are the six branchial cells, with their internal communications with the central canal ; on the outside each cell communicates with a tube that is common to the six cells on that side, which, passing downward, ends at the external aperture below. Beneath this is the stomach and intestinal canal, which, as in the Lamprey, is straight ; the rounded marks along the margin on both sides from end to end show the numerous mucous glands that have already been referred to. It is impossible to dissect a Myxine, and attend to the structure and substance of its investing skin, without being forcibly reminded of its great resemblance to the investing mantle of the Cephalopods.

The relations of structure in the Myxine to the Lampreys on the one hand, and the first class of mollusca, the Cephalopods, on the other, appear to prove that the situation claimed for this fish by Bloch, and systematic authors since his time, is the natural one. The relation to the Lampreys is shown in the elongated, cylindrical form of the body ; the single spiracle on the head ; the general similarity in the parts of the mouth ; the character of the branchial cells, and the viscera.

The relation to the Cephalopods is apparent in the eight

tentacula or feelers about the head, the horny but flexible
nature of the columnar support of the body, the character of
its external covering, and by the power of ejecting a copious
secretion whenever it considers itself in danger.

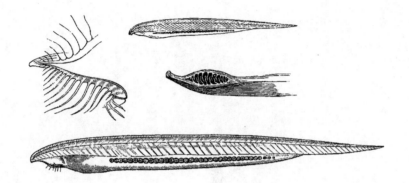

THE LANCELET.

Amphioxus lanceolatus, YARRELL.
Limax   ,,   PALLAS, Spic. Zool. X. p. 19, t. i. fig. 11.

Generic Characters.—Body compressed, the surface without scales, both ends pointed ; a single dorsal fin extending the whole length of the back ; no pectoral, ventral, anal, or caudal fins ; mouth on the under part of the head narrow, elongated, each lateral margin furnished with a row of slender filaments.

THE singular little animal here figured of the natural size, although one of the smallest, as well as the last, among British fishes, is by no means deficient in interest. The specimen, the only one I ever saw, and which is probably also the only one that has been taken for many years, was sent to me by Mr. Couch, who found it himself on the shore near Polperro. A portion of the tail of this little fish was sticking out from underneath a stone in a small pool left by the tide. Mr. Couch perceiving it, took it up with some water in the hollow of his hands. It was alive, very active, and so transparent that the viscera were perceivable through the external covering. It was taken home by Mr. Couch, who made a drawing of its appearance under a microscope.

The only notice of this little animal on record that I have become acquainted with is that by Pallas, in his *Spicilegia Zoologica*, already quoted ; and I insert at the foot of the page, as a note, the Latin description of Pallas,* believing that the reader will then have before him all that has been published of this very rare little animal, of which, at least as far as I am aware, possibly no other specimen has been found or noticed since that to which Pallas refers, and which, it is not a little singular, was also obtained from Cornwall.

Of the specimen in his possession Pallas says, " Quod nunquam vivum vidi, sed liquore servatum e mari Cornubiam adluente accepi olim, quodque prima facie refert piscem Leptocephalum Gronovii."

At first sight this little fish has somewhat the appearance of a *Leptocephalus*, a British fish first sent to Gronovius by our countryman and zoologist Pennant ; it more particularly resembles it in the arrangement of the striæ on the flattened sides : but *Leptocephalus*, as will appear by a reference to the figure of it in this volume at page 311, has a perfect head, though a small one, with jaws, teeth,

* " Limax lanceolatus. Corpus anceps, planum, lineari lanceolatum, utrinque acutissimum. Margo undique limbo membranaceo auctus ; subtus vero ad duas tertias longitudinis margo bilabiatus est, sulcatusque, ut sit quasi pes limacinus angustissimus. Tentacula plane nulla. Latera striis obsoletis, antrorsum obliquatis prope dorsum angulo recurvatis, ut quasi latus pisciculi desquamatum referant."

eyes, and gill-covers ; while the fish under consideration has neither eyes nor gill-covers, nor any fins except one along the back.

Supported by the opinions of three or four zoological friends, I have placed this little animal in this family, near the cyclostomous fishes, believing it to be, as far as at present known, the lowest in organization among this class ; and although I am unwilling to mutilate entirely by my rough dissection the only specimen probably I shall ever possess, and which is perhaps unique, I shall yet be able to show, by the figures given and some further description, that this animal is entitled to a place at the end of the present family.

The form of the fish is compressed ; the head pointed, without any trace of eyes ; the nose rather produced : the mouth on the under edge, in shape an elongated fissure, the sides of which are flexible ; from the inner margin extend various slender filaments, regularly disposed, which cross and intermingle with those of the opposite side. Along the sides of the body the muscles are arranged in regular order, diverging from a central line, one series passing obliquely upward and backward, the other series as obliquely downward and backward : the anal aperture is situated one-fourth of the whole length of the fish, in advance of the end of the tail ; the tail itself pointed : from the nose to the end of the tail a delicate membranous dorsal fin extends the whole length of the back, supported by very numerous and minute soft rays ; the surface of the body smooth.

The body is strengthened and supported internally throughout its length by a flexible cartilaginous column, from which the numerous muscles diverge ; the cavity of the abdomen is comparatively large ; the intestine a canal of considerable calibre, without convolution ; above it a double row of flattened globular bodies, which have all the appearance of ova. The figure at the top of the illustration represents this fish of the natural size. The right-hand figure in the middle line is an enlarged representation of the mouth as seen from below, with the filaments from each side stretching across the opening ; the outline on the left of the middle is a magnified view of the two portions of the hyoid or lingual bone, to which the filaments are attached, one branch of which bone is divided, and the cut portions turned up and down to expose the other perfect side ; the figure at the base is a magnified view of the appearance of the whole fish.

Several relations in structure to the Lampreys and Myxine are observable in the fringed mouth, the armed lingual bone, the absence of eyes, and the want of pectoral and ventral fins. Of its habits, that only which has been stated is known : it is extremely active when in water, and its food is probably some of the most minute among the thin-skinned crustacea, or decomposing animal matter.

It may perhaps be expected that I should state on what grounds I have ventured to differ from such a naturalist as Pallas in considering this animal a fish, and not a *Limax.* It is distinguished from the *Limaces* by the absence of the ventral muscular disk for locomotion ; and from every other molluscous genus, in the position of the anal aperture, which is unconnected with the respiratory cavity. On the other hand, the dorsal fin, and regular oblique strata of muscular fibres clothing the sides of the body and having their points of origin attached to a firm dorsal internal axis,—with the existence of a lengthened internal vertebral column, although in a soft cartilaginous state, as in the Myxine,—are sufficient to determine the primary division of animals to which the *Amphioxus* belongs.

The vignette closing this second volume, and the History of British Fishes, represents the New Hall and the Barge of the Company of Fishmongers of London.

The arms on the title-page of the first volume are also those of the Company.

LONDON :
PRINTED BY SAMUEL BENTLEY,
Dorset Street, Fleet Street.

Frederick A. Heath, Engraver.   Maull & Polyblank, Photo.rs

Yours sincerely
Wm Yarrell

London; John Van Voorst, Paternoster Row, 1859.

SECOND SUPPLEMENT

TO

THE FIRST EDITION

OF THE

HISTORY

OF

BRITISH FISHES,

BY

THE LATE WILLIAM YARRELL, V.P.L.S., F.Z.S.

BEING ALSO A FIRST SUPPLEMENT TO THE SECOND EDITION.

ILLUSTRATED WITH WOODCUTS.

EDITED BY SIR JOHN RICHARDSON, C.B.

LONDON:
JOHN VAN VOORST, PATERNOSTER ROW.
M.DCCC.LX.

PREFACE

TO THE

SUPPLEMENT TO THE SECOND EDITION.

———

This Supplement is published for the use of the purchasers of Mr. Yarrell's First and Second editions of his History of British Fishes, and contains all the species that have been discovered in the British Seas since the year 1839, as far as they have come to the knowledge of the Editor.

Lancrigg, Grasmere, 1859.

NAMES OF FISHES IN THIS SUPPLEMENT.

FIRST VOLUME.

SECOND VOLUME.

MEMOIR OF WILLIAM YARRELL.

WILLIAM YARRELL was a man rather below the middle
height, having a robust, well-knit frame, a sagacious
and pleasing countenance, and frank and agreeable manners. His aspect was that of a stout yeoman, such as those
who in times past have contributed with head and hand
to elevate their native England to its present rank among
the nations; or rather his demeanour may be said to have
indicated exactly what he was in fact—a citizen who had
thriven in the greatest of commercial cities, but who,
strong in native honesty and self-respect, had passed un-
scathed through the perils of money-making, his cheerful
countenance bearing no lines traced by the thirst of gain
or the debasing passion for hoarding: on the contrary, his
mild but fearless eye, and his open forehead, showed,
even to a stranger, a man at peace with himself and with
his fellow men.

He was born on the 3rd of June, 1784, in the parish
of St. James's, where his home continued to be for the
seventy-two years of his life. In Duke Street his
father and uncle carried on in partnership the business
of newspaper agents. On the death of his father, his
mother removed to a private residence in Great Ryder

Street, and there the son lived with her, and during that time was joined in trade with his cousin, then carrying on the business of their late fathers at the north-east corner of Little Ryder Street, to which house it had been removed; and whither, on Edward Jones ceasing to reside, William Yarrell went, and continued to dwell, till death.* A domicile so permanent offers no field for stirring incident, but it is salutary to contemplate the career of a man, who, possessing the ability, judgment and industry that lead to success, and placed by the accidents of birth and connection among the busy throng of the metropolitan worshippers of wealth, deliberately chose the safer middle path of competency,— in an age when money has power to raise its possessor to a seat among the law-givers of the land, and the art of acquiring it is considered in the social estimate of the day as equivalent to high breeding, education and virtue, —when, in short, the cry " get money, *per fas aut nefas*," has gone far towards sapping the national character for honesty, and the vaunted good faith of the British merchant is in danger of becoming a myth.

The following brief narrative is compiled from obituary notices published immediately after Mr. Yarrell's death by several of his intimate and attached friends— Professor Bell, President of the Linnean Society, Dr. R. G. Latham, Edward Newman and Lovell Reeve, Esqs. These gentlemen have referred mainly to Mr. Yarrell's scientific pursuits, and have mentioned few or no particulars of his private life, nor is the compiler of this memoir able to supply the deficiency. But he, who attained the length of days usually allotted to man, and survived all his brothers and sisters as well as father and

---

* A year before that event, he had ceased to have any connection with the business, having retired in favour of Messrs. Joseph and Charles Clifford.

mother, though he never married, must have had the
depths of his sensitive nature often stirred by the breaches
made by death in the circle of his relatives and friends,
even should no tenderer tie have been untimely snapt
asunder. That such was the case may be inferred
from the feeling which only two years before his death
prompted him to transfer to the album of his relatives,
the Misses Pallett of Dover, the subjoined lines from
Wordsworth :—

> " first and last,
> The earliest summoned and the longest spared,
> Are here deposited."

The following is the marriage certificate of his parents :—

"At Bermondsey Parish Church, Surrey, Francis
Yerrall, of this Parish, Bachelor, to Sarah Blane, of this
Parish, Spinster. By Banns, 26 June, 1772.

Present, William Hawkins,
John Beszant."

Subsequently his father transposed the e and a in
writing his surname, as appears by this register of
birth :—

"St. James's, Westminster, June 27, 1784. William
Yarrell, son of Francis and Sarah, born June 3rd."

Of his father's origin, except that he was born the
10th of February, 1749, married the 26th of June, 1772,
died the 25th of March, 1794, was the eldest of seven
brothers and sisters, the children of Francis Yerrall, born in 1727, died the 5th of January, 1786, and
of Sarah his wife, born in 1719, died the 12th of December, 1800,—nothing can now be ascertained ; and it is
believed that the son never knew his father's native
place exactly, though he used to think that he came
from Bedfordshire, where the surname is a common

b 2

one, but is spelt in various ways. The second Francis
Yerrall is reported to have been a proud man. Sarah
Blane is said by a relative to have been born of parents
who were small farmers at Bayford in Herts, and to
have been remarkable for nothing but a tartness of
temper, wholly unlike to that of her distinguished
son. If the dispositions of the mind are, as has been
supposed, like the constitution of the body and the
lineaments of the countenance, in some degree here-
ditary, and that consequently pride and quickness of
temper descended to the offspring of Francis and Sarah
Yarrell, William, the ninth-born child, was fortunately
endowed at the same time with so much firmness and
good sense as to be able to keep his passions under control and to become remarkable in after-life for modesty
and urbanity.

In his boyhood William Yarrell occasionally visited
his maternal relatives at Claypits Farm, Bayford, and
there, doubtless, his love of rural scenery originated;
but his earliest tastes for Natural History seem to have
been fostered by his mother, who took him with her in
the frequent excursions she made to Margate, then a
favourite resort of Londoners. Their conveyance was
the usual one of the time, the Margate Hoy, and young
Yarrell found amusement on the sands in picking up
sea-weeds, which he and his sister afterwards laid out
on paper. He also collected shells and other marine
productions.

His school days were passed at the large scholastic
establishment kept by Dr. Nicholas at Ealing, where he
acquired the character of a quiet, studious boy. The
late General Sale, G. S. Heales, Esq., of Doctors' Commons, who survived him but a few months, and Mr.
Edward Jones, were his fellow pupils, the last-mentioned

being his cousin, the son of his father's partner, and his own future associate in the business carried on in Ryder Street. He had also for playmates his relatives, Mr. Bird and Mr. Goldsmith.

In 1802, being then in the eighteenth year of his age, he entered the banking-house of Herries, Farquhar and Co., as a clerk, but shortly afterwards left that employ, and returned to his father's business. Previous to this event his love of angling had made him acquainted with the streams in the vicinity of London, and the perusal of Izaac Walton's fascinating colloquies had taught him to combine practical philosophy with that pastime. In the course of this pursuit he afterwards often associated with a Londoner of maturer years, an old sportsman named Adams, with whom he was wont to angle under Putney bridge and in other parts of the Thames when the calls of business did not press. Under the guidance of this early friend he acquired the art of shooting, and as it was his custom throughout life to pursue zealously whatever he undertook, he became a proficient in the management of the gun. This led to an intimacy with George Manton, the well-known gun-maker of Bond Street, and with Shoobridge, the hatter of Bond Street, known among sporting men as an unerring shot. Yarrell, who was thought by some to be the better shot of the two, became a member of the Old Hats Club, and was a successful competitor at shooting matches near London. He was constantly in friendly consultation with George Manton when any new form of breech or lock in a fowling-piece was to be tried. At a later time he shot game in Hertfordshire and Cambridgeshire over different manors which he rented in conjunction with his friend Wortham. His exploits with the gun are still remembered in the neighbourhood of Royston, and the same healthful exercise led him also frequently into other localities. His tastes, says one of his friends, were those of a Londoner, whom the *rus in urbe* suited better, perhaps, than the unmixed country. They were those of Izaac Walton, citizen and angler, rather than those of the full and perfect yeoman.

These amusements of his earlier life led to his acquiring an intimate knowledge of the habits of our native birds and fishes, their food and migrations, his observation of the objects that engaged his attention being as accurate as it was keen. They were not, however, the only occupations in which he sought relief from the monotony of business, for in 1817 he studied Chemistry at the Royal Institution. Before he attained middle life he engaged in the systematic study of Zoology, and pursuing it in the intervals of business with his accustomed application, he gradually gave up field sports, and it is believed that for thirty years before his death he handled neither rod nor gun.

In 1823 he commenced noting the appearance of strange and rare birds, and in 1825 he lent his aid to Bewick by sending him scarce British birds to figure. He also presented a collection of the tracheæ of water-birds to the Royal College of Surgeons. His own museum at this time contained a series of British Birds and their eggs, and he now cultivated the society of scientific men, among whom he had made the acquaintance of Sir William Jardine, Bart., and P. J. Selby, Esq., of Twizel House, who were then engaged in publishing their respective works on British Ornithology. In November of the same year he was admitted a Fellow of the Linnean Society, and in 1826 he became one of the original members or founders of the Zoological Society. Next year he was chosen to be one of the

Council of the Medico-Botanical Society, and henceforward his readiness to oblige, the clearness of his understanding, and his business habits, coming to be known, his services in the management of the societies to which he belonged were in constant requisition. " It was only with reluctance, and in compliance with established rules," says Professor Bell, " that his name was omitted from the council lists of either the Linnean or Zoological Societies." Of the latter he was occasionally Auditor, for a time Secretary, and frequently one of its Vice-Presidents. He was also a warm supporter, and for a long time treasurer, of the Entomological Society. On the death of Mr. Forster, in 1849, he was elected Treasurer of the Linnean Society, and continued to fill that office and to be one of its Vice-Presidents until his death. With respect to the Royal Society, the following is the statement of Professor Bell, who was fully cognizant of the circumstances : " Many years since, long before the present plan of selecting a certain number of candidates by the Council was adopted, Mr. Yarrell was proposed as a Fellow, and his certificate signed and suspended. At that time the Council had nothing whatever to do with the election nor with the recommendation of the candidates. Mr. Yarrell's scientific character was not so well known and appreciated as it has since become. A gentleman, long since deceased, who would afterwards have gladly recalled the act, expressed some objection to his being elected, and his certificate was, from a feeling of delicacy on Mr. Yarrell's part, withdrawn ; but subsequently, since the present system has been in action, the writer of this notice, with the full concurrence of many members of the Council, who were most desirous of his election, drew up a certificate in his favour, and obtained some signatures before he men-

tioned the subject to Mr. Yarrell, hoping that when he
knew such a step had been taken he would consent to be
put in nomination.   On being informed, however, of this
movement, which there is no doubt would have met with
the unanimous approval of the Council, he declined the
honour solely on the ground of advancing age, and his
increasing inability to avail himself of the advantages of
the position."

The subjoined list of Mr. Yarrell's publications affords
evidence of his industry and the variety of his zoological
studies.   He became an author evidently from the love
of his subjects, and being in no haste to publish until he
had duly investigated the matter in hand and brought his
clear judgment to bear on the evidence before him, his
writings on Natural History soon acquired that value
among scientific men which simple and truthful narrative always commands.   His great works on the Birds
and Fishes of Britain are quoted as authorities in all the
scientific circles of Europe and America, and are models
of local Faunæ, both on account of the strictness with
which doubtful species are noted or rejected, as well as
for the completeness of the lists gained by unwearied
diligence and inquiry in every direction.   The synonymy
is elaborated with care and skill, and the illustrations,
liberally provided by his friend and publisher, are worthy
of the works, which is no mean praise.   Mr. Yarrell and
Mr. Van Voorst were first brought together by their
mutual friend, the late Mr. Martin, the librarian of the
Duke of Bedford, and the conjunction was a most fortunate one for the progress of British Zoology, a series of
unrivalled illustrated monographs having originated therefrom.   Mr. Yarrell's inquiries into the changes of plumage of hen Pheasants and of birds generally, his dis-
sertations on the horny tip of the bill of young chickens,

on the production and migration of Eels, on the gestation of eggs by the male Pipe-fish, his investigations into the route pursued by American birds in their casual visits to England, and numerous other passages of his works, show much originality of thought and a careful examination of facts.

One of his friends * says, "There was one trait in the character of Mr. Yarrell which must not be passed over in silence, a trait which no one was better acquainted with than myself, and that was his extreme readiness to afford information. Often have I had occasion to appeal to him in difficulties about specific character or points of economy, and from the very moment of mentioning the doubt or the object of inquiry, his whole attention was absorbed by it; books, specimens, memory, every auxiliary was at his fingerends; and no sacrifice of time or trouble was too great for him to make; neither was the subject ever left undecided while diligence or a disposition to teach could throw on it a single ray of light. No other subject seemed to occur to him during the investigation; he had no other occupation; that one inquiry was, for the time, the object of his life. His power of concentrating his attention on a single subject was most extraordinary, and more extraordinary still was the facility with which that concentrated attention was turned to *any* subject; he used it after the fashion of a burning glass, casting the focus wherever he pleased. This faculty was at the service of all; and the attention of which I speak thus gratefully from personal experience was given to every truth-seeking inquirer."—*Zoologist*, 5258.

Another friend † writes as follows: "Mr. Yarrell's

* Edward Newman, Esq., Editor of the Zoologist.
† Dr. R. G. Latham.

purely intellectual character is seen in his works. The part which the author himself always took most credit for, was the geographical distribution of birds. He considered that in treating it as he had done, he smuggled in a certain amount of geography under the garb of ornithology. For the high qualities of accuracy, terseness of description, and felicity of illustration, they speak for themselves."—*Edinburgh New Philosophical Journal,* 1856.

So much for the excellence of his works on Zoology, of which the best tests are the continually-increasing demand for them at home after twenty-two years' circulation, and the frequency and confidence with which they are quoted by naturalists abroad. With regard to Mr. Yarrell's character as a man, the following extracts from the obituary notices referred to, will show the estimation in which he was held by those who were most intimately acquainted with his conduct in private and public life. " Strong social instincts," says a keen observer of men and manners, " geniality of temper, warmth of heart (exhibited in an extreme fondness for children), made him loved, even as his simple and straightforward independence of character made him respected. His advice, too, was always valued, freely asked and freely given, for his mind was observant, active, practical, and wholly unclouded by fancies or prejudices; his knowledge varied and accurate. Indeed he was essentially a reliable man, knowing what he knew well, and caring to undertake nothing that he was likely to fail in. For this a strong will and perseverance is needed. It was strong enough to keep a warm temper in thorough control; for Mr. Yarrell, knowing what was due to himself, knew also what was due to others. He helped many, not only with his advice but by his purse, ever valuing money for its uses only, never for its own sake; moderate (as a man of business) in his aims, though attentive to what he undertook; hating waste, yet never ambitious of accumulation." . . . . "For many years his house was familiar to all naturalists, and to visitors of every rank from the country, not to mention foreigners, to whom the reputation of one of the soundest of living zoologists was well known, and who never visited it without being struck by the kind and communicative manners of its hospitable inmate." . . . . "His habits, angler and ornithologist as he was, were eminently those of a Londoner. He loved glees, and sung them well, and at one time of his life was a frequent attendant at the theatres."—*Dr. R. G. Latham.*

The testimony of Professor Bell, who knew him well, is as follows:—"In speaking of Mr. Yarrell's intellectual and social qualities, it is difficult to do them justice without danger of appearing hyperbolical. His judgment was clear and sound, his appreciation of the value of facts and of evidence most accurate, his advice always practical and thoughtful. His truthfulness and simpleheartedness were even child-like, his temper gentle, his heart loving and affectionate, and he was liberal and charitable almost to the verge of imprudence. A kindlier spirit never lived. His friendships were sincere and lasting, and only changeable on discovery of the worthlessness of the subject, and then how hard was he to believe the painful truth! If ever man realized the beautiful apostolical definitions of Charity, it was William Yarrell. There were, indeed, in Mr. Yarrell's character many points of resemblance to that of Izaac Walton and of Gilbert White. The same charming *bonhomie* and truthfulness and simplicity and elegant taste as in the former; and the close and accurate obser-

vation and clear and graphic description which charac-
terize the writings of the latter."

A third friend (Mr. Lovell Reeve) mentions the following traits of character :—" Notwithstanding his retired
manners, Mr. Yarrell was a frequent diner-out, and a
jovial companion at table. He sang a capital song, and
was a constant attendant at the theatre, generally selecting, with the gusto of a dilettante, the front row of
the pit. In the days of the elder Mathews, he would
manage to get the songs of the great mimic, in spite
of the rapidity of their utterance, by taking down the
alternate lines one night, and filling in the others on
the next. A song of Dibdin's we heard him sing only
recently, with admirable spirit and pathos. He seldom
missed the Linnean Club dinners and country excursions, and was at all times the liveliest of the
party."

By the methodical distribution of his time Mr.
Yarrell was enabled, without neglecting his business
concerns, to assist in the management of the scientific
societies of which he was a member, and to carry on his
zoological inquiries and publications. His enjoyment
of social life was combined with temperance ; and being
blessed with a sound constitution he possessed continuous good health up to the year 1853, when some
premonitory symptoms of indisposition began to appear,
without, however, affecting the activity of his intellect
or the cheerfulness of his manners. On the 3rd of
August, 1856, as he was returning from St. James's
Church, which for some years he had constantly attended, a slight giddiness seized him, his steps became
uncertain, and he felt for a moment unable to proceed.
After a short rest he reached home without assistance.
This attack proved to be a slight paralysis, from which

he so far recovered as to able to give his uninterrupted attention to matters of business. On Monday, the 25th of August, he attended a Council of the Linnean Society, and was as cheerful, and apparently nearly as well as usual. In answer to a wish expressed by his intimate and attached friend the President of the Society, that he would soon be able to pay him a quiet visit, he said that though pretty well he felt a "wooliness" in the brain, and that he was restricted in his diet. On the following Saturday, however, he felt himself well enough to take charge of an invalid friend in a voyage by sea to Yarmouth, and thus the very last act of his life was one of kindness. He enjoyed the voyage, took a moderate dinner at the Royal Hotel with appetite, and retired to bed anticipating a good night's rest. But scarcely had he lain down before he felt a difficulty of breathing, and fearing, as he said, that "he might die and no one know it," he got up, unlocked the door, and rang the bell. The attentive landlady was speedily at his bedside, medical assistance was procured without delay, but nothing availed, and he expired calmly at half-past twelve on Monday morning the 1st of September, in the seventy-third year of his age. He experienced no pain, and remained perfectly conscious until within a few minutes of his entering the unseen world. The immediate cause of his death was judged to be disease of the heart, with which the previous symptoms of apoplexy were but indirectly connected. No autopsy was made. On the following Monday he was buried at Bayford in Hertfordshire (where a great many of his maternal kinsfolk and ancestors lie), in a spot selected by himself, his body being attended to the grave by the President and other office-bearers of the Linnean Society, as well as by his executors and surviving relatives.

A tombstone erected to his memory bears the following inscription :—

HERE LIE THE REMAINS

OF

WILLIAM YARRELL, V.P.L.S., F.Z.S.,

of St. James's, Westminster,

Author of a History of British Birds, and of a History of British Fishes.

BORN, JUNE 3rd, MDCCLXXXIV.

DIED, SEPT. 1st, MDCCCLVI.

He was the survivor of twelve Brothers and Sisters, who, with their Father and Mother, are placed close to this spot.

"first and last,
The earliest summoned and the longest spared—
Are here deposited."—WORDSWORTH.

His executors were his relative Mr. Bird, and his friend and publisher Mr. Van Voorst, and the property administered to amounted to about 17,000*l.* After his death his extensive library of Natural History books and his valuable collections of British Birds and Fishes were sold by auction, at which the Fishes were purchased for the British Museum.

A portrait of him, painted in 1839 by Mrs. Carpenter, is suspended in the hall of Burlington House, the expense having been defrayed by forty Fellows of the Linnean Society ; and a Medallion Tablet executed by Mr. Neville Burnard, has been affixed in St. James's Church, at the west end of the north aisle.

The portrait facing the title-page of Vol. I., engraved by Mr. Frederick A. Heath, is from a photograph by Messrs. Maull and Polyblank, taken in 1855.

LIST OF MR. YARRELL'S WRITINGS.

1. Notices of the occurrence of some rare British Birds observed during the years 1823, 1824, and 1825.—*Zool. Journ.*, ii. p. 24, March, 1825.

2. Ditto, second communication.—*Zool. Journ.*, iii. p. 85, October, 1826.

3. Ditto, third communication.—*Zool. Journ.*, iii. p. 497.

4. Some observations on the anatomy of the British Birds of Prey.—*Zool. Journ.*, iii. p. 181, October, 1826.

5. On the small horny appendage to the upper mandible in very young chickens.—*Zool. Journ.*, ii. p. 443, written 17th October, 1825, published 1826.

6. Notice of the occurrence of a species of duck (*Anas rufina*) new to the British Fauna.—*Zool. Journ.*, ii. p. 492, 1826.

7. Observations on the tracheæ of Birds, with descriptions and representations of several not hitherto figured.—*Linn. Trans.*, xv. p. 378, 1827. Read February 6th, 1827.

8. On the change of plumage of some Hen Pheasants.—*Phil. Trans.*, written February, 1827. Read in May, 1827.

9. On the osteology of the Fennec (*Canis cerdo*).—*Zool. Journ.*, iii. p. 401, 1827.

10. On the osteology of the *Chlamyphorus truncatus* of Dr. Harlan, March, 1828.—*Zool. Journ.*, iii. p. 544.

11. Some remarks on the habits of the Kingfisher, March, 1828. —*Loudon's Mag. of Nat. Hist. and Journ. of Zool.*, &c., i. p. 23, 1828.

12. Description of a species of Tringa (*T. rufescens*) killed in Cambridgeshire, new to England and Europe.—*Linn. Trans.*, xvi. p. 109. Read June 17th, 1828.

13. On the supposed identity of Whitebait and Shad, August, 1828.—*Zool. Journ.*, iv. p. 137.

14. Observations on the Tapir of America. — *Zool. Journ.*, iv. p. 210.

15. On the use of the Xiphoid bone and its muscles in the Cormorant (*Pelecanus carbo*, L.), August, 1828.—*Zool. Journ.*, iv. p. 234.

16. Notes on the internal appearance of several animals examined after death, in the collection of the Zoological Society (Otter, Paradoxure, Ocelot, Chinchilla, Agouti, Porpoise, Touraco, Javanese Peacock, Silver Pheasant, Hybrid Pheasant, White Stork, Common Bittern, Crested Grebe, Red-throated Diver, Tame Swan, Wild Swan, Black Swan, Canada Goose, White-fronted Goose, Indian Tortoise, Active Gibbon, Diana Monkey, Weeper Monkey, Mexican Dog, Jerboa, Bobac, Malabar Squirrel, Crested Porcupine, Alpine Hare).—*Zool. Journ.*, iv. pp. 314-322.

17. On the structure of the beak and its muscles in the Cross-bill (*Loxia curvirostra*).—*Zool. Journ.*, iv. p. 459.

18. Remarks on some English Fishes, with notices of three species new to the British Fauna (*Solea pegusa, Cottus bubalis, Anguilla*). —*Zool. Journ.*, iv. p. 465.

19. Descriptive and Historical Notice of British Snipes.—*Loudon's Mag. of Nat. Hist., &c.*, ii. p. 143, 1829.

20. Supplement to ditto.—*Loudon's Mag. of Nat. Hist.*, iii. p. 27, 1830.

21. On the organs of voice in Birds.—*Linn. Trans.*, xvi. p. 305. Read June, 1829.

22. On a new species of Wild Swan (*Cygnus Bewickii*) taken in England, and hitherto confounded with the Hooper.—*Linn. Trans.*, xvi. p. 445. Read January, 1830.

23. Reply to the statement respecting the Discovery of *Cygnus Bewickii*, published in the Philosophical Magazine and Annals for August.—*Richard Taylor's Philos. Mag. and Annals.*

24. On the occurrence of the *Sylvia Tithys* of Scopoli in England.—*Proceed. of Com. of Science of Zool. Soc.*, i. p. 18, 1830.

25. On the assumption of the male plumage by the female of the common Game Fowl.—*Ibid.*, i. p. 22, 1830.

26. On the anatomy of the *Cereopsis Novæ Hollandiæ*, Lath., and on the relations between the *Natatores* and *Grallatores.—Ibid.*, i. p. 25, 1830.

27. On the sexual organs of the hybrid Pheasant.—*Ibid.*, i. p. 27.

28. On the specific identity of the Gardenian and Night Herons (*Ardea Gardenii* and *nycticorax*).—*Ibid.*, i. p. 27.

29. On the Anatomy of the *Chinchilla lanigera.—Ibid.*, i. p. 32.

29*. On the trachea of the Red-knobbed Curassow (*Crax Yarrellii*, Benn.).—*Ibid.*, i. p. 33.

30. Characters of a new species of Herring (*Clupea*, L.).—*Ibid.*, i. p. 34.

31. On the occurrence of several North American Birds in England.—*Ibid.*, i. p. 35.

32. On the anatomy of the Lesser American Flying Squirrel (*Pteromys volucella*).—*Ibid.*, i. p. 38.

33. On the anatomy of the *Ctenodactylus Massonii*, Gray.— *Ibid.*, i. p. 48.

34. On the sterno-tracheal muscles of the Razor-billed Curassow (*Ourax mitu*, Cuv.).—*Ibid.*, i. p. 59.

35. On the distinctive characters of the *Tetrao medius*, Temm.— *Ibid.*, i. p. 73.

36. On two species of Entozoa in the Eel.—*Ibid.*, i. p. 132.

37. On the generation of Eels and Lampreys.—*Ibid.*, i. p. 132.

38. On the Brown-headed Gull (*Larus capistratus*, Temm.).— *Ibid.*, i. p. 151.

39. On the anatomy of the Conger Eel (*Conger vulgaris*), and on the differences between the Conger and the Fresh-water Eels.— *Ibid.*, i. p. 159.

40. Additions to the British Fauna, Class Fishes, September, 1830. —*Loudon's Mag. of Nat. Hist.*, iii. p. 521.

41. Specific characters of *Cygnus Bewickii* and *C. ferus*.—*Taylor's Phil. Mag. and Annals*, vii. p. 194.

42. Additions to the catalogue of British Birds, with notices of the occurrence of several rare species, January, 1831.—*Loudon's Mag. of Nat. Hist.*, iv. p. 116.

43. On a hybrid between a Muscovy Duck (*Anas moschata*) and a Common Duck (*Anas boschas*).—*Proc. Com. of Sc. of Zool. Soc.*, ii. p. 100, 1832.

44. On two species of Mammalia new to Britain, one of them (*Sorex remifer*) new to science.—*Ibid.*, ii. p. 429, 1832.

45. Description of the organs of voice in a new species of Wild Swan (*Cygnus buccinator* of Richardson).—*Linn. Trans.*, xvii. p. 1. Read 20th March, 1832.

46. Description of three British species of Fresh-water Fishes belonging to the genus *Leuciscus* of Klein.—*Ibid.*, xvii. p. 5. Read June, 1832.

47. Additions to the British Fauna, Class Mammalia (*Arvicola riparia, Sorex remifer*), August, 1832.—*Loudon's Mag. of Nat. Hist.*, v. p. 598.

48. Notice of a new species of Herring.—*Zool. Journ.*, v. p. 277.

49. Observations on the laws which appear to influence the assumption and changes of plumage in Birds.—*Proceed. Zool. Soc.*,

VOL. I.   *(2nd Supp.)*   c

i. pp. 9, 56; *Trans. Zool. Soc.*, i. p. 13. Read February and April, 1833.

50. Description, with some additional particulars, of *Apteryx australis* of Shaw.—*Proc. Zool. Soc.*, i. p. 24; *Trans. Zool. Soc.*, i. p. 71. Read June, 1833.

51. On the tracheæ of the *Penelope gouan*, Temm., and the *Anas magellanica*, Auct.—*Proceed. Zool. Soc.*, i. p. 3.

52. On the Woolly and Hairy Penguins (*Aptenodytes*).—*Proceed. Zool. Soc.*, i. pp. 24, 80; *Trans. Zool. Soc.*, i. p. 13, 1833.

53. On the identity of the Woolly Penguin of Latham with the *Aptenodytes patachonica* of Gmelin.—*Proceed. Zool. Soc.*, i. p. 33, 1833.

54. Characters of the Irish Hare, a new species of *Lepus*.—*Ibid.*, i. p. 88.

55. On the deficiency of teeth in the hairless Egyptian variety of the dog.—*Ibid.*, i. p. 113.

56. Notice of the occurrence of *Squilla Desmarestii* on the British shores.—*Loudon's Mag. of Nat. Hist., &c.*, vi. p. 230.

57. On the reproduction of the Eel.—*Report of Brit. Assoc.*, 1833, p. 446.

58. On the anal pouch of the male fishes in certain species of *Syngnathus*.—*Proceed. Zool. Soc.*, ii. p. 118.

59. Observations on the economy of an insect destructive to turnips (*Athalia centifolia*).—*Trans. Zool. Soc.*, ii. p. 67. Read November, 1835.

60. On the mode of union after fracture of the processes of the vertebræ of a Sole (*Solea vulgaris*, Cuv.).—*Proceed. Zool. Soc.*, iii. p. 57, 1835.

61. On the trachea of the Stanley Crane (*Anthropoides paradiseus*, Besch.).—*Ibid.*, iii. p. 183.

62. On the fœtal pouch of the male Needle Pipe-fish (*Syngnathus acus*, L.).—*Ibid.*, iii. p. 183, 1835.

63. A History of British Fishes. Van Voorst, London, 1836, 2 vols., 8vo., (published in parts, finished in 1836, containing vol. i. pp. 408, vol. ii. pp. 472).

64. Supplement to ditto, March, 1839 (vol. i. pp. 48, vol. ii. pp. 78, containing 27 new species).

64*. On an interwoven mass of filaments of *Conferva fluviatilis* of extraordinary size.—*Proceed. Linn. Soc.*, i. p. 65, 1838.

65. A History of British Fishes. Van Voorst, London, 1841, 2 vols., 8vo., Second Edition (vol. i. pp. 464; vol. ii. pp. 628, containing 263 species, and 500 figures).

66. A History of British Birds. Van Voorst, London, 1843, 3

vols., 8vo. (published in parts at intervals of two months, the first one in July, 1837, and the last one in May, 1843. Vol. i. pp. 525; vol. ii. pp. 669; vol. iii. pp. 528).

67. Supplement to ditto, October, 1845 (number of species in the first edition and supplement 354).

68. A History of British Birds. Van Voorst, London, 1845, 3 vols., 8vo., Second Edition.

69. On a new species of Swan (*Cygnus immutabilis*).—*Proceed. Zool. Soc.*, ix. p. 70, 1841.

70. On the trachea of a male Spur-winged Goose (*Anser gambensis*).—*Ibid.*, ix. p. 70, 1841.

71. On a new species of Smelt from the isle of Bute (*Osmerus hebridicus*).—*Report of Brit. Assoc. for* 1838, p. 108.

72. On the preservation of *Crustacea.*—*Entom. Mag.*, vi. p. 421.

73. Remarks on some species of *Syngnathus.*—*Annals of Nat. Hist. and Mag. of Zool.*, Jardine, &c., iii. p. 81.

74. Growth of Salmon in fresh water.—*Ibid.*, iv. p. 334.

75. On *Motacilla alba*, L.—*Ibid.*, vii. p. 350.

76. Description of the eggs of some of the birds of Chile.—*Zool. Proceed.*, 1847.

77. Occurrence of a Petrel new to Britain on the coast of Ireland, June, 1853.—*Zoologist*, 3947.

78. On birds lately ascertained to be British, p. 79; and on rare English fishes, p. 85.—*Ibid.*, 79. 1843.

79. On the influence of the sexual organ in modifying external character.—*Journ. Linn. Soc.*, June, 1856, i. p. 76.

80. On Mucor observed by Colonel Montagu in the air-cells of a bird.—*Annals and Mag. of Nat. Hist.*, &c., ix. p. 131.

81. Chapter VIII. in the Third Edition of Harvey's Sea-side Book "On Marine Fishes," pp. 237–269.

SECOND SUPPLEMENT

TO THE FIRST VOLUME OF

THE HISTORY OF BRITISH FISHES.

ACANTHOPTERI.   *SCLEROGENIDÆ.*

FABRICIUS'S SEA-BULLHEAD.

KANIOK, *Greenland Eskimos.*

Acanthocottus groenlandicus, GIRARD.
Cottus scorpius, FABRICIUS, F. Groenl. p. 156 (excl. syn.).
 ,, *groenlandicus,* Cuv. et. VALENC. Poiss. vol. iv. p. 185.
 ,,   ,,   RICHARDSON, F. Bor. Am. iii. pp. 46, 297, 314, pl. 95.
 ,,   ,,   *Greenl. Bullhead,* THOMPS. Nat. Hist. of Irel. iii. p. 81.

An example of this species was captured in Dingle
Harbour in February 1850, and exhibited in the Dublin
Natural History Society, by Mr. William Andrews. As
yet, this and one seen by Dr. Ball, are the only recorded
examples of the fish that have been met with on our
shores. The figure at the head of the article, which
corresponds closely with the one published by Mr.
Andrews, and the vignette at the end, are borrowed from
the *Fauna Boreali Americana,* where a Newfoundland
specimen is described at length.

 This, Fabricius says, is a most voracious fish, and very

VOL. I.   *(2nd Supp.)*   **B**

destructive of the fry of Blennies, Salmon, Herrings, and Haddocks. It even attacks larger fish, does not spare its own species, devours crabs and worms, and in fact pursues every living thing that it can master.* It is bold, lively, and incautious; but habitually keeps at the bottom of the sea, coming to the surface only when it is led thither in pursuit of its prey. It spawns in December and January, depositing its roe on sea-weeds. It is prized as an article of food by the Greenland Eskimos, who eat it daily both boiled and dried, and find it agreeable and wholesome for the sick. Many of them eat its eggs raw; and some even consume the fish itself in that condition. They capture it with lines armed with four hooks, disposed crosswise, and with no other bait than something coloured or shining placed above the hooks. Sometimes they spear it.

The female, Fabricius states, is larger than a male of the same age, and may be distinguished at once by its white belly, which appears yellow in the water and is spotted. The posterior cranial tubercles are nearer to each other in the males than in the females. There are four of these tubercles on the upper aspect of the head, one at each corner of an area, which in the female is nearly square and flat. There are besides eight spines on each half of the head and shoulders, viz. a nasal, opercular, subopercular, scapular, and humeral one, with three preopercular ones. The principal spine is the one at the angle of the preoperculum. Its tip falls about its own length short of the point of the opercular spine. The interval between the orbits is much depressed, and

---

* The omnivorous appetite ascribed to this Bullhead by Fabricius was proved by an examination of the contents of the stomachs of several Newfoundland specimens, which consisted of the vertebral columns of several small fishes, some entire crabs, the peelings of potatoes, and other substances. These Bullheads were caught off the end of a landing jetty.

is bounded anteriorly by the two nasal spines and the prominent ends of the premaxillary pedicles. There are no serratures on any of the spines or bones of the head or shoulder, in which respects this species differs from the Father-Lasher.

The top of the head is sprinkled with soft conical pimples, and the skin generally is naked and smooth, but some small, circular, minutely-spiniferous scales exist on the back and posterior surfaces of the pectoral rays.

Br. 6: D. 10—17 or 18: A. 12 or 13: P. 17: V. 1+3: C. 11¾.

Colours, after the specimen had been kept in spirits, dark brown on the dorsal aspect, mixed with clay-coloured patches on the head, and crimson blotches on the gill-covers, nape and pectorals. The sides, belly, pectoral fins, and ventrals, are ornamented with circular spots of dead white, each surrounded by a dark rim. The liver has a bright red colour in the spirits.

FRONT OF COTTUS GROENLANDICUS.

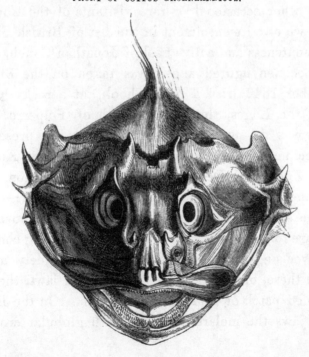

B 2

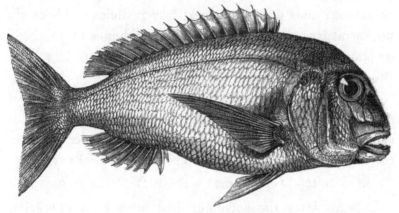

COUCH'S SEA-BREAM.

Pagrus orphus, Le Pagre Orphe, Cuv. et Valenc. vi. p. 150, pl. 149.
" *Aurata orphus* " ,, ,, Risso, 2ᵉ edit. p. 356.
Couch's Sea-Bream, ,, Couch. (Ion.) Zool. for 1843, p. 81.
Pagellus Rondeletii, ,, Couch. (R.Q.) Zool. for 1846, p. 1406.

ONLY one example of this fish is known to have been captured on the English coast, and, as is the case with several other occasional or rare visitants of the Cornish shores, we owe its enrolment in the list of British Fishes to the acuteness and active zeal of Jonathan Couch, Esq. The specimen figured above was taken on the 8th of November 1842 with a baited hook, at a rocky place termed the Edges, three miles south of Polperro. Its weight was six pounds. Mr. Couch having presented the specimen to the British Museum, Dr. Gray, Keeper of the Zoological Department of that Institution, has most kindly furnished the following account of it :—

" The specimen is stuffed. The front teeth above and below are four on each side, the upper ones being conical, the lower ones elongato-conical, and set widely apart. Behind these, but in the front part of both jaws, there is a crowded patch of small subulate teeth. On the limbs of the jaws the molars are large with globular crowns, and rounded teeth of unequal size cover the roof of the mouth. The fish is moderately like fig. 149, in the *Histoire des Poissons*, but that figure does not show space enough between the tall conical teeth in front of the mandible, and the flat molars on the limb of the bone. The specimen also has a higher front than the figure referred to, with more resemblance in profile to *Pagellus calamus*, fig. 152, of the same work. This elevation of the face may be owing to age, for the specimen figured in the *Histoire des Poissons* was only eight inches long, while the one in the British Museum measures above twenty. (For a side view of the mouth see p. 36.)

D. 12+11 : A. 3+9 : V. 1+5 : P. 15 : C. 29.

The last two rays of the dorsal and anal are contiguous at the base, and the last ventral ray is also divided to the bottom." — *Gray.*

TEETH OF COUCH'S SEA-BREAM.

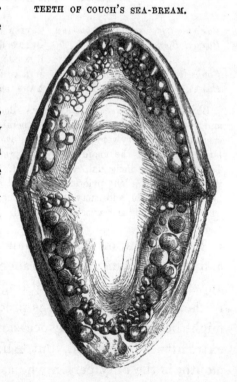

Mr. Couch says, that the body is not unlike the *Pagellus centrodontus*, but is rather deeper and more stout. The head is thick, and the snout remarkably so. The back rises high above the head. The colour of the front and top of the head was a brownish-red, that of the back and fins between lake and vermilion, or like the Becker, except the anal, which was pale-yellow: the sides being pale-red and the belly whitish.—*Couch.*

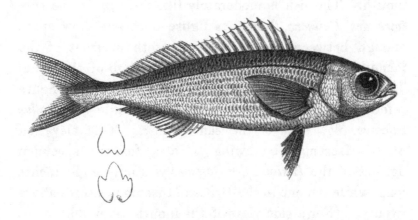

THE BOGUE.

BOGA, *Provence, Madeira.*—BOBBA, *Venice.*—UOPPA,
Messina.—BALAIJOLA, *Catania.*

Box vulgaris, Le Bogue commun,	CUVIER et VALENC. vi. p. 348, pl. 161.
Box vel Boops, ,, ,,	BELON. de Aq. p. 230.
., ,, ,, ,,	RONDELET, Pisc. p. 136.
Boops Rondeletii primus,	WILLUGHBY, 317, t. U. 8. f. 1.
Box vulgaris, Common Bogue,	YARRELL, Zool. for 1843, p. 85.

Box or Boops. *Generic Character.*—Body elongated, rounded, the dorsal and ventral profiles alike, and the general aspect peculiarly trim. Mouth small; premaxillary very little movable, overlapped together with the maxillary and edge of the mandible by the preorbitar when the mouth is shut. Teeth uniserial, incisorial, chisel-shaped, lobed, and crenated. Stomach very small, with a long pyloric branch, and about five pancreatic cæca. Air-bladder large, thin, and nacry, with two long, posterior, horn-like appendages, which enter among the muscles of the tail.

THE BOGUE, or BOGA, abounds in the Mediterranean, and as it has an Atlantic range southwards to Madeira and the Canaries, and according to Cornide, northwards to the coasts of Galicia, it is precisely one of the fish that might be expected to pay occasional visits to the southern extremity of England, but Alfred Fox, Esq., of Falmouth, is the only person who as yet has had the fortune to recognise and secure an English example of the species. That specimen was caught at St. Mawes, in a ground seine, early in October 1843, and deposited by Mr. Fox in the Museum at Truro, where it is preserved, stuffed, and varnished. Through the kindness of Dr. Barham, Senior Physician of the Cornwall Infirmary, the specimen has been lent that the subjoined description might be taken of it, but the figure on the preceding page is copied from one in the *Histoire des Poissons*, which was drawn from a fresh specimen, rather than from the Truro one, which has suffered mutilation in the fin-rays.

The genera *Box, Oblata, Boxaodon* (Guich.), *Scatharus,* and *Crenidens,* form the fourth Sparoid tribe of Cuvier, and are characterized by simple, lobed or serrated, trenchant teeth set closely side by side on the edges of the jaws; sometimes with villiform teeth behind them, or more often with many-crowded rows of minute teeth having a villiform appearance to the naked eye, but being in fact similar in form to the large incisorial ones that constitute the exterior row, and destined to succeed them as they wear away and drop out. In this tribe there are no rounded molars on the limbs of the jaws, which, consequently, are neither so strong nor so thick as in the members of the first tribe, which have broad molars, that necessarily require space. In accordance with the dentition the mouth of *Box* and its allies is small, and the neat head is very unlike that of the bull-headed *Chrysophrydes* and *Pagri.*

The Bogue, according to the *Histoire des Poissons,* spawns twice in the year, and at these times it approaches the shore in large sculls. The fishermen of Provence and Nice take it in nets of a peculiar kind, named by them *bughicra,* and to render the fishery more prosperous, they adorn their boats with small figures of the Bogue cut in silver. Rondelet says that the flesh of the Bogue is easily digested, and on that account is a wholesome aliment for invalids, but Cornide, who speaks of it as he observed it on the Atlantic coasts of Spain, states that it has a disagreeable taste, and is consumed chiefly by poor people. Mr. Lowe tells us that it is exceedingly common at Madeira, and though he says nothing of its qualities as an article of food, he remarks of another species of the same genus that it is one of the handsomest and most worthless of fishes.

The Bogue has an elegant, moderately-compressed, elliptical form, the curves of the dorsal and ventral profiles meeting at the terminal mouth which is slightly obtuse. The greatest height of the body is at the fifth or sixth dorsal spinous ray, and is contained thrice and one-half times in the length, excluding the caudal fin. The head makes a fourth of the same distance, or a fifth of the entire length of the fish, including the caudal fin. The face forms part of the general dorsal curve without inequalities, and the crown of the head is moderately rounded transversely, the width at the posterior angles of the orbits being a little more than a diameter of the eye, but at the anterior angles a little less. The nostrils are small pore-like openings in a membrane near the anterior angle of the orbit, and close to the upper end of the preorbitar.

The mouth is very small, and is armed above and below with a single close-set series of incisorial teeth, which are channelled in front, bevelled and crenated on the edges. The minute crenatures of each of the upper teeth number about seven or eight, the cutting edges being otherwise nearly straight; but the lower teeth have a convexly-curved edge, and from the depth of the lateral crenatures are more or less lobed; in the St. Mawes specimen none of the under teeth have the strongly-projecting middle lobe represented in the *Histoire des Poissons;* and if it existed in the younger fish, it has worn down in the older specimen before us. Cuvier enumerates twenty-four teeth in the upper jaw, but the jaws being only half open in the example we are describing, we cannot reckon beyond nine or ten on each premaxillary or limb of the mandible.

The preorbitar is highest anteriorly, and narrows gradually towards its termination under the centre of the pupil, its length being about twice its greatest height; the rest of the suborbitar chain is narrow, the whole forming a half circle close beneath the eye, with a silvery lustre and many pores. The mandible has the same kind of porous nacry surface on its under aspect, and all the naked parts of the head seem to be copiously mucigenous. The upper edges of the mandible, as far back as the articulation of the jaw, are received under the preorbitars, which also wholly cover the maxillaries and all the lateral portions of the premaxillaries. A crescentic band of scales, five deep in the centre of the crescent, covers the cheek entirely between the suborbitar chain and the naked preopercular disk, which has a perfectly even hyperbolically-curved edge.

Four rows of smaller scales cover the interoperculum, which, when the jaws are closed, touches its fellow, and conceals the branchiostegous membrane. The posterior margin of the gill-cover is a small segment of a circle, of which the suboperculum constitutes about two-thirds. A small shallow obtuse notch, with rounded corners, terminates the bony edge of the operculum, above the level of the pectoral fin : with the lower corner of the notch the point of the suboperculum coincides exactly so that there is no projection, and neither bone nor notch would be perceptible in a recent specimen. The membranous edge of the gill-cover is very narrow, and the disk is covered with six rows of scales nearly as large as those on the body, but diminishing to five and four rows as they descend over the suboperculum whose junction with the operculum they wholly conceal. From opposite the upper corner of the opercular notch, a strip of scaleless very porous integument curves upwards and forwards to the mesial line of the occiput opposite the posterior angles of the orbits where it meets its fellow: it includes the porous disk of the suprascapula, which looks like a scale, and is bounded posteriorly by eight or nine scales, being the commencement of those on the body, but appearing larger from their whole disks being exposed. On the temples between the naked border of the orbit and the humero-nuchal arc of integument, there is a short isolated patch of scales ranged in four rows. With this exception, the top of the head, the snout, and jaws are destitute of scales.

The lateral line bounds the upper fourth part of the height, having a rather flatter curve than the back: it is composed of seventy-eight scales, exclusive of the small scales on the base of the caudal, where the line cannot be traced in the dried specimen. Where the body is highest there are six rows of scales above the row which forms the lateral line, and about twelve below, all ranged so as to form nearly a semicircular curve between the dorsal and ventral profiles, and having a Sciænoid aspect, with more or less obliquity. The free border has a smooth nacry surface, with many little pits, producing the same appearance of frosted silver which the naked parts of the head exhibit. A detached scale has a straight base, impressed with six, eight, ten, or more furrows, separated by ridges that diverge, like the rays of a fan, from a point situated in the posterior third of the disk: the sides are also straight, and the free margin is curved, and smooth when its nacry epidermis is entire; but the adjoining half of the disk is composed of microscopical polygonal areas, like denticles worn down, and the exterior row of these denticulate the margin when the epidermis is removed: the lines of structure run parallel to the free edge and sides, bending at the angles to do so.

(D. 14+15: A. 3+16: P. 18: V. 1+5: C. 15⅔.—(*Hist. des Poiss.*)

The coloured drawing of the British specimen sent by Mr. Fox to Mr. Yarrell is not among the papers handed to the Editor of the present edition of British Fishes, but the following tints are enumerated by Cuvier, who describes a fresh specimen. "The back is yellowish-olive, and the belly silvery. Three or four bright golden lines traverse the sides." Even in the dried specimen the course of these lines can be traced.

The total length of the specimen is about ten inches.

MULL OF CANTYRE.

The barrier of that iron shore.—Scott.

THE DOTTED MACKEREL.

Scomber punctatus, Couch, Zool. 1849, p. xxix. App. fig.
    ,,   ,,   Id. Rep. to Penz. Nat. Hist. Soc. for 1848, pl. iii. f. 1.
    ,,   ,,   White, Cat. Brit. Mus. p. 30.

THIS fish was taken in a Mackerel Seine at Looe, in
Cornwall, on the 6th of July, 1848, and fortunately fell
into the hands of Jonathan Couch, Esq., the able and
industrious cultivator of Cornish ichthylogy. As no
second example has as yet been met with, and the chief
peculiarities of the Dotted Mackerel are its colours
and markings, its specific rank may remain a question,
until the acquisition of other specimens furnish the
means of investigating its internal structure.   In the
meanwhile Mr. Couch's description is quoted from the
Zoologist.   The figure is from a drawing by him.

" The length of the specimen was fifteen inches and a
half, and the general proportions were those of the
Common Mackerel.   Conspicuous scales, marked by
minute transverse lines, cover the sides and belly, where
none are distinguishable in the common species.   There
was no corselet, but there was some appearance of it in
a line of denser scales above the pectoral fin which

vanished below that fin. The dorsals were three inches apart.

D. 12—11, V: P. 20.

" The tail at the setting on of the caudal fin is depressed and square. Lateral line waved. The colour afforded a marked distinction from the Common Mackerel, being of an uniform dark neutral tint, or bluish-olive, green on the head and back without any coloured bands or variations, but with green reflections on the sides : round, well-defined spots, of the size of a small pea, cover the sides thickly from head to tail; on the summit of the back they are a little larger, and are transversely elongated ; they end a little below the lateral line, the belly being pure white. Between the caudal crests the surface is a bronzed yellow. The specimen was a female, and had an air-bladder."—*Couch, l. c.*

WHEEL AND SPINDLE, ST. ANDREW'S.

"Prima diocœsis et antiquissima regni
Patroni Andreæ nobile nomen habet."
(*Carmen de Fifa*, SIBBALD.)

"St. Rule, a monk of Patræ, in Achaia, warned by a vision, A.D. 370, is said to have sailed westward till he landed at St. Andrew's, where he founded a chapel and tower."—SCOTT, *Marmion,* i. *notes.*

THE GERMON.

Thynnus alalonga, Le Germon, Cuvier et Valenc. viii. p. 120, t. 215.
Germon,   Barbot, Churchill's Voy. v. pl. 29 (1732).
Alilonghi,   Duhamel, Peches, pp. 203, 207.
Ala-longa,   Cetti, Hist. Nat. Sard. iii. p. 191.
Orcynus alalonga, Couch (Jon.), MSS. fig.
Long-finned Tunny, Couch (R. Q.), Zool. 1413, with fig.

Cuvier considers it to be one of the most remarkable facts in the history of ichthyology, that this fish of great size, very distinct in its characters, excellent as an article of food, and the subject of productive fisheries on the coasts of Europe, should have remained unnoticed by ichthyologists until a recent period. Though it is captured in abundance on the north coasts of Spain, facing the Bay of Biscay, and appears to be not uncommon on the French Atlantic coasts as high as Rochelle, it either rarely enters the English Channel, or it has been overlooked by British naturalists as much as it had been by those of Spain and France. It is to the Messieurs Couch, father and son, that we owe its introduction into the list of English fishes. Mr. R. Q. Couch informs us in the Zoologist for 1846 that two

specimens have been taken in Mount's Bay by fishermen who have spread their seines for Mackerel. One
of them in the year 1846, whereof the published figure is
quoted above, and the other, which was captured several
years previously, was then deposited in the Penzance
Museum of Natural History.

This fish ought to interest Englishmen peculiarly,
since its appellation of *Germon*, by which it was first made
known to science, is supposed to be a corruption of the
word War-man, in use at the Ile d'Yeu, when the English were masters of Guienne and Poitou. The Basques
name it *hegalalonchia*, which signifies long-winged, and the
French mariners also, with reference to the length of its
pectorals, call it *long-oreille* (long-ear). Cuvier had not the
means of comparing Mediterranean with Atlantic specimens of this fish ; their identity, therefore, rests on the
accuracy of the details given in books. His description
was drawn up from a specimen procured from Rochelle,
and ought to accord with the British fish.

M. Noel de la Moriniere has given the best account
of the fisheries of the Germon on the French Atlantic
coasts. The fishermen of Ile d'Yeu begin the fishery in
the south of the Bay of Biscay opposite St. Sebastians,
follow the fish in their movements to the north of
Belleisle ; and the numbers they capture in a season
average 13,000 or 14,000. They use lines of eighty
fathoms in length, and bait their hooks with salted eel,
but the Germon being very voracious, a piece of white
or blue cloth or some shining piece of earthenware, or
tin cut into the form of a Pilchard, often serves the purpose.

The Germons arrive in the Bay of Biscay in numerous
bands about the middle of June, sometimes a few come
as early as May, and they continue to be met with as

late as October. Their fishery is generally two months later than that of the Tunny. The Germons prey on Mullets, Pilchards, Anchovies, and other fishes that assemble in sculls, and they also pursue the Flying-fishes. When the Germons come to the surface of the water, the fishermen take few, and large captures are only made at great depths. Experience alone points out the places where they may be sought with success, and when once the fishermen fall in with a scull of these fish, they pursue it till the end of the season. A cloudy sky, a fresh north-west or south-west wind, and a gently-agitated sea, are favourable for this fishery. When in full season, that is, in July and August, the meat of the Germon is whiter and more delicate than that of the Tunny, and fetches a better price, but before and after these months it is inferior. These details are borrowed from the *Histoire des Poissons,* wherein the history of the species is carried to a much greater length.

The specimen described in the Zoologist by Mr. R. Q. Couch was eighteen inches long and five high, excluding the vertical fins. The Germon has the usual form of the Tunnies, and a thickness equal to about two-thirds of its height. The falcate pectoral reaches as far towards the tail as the middle of the anal fin. The corselet, composed of larger scales, commencing on the humeral chain, embraces the base of the pectoral, and extending as far as that fin reaches, forms a recess in which the fin lies when it is laid to the side of the fish. The formula for the fin-rays is—

P. 37: D. 14—3+12, VIII. : A. 3+12, VIII. : V. 1+5: C. 40.

There are three graduated spines buried in the front of the soft dorsal and anal, and eight detached finlets follow each of these fins. The ventrals are closely approximated to each other, and between them there is a slender scale which looks like an additional ray. The caudal fin is widely crescentic with very short rays towards the middle. The mouth is small, and the mandible is longer than the upper jaw. The teeth are small, and not thickly set on the jaws. On the palatines and tongue they are very short and densely villiform. The colour of the specimen was blackish-blue or deep mackerel tint on the dorsal aspect, fading on the ventral surface into pale blue, yellow, and white. These particulars are chiefly from Mr. Couch.

THE TUMMEL AT BONSKIED.

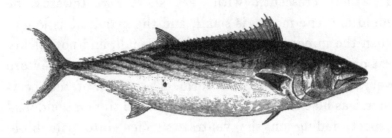

THE PELAMID.

Pelamys sarda, La Pélamide,	Cuv. et Val. t. viii. p. 149, tab. 217.
,, *vera Aristotelis,*	Rondelet, 238. A.D. 1554.
Pelamis,	Salvian, 123. A.D. 1554.
Thunnus,	Aldrovand, 213. A.D. 1640.
Pelamys Belloni,	Willughby, 180. A.D. 1686.
Scomber ponticus,	Pallas, Zoogr. vol. iii. p. 217. A.D. 1831.
,, ,,	Bloch, 334.
Pelamys sarda, La Pélamide,	Webb et Berth. Can. Poiss. p. 50.
Scomber sarda, Bonetta,	Mitchill, New York Trans. vol. i. p. 428, No. 8.

Pelamys.   *Generic Characters.*—The general shape of the members of this group is fusiform, and they have a cutaneous keel on each side of the slender part of the tail.   On the coracoidal or pectoral region, scales of larger size form a corselet ; elsewhere the scales are small and tender, passing, on the belly, into soft nacry integument.   The dorsals are contiguous ; and the first one has its rays, which are spinous, connected by a continuous membrane : behind the second dorsal there are numerous detached finlets, and one or two fewer behind the anal.   The branchiostegals are seven.   These, and other characters, they have in common with the Tunnies (*Thynni*) ; but they are distinguished by having longer and stronger subulate teeth on the jaws, widely set.   The head is conical, with a rather fine apex formed by the symphyses of the equally long jaws.

This fine fish has a wide distribution, having been taken of full size on the Russian coasts of the Black Sea, in all districts of the Mediterranean, and on both sides of the Atlantic—on the east side from the Cape Verds and Canaries, northwards along the coast of Spain, and on the west side off Connecticut, at New York, and on the Brazil coast. Ichthyologists might naturally have looked in the British seas for this active and wide-travelling fish, especially on the Cornish or Irish coasts; but the first of our naturalists who has had the fortune to procure a British example, or at least to recognise the species, is William Beattie, Esq., Honorary Secretary of the Montrose Natural History and Antiquarian Society. The specimen was captured in a salmon-net set at the mouth of the North Esk, which falls into the North Sea in latitude $56\frac{3}{4}°$, and fortunately came into the possession of a gentleman competent to understand the value ichthyologists set on such a discovery. Before intelligence of this fish reached us the entire impression of the third edition of British Fishes had been printed off; but as there had been no issue, we are enabled to interpolate this notice in the place that the species ought to occupy in the volume: and we beg to tender many thanks to Mr. Beattie and Dr. Gray for their communications; and to the directors of the Montrose Society for their liberality in lending the specimen.

According to Pallas, the Black Sea specimens attain the length of an ell; Webb and Berthollet's Canary example was twenty-five inches long; Storer quotes the dimensions of the New England ones at two feet; and Mr. Beattie's Forfar one measures twenty-two inches and three-quarters. These dimensions approach those of the Tunnies, and sailors very commonly confound the Striped Thynni and the Pelamids with each other under the general term of Bonitos; they also give them the name of Skip-jacks, expressive of the habit which many of the large Scomberoids have of skimming the surface of the sea, and springing occasionally into the air.

Pallas describes the Black Sea Pelamids as being vari-

c 2

ously clouded, on the upper parts, with brown and blue bands, while the under parts are silvery white and highly polished, and he adds that a blue stripe runs along under the lateral line. The branchiostegous membrane and the first dorsal are black, the pectoral fins azure-coloured— the purity of the colours and elegant form of the fish rendering it a very beautiful object.

The British Scomberoid to which the Pelamid has the nearest resemblance in external form is the Belted Bonito, which has been detected in our seas by Mr. Couch alone: but that Thynnus has shorter teeth, and only thirteen spinous rays, in the first dorsal. In the Forfarshire specimen the following is the formula of the rays—

Br. 7: D. 21—1 + 13—viii.: A. 4 + 12—vi.: P. 24: V. 1 + 5: C. 21—20.

The spinous rays of the first dorsal are slender, and the third is the tallest, while the first is not above a sixth or a seventh part shorter. The figure shows the form of the fin, and how it falls off posteriorly. There is, perhaps, a short incumbent ray on the base of the second dorsal spine, but its existence cannot be proved without dissection, and it may be that the spine is merely thickened at the base. The numbers of the detached finlets behind the dorsal and anal vary with the age of the fish in the Scomberoids, the membrane being more continuous in the young, and including more of them. Four slender, graduated, jointless rays commence the anal. The pectoral is triangular, and when in repose, fits into a depression of the corselet. Its tip, when laid back, just passes the eighth ray of the first dorsal: and the ventrals, which also recline in a cutaneous depression, have their origin opposite to the base of the first pectoral ray. The corselet composed of scales, larger and somewhat more conspicuous than the others, covers a triangular area on each side, which extends from the supra-scapula to a little beyond the point of the pectoral, where it ends rather obtusely. Its inferior edge is straight, and running along and near the under margin of the pectoral, joins the coracoid above the curve of the gill-cover. On the back the scales are very small, but sufficiently visible to the naked eye by reflected light, particularly a row or two under the spinous dorsal. They become gradually imperceptible on the sides, and are lost on the belly in the smooth nacry integument.

In the supra-scapular region, the lateral line bounds the corselet, receding from it over the proximal third of the pectoral, in a small curve convex upwards, then, before it passes the posterior third of the pectoral, resuming a straight course parallel to the back and nearer to its profile than to that of the belly. It makes, however, some slight undulations before reaching the region of the vent. Opposite the penultimate dorsal and anal free finlets, the lateral line is replaced by a callous cutaneous crest, which terminates at the base of the caudal. There are no oblique crests on the bases of the fin just named, such as the Common Mackerel possesses.

Both jaws are armed by conico-subulate teeth, rather widely set, with smaller ones springing up in some of the intervals. Most of these teeth are moderately curved, and the tallest ones arm the sides of the mandible; a pair, equally tall, however, stands on each side of the point of that bone, and rather more interiorly than the general row. On the premaxillaries the teeth are smaller and closer. The palatine ones are strongly curved and rather crowded: there are none on the vomer.

Ten dark bars traverse the back and upper half of

the sides, descending below the lateral line. They run obliquely, the longest one extending from between the first and second free dorsal finlets to the apex of the corselet; the others lie parallel to it and at equal distances, and consequently, owing to the curve of the back, decrease in length the further they are situated from the one above mentioned. In the dried specimen the cheek is impressed by brownish grooves or wrinkles, intercepting elliptical areas, and similar depressions exist on the integument covering the coracoids. On the head generally, and especially on the jaws and gill-covers, the skin is very smooth, even, and nacry, without scales. The branchiostegous membrane and the isthmus of the gills are bluish-black. The first dorsal also appears to have been blackish. Including the caudal, the specimen measures twenty-one and three-quarter inches in length.

In the Zoologist for 1859 (p. 6731) mention is made of the capture, in a herring-net set off the coast of Banff, of an example of the Plain Bonito (*Auxis vulgaris* of the second edition)—a fish which has hitherto been but seldom recognised on our coasts.

In the warmer districts of the Atlantic, Bonitos, Pelamids, and other large Scomberoids, are fished for with tackle rigged like a Mackerel line, but considerably stronger. The bait is a piece of bright tin, shaped like a Flying-fish, or a slice of the skin of pork, or of the tail of a Mackerel. The hook is weighted so as to sink a little beneath the surface of the water, and produces most sport when it is dragged at the rate of five miles an hour, or thereabouts.

ACANTHOPTERI. *SCOMBRIDÆ.*

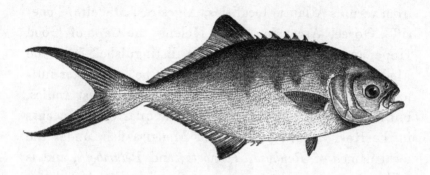

THE DERBIO.

Lichia glaucus, Cuv. et Valenc. viii. p. 558, pl. 234.
Premier glaucus, Rondelet, p. 252.
Lampuge des Marseillais, Belon, p. 155.
Scomber glaucus, Linnæus.
Gasterosteus glaucus, Forster, Des. An. p. 5.
Centronotus vel *Lichia glaycos,* Risso, 2me Edit.
Centronotus binotatus, Rafinesque.
Albacore, Couch, Linn. Tr. xiv. p. 82. Jenyns, Man. p. 366.

LICHIA. *Generic Characters.*—Form oval, compressed, covered with leathery scales, without keel or lateral ridges on the tail; head small; teeth minute. Dorsal spines low, isolated, each with an axillary membrane, and, in front of all, a recumbent spine; two preanal spines; second dorsal and anal long, similar to one another; no spurious fins. Seven branchiostegals. A large air-bladder, expanded posteriorly. Five conspicuous cranial ridges, the median one being the longest and highest.

FOUR species of this genus are described in the *Histoire des Poissons,* three of them inhabitants of the Mediterranean, but together with the fourth, ranging also along the western coast of Africa, some of them as far as the Cape of Good Hope, where the Dutch colonists call them *lyre-vish.* The species which we have to describe is the one which Rondelet says is known at Montpellier, by the name of *Derbio,* but which is called *La liche* and *La cabrolle* by the Provençals, and *La lechia* by the Sards and Romans. At Nice its name is *lecco*, and, according to Rafinesque, its Sicilian appellations are *cionana*, *ciodena* and *ciodera*. Cuvier received specimens from various Atlantic localities, Algesiras, Madeira, Teneriffe, Goree, Ascension, Saint Helena, the Cape of Good Hope, and from Brazil, not to be distinguished from the Mediterranean ones. It may possibly be, as Forster intimates, the *Sea-blueling* or *Silver-fish* of the West Indies, but we have seen no example from that quarter. It occurs in the Rev. R. T. Lowe's list of Madeira fish, under the local names of *Ranhosa, Toonbeta*, and *Pelumbeta*, and is said to be extremely common at that island. It belongs to the same tribe of Scomberoids, with free dorsal spines, as *Naucrates*, that is, to the *Centronoti* of Lacépède.

Notwithstanding its extensive southern range, it seems to wander rarely into the more northern parts of the Atlantic. It is not mentioned by French ichthyologists as having been captured on the western coasts of their country; and Mr. Couch, to whose industry and acute discrimination British Ichthyology owes so much, is the only person who has procured an English specimen. That solitary example is carefully preserved in the Museum of the Natural History Society of Penzance, and we have not had an opportunity of seeing it, but through the kindness of Dr. Gray, of the British Museum, we have been enabled to compare two excellent photographs of the specimen with the figures given in the *Histoire des Poissons*, and in Webb and Bertholet's *Histoire Naturelle des Isles Canaries*. (Poiss. pl. 13.) With the latter the photograph agrees so closely as to leave no doubt of the specific identity of the fish they represent, and the former differs merely in the lateral line, being a little more undulated anteriorly. The lateral spots are not exhibited in the photograph, its prototype being, probably, too young for the development of these markings. Cuvier regards the *Lichia tetracantha* discovered by Mr. Bowdich near the Gambia as merely a variety of this species, but from a drawing of *tetracantha* made at Sierra Leone by Dr. Mitchill, Surgeon in the Royal Navy, it appears to be a considerably more oblong species, and instead of about four spots on the fore part of the sides, there is a series of ten smaller ones, extending nearly to the base of the caudal. Its colour is bright ultramarine blue, and silvery white below, the lateral spots being darker blue, and the tips of the fins blackish-blue, as in *glaucus*. The true *glaucus* was also obtained by Dr. Mitchill, off the Niger in the Bight of Benin, and his drawing represents it as of a darker blue than *tetracantha*, and of a considerably deeper oval form. The following description of the Derbio is drawn up from the photographs, with additions furnished by Mr. Couch.

This gentleman states the length of the specimen to have been thirteen inches and a half, and its height three inches and seven-eighths. The comparative length, omitting the acute caudal lobes, is thrice the height; and the head forms a sixth part of the total length, including the whole caudal fin. The scales are small, and not strong. The cheeks nacry and scaleless. The lateral line descends obliquely without an abrupt curve from the suprascapular, till it comes nearly over the first free anal spine, and a little above mid-height, and from thence runs straight to the central rays of the caudal fin without any keeling or armature perceptible in the photograph.

The fin formula is—

D. VI.—1 + 22 : A. II.—1 + 23 or 24.

The ventrals and pectorals are both small. There is a couchant spine pointing forwards before the first dorsal, which is composed of six detached spines, all nearly of equal height, and each with a triangular membrane in its axilla. The soft dorsal and anal are alike, each being higher in front, but not decidedly falcate, and each having a short spine incumbent on the base of the first articulated ray. Two detached spines stand in front of the anal similar to those of the first dorsal, and midway between them and the ventrals is the vent.

The caudal is deeply swallow-tailed.

There are several vertical oval spots or bars in a row just above the lateral line, and touching it. These are said in the *Histoire des Poissons* to be peculiar to the adult fish. Mr. Couch says, that the dorsal aspect and the lateral bars were of rather a dark blue; the ventral aspect from the mandible to the caudal, and including the eye, was pale yellow, and the dark blotches on the anterior tips of the soft dorsal and anal were well marked. Cuvier describes the air-bladder as forked posteriorly, its long points entering among the muscles of the tail on each side of the anal interspinal bones : and he considers the course of the lateral line without a decided elbow as a mark by which the Derbio may be readily distinguished from the *Lichia amia*, which is the second *glaucus* of Rondelet, the *amia* of Salviani, and the *cerviola* of the Sicilians. In this second species the lateral line has a strong curvature in form of the letter ɷ. Willughby introduced much confusion into the history of these two species, having mistaken the Derbio or the first *glaucus* of Rondelet for the second one; and Ray, Artedi, and Linnæus have all, while correct on some points, fallen into error on others in their accounts of the several species of Lichia.

ACANTHOPTERI. *TÆNIÆDÆ.*

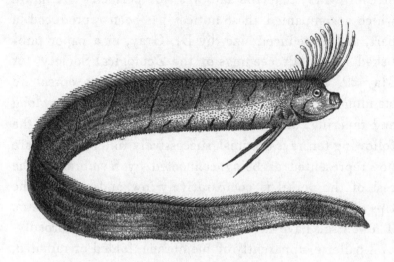

BANKS'S OAR-FISH.

Regalecus Banksii, GRAY, Pr. Zool. Soc. for 1849, p. 80.
Gymnetrus Banksii, CUV. et VALENC. Poiss. x. p. 365.

REGALECUS. *Generic Characters.*—Greatly compressed and elongated sword-shaped fishes. Teeth minute or none. Dorsal fin rising on the occiput like a plume. Caudal said to be continuous with the dorsal, and to embrace the point of the tail, but seldom seen entire, and of doubtful form in most species. Ventrals uni-radiate and very long, edged with membrane which expands at the end. Branchiostegals seven. A very long slender tapering stomach, of which three-fourths is cæcal; pancreatic cæca simple and very numerous. Scales microscopical in the nacry epidermis, also scattered osteoid tubercles in the skin. No air-bladder.

IN the Banksian library at the end of a quarto copy of Pennant's British Zoology, published in 1776, is the following manuscript note:—" On Saturday, the 23rd day of February 1788, was caught near Newlyn Quay, on the sand at ebb-tide, a fish which measured in length eight feet four inches, in breadth ten inches, and thickness two inches and a half; its weight was forty pounds." Another marginal note states further that " a gentleman

who saw this fish informed Capt. Chemming (Chelwyn?
or Chirgwin?) that the tail was not perfect." A figure
which accompanied these notices has been reproduced in
part, and of reduced size, by Dr. Gray, in a paper published in the Proceedings of the Zoological Society, for
May 29, 1849. In this the front ray of the dorsal fin
standing on the forehead between the eyes is very long
and tapering, and curves forward before the face : the
following ten rays diminish successively in length, and are
not represented as being connected by membrane : the
rest of the dorsal is comparatively low, and has only the
tips of the rays rising above the continuous membrane.
The ventrals have each one long ray dilated into a broadly-
oval pallette, apparently of membrane, folded or radiated.

Pasted into the same copy of Pennant there is also a
paragraph cut from the York Chronicle, stating that on
the 18th of March 1796 four women picked up a curious
and uncommon fish, which came ashore in Filey Bay.
They sold it to a man who carried it to York. Sir Joseph
Banks's correspondent sent him a tracing of a drawing of
the fish by Dr. Burgh, together with observations which
are here abridged from Dr. Gray's paper :— " Thirteen
feet long, one foot deep, three inches thick, head seven
inches long, eye one inch three-eighths in diameter. The
dorsal fin runs from the head to the other end, at which
there is no tail; it has 290 and 13 rays, and is red like
that of a roach or perch ; the pectoral has twelve ; the ventral one ; no anal. Branchiostegous rays six. No teeth,
a soft tongue. Anus, four feet nine inches from the head.
The face, and inside of the mouth, black ; the irides silvery white. Though there was no caudal fin when I saw
it, it is not clear that he never had one, for there was an
appearance of mutilation in its place.—W. B."

Notices of the capture on our coasts of similar fish,

but mostly too imperfect for the identification of species, occur in various periodicals. The Annual Register records the taking of one of these at Whitby on the 22nd of January 1759; and Mr. Stanton of Newcastle informed Messrs. Hancock and Embleton, that about the end of the 18th century he recollected the exhibition of a similar fish in Newcastle. It was ten feet long, and two inches thick. A sketch was made of it by Bewick the celebrated wood-engraver, which has been unfortunately mislaid. The same gentlemen were told by John Blackett Anderson, of Walker, near Newcastle, that he recollects the capture of two fishes about the year 1800, in a shallow pool at the outer Fern Islands. The larger was eighteen feet long, about a foot deep, and of a silvery colour. In 1796 one was got at Cullercoats, near Newcastle, as mentioned in a pamphlet published in 1849 by John Such, of that town. On the 19th of March 1844, one was stranded, after a severe north-east gale, at the village of Crovie, in the estuary of the Doveran, near Macduff in Banffshire, and was afterwards exhibited in the Town-hall of Elgin. From the correspondence of Mr. John Martin of the "Elgin Institution," with the late Dr. Johnston of Berwick, and the sketches he sent, Mr. Yarrell entertained no doubt of its specific identity with the *Regalecus* to be fully described below, and whose portrait is placed at the commencement of this article. Mr. Martin states the measurements of the Crovie specimen to be, total length twelve feet; depth one foot; thickness two inches and three-quarters; height of the dorsal fin two inches and a half, length of the ventral rays three feet; length of the pectorals two inches and a half. The head measured nine inches from the symphysis of the mandible to the end of the gill-cover; and from thence to the vent the distance was forty-six inches. There was no caudal fin. The shaft of each ventral was about the thickness of a goose-quill, was fringed on each side by membrane, and broken short off at the extremity; the ends having been thrown away at the sea-side, their original length could not be ascertained. The dorsal fin contained two hundred and seventy-nine rays, of which fifteen standing on the head were very tall, but were connected at the base by membrane. The pectorals were supported by twelve rays. The lateral line was straight, and about one-third of the height above the ventral profile, except where it rose gradually over the pectorals. The body is described, and the drawings represent it, as having a slightly-tapering profile from the operculum to the end, which is rather abrupt, with a spur at its lower corner, and without any indication of a caudal fin. If one existed, Mr. Martin thinks that it must have been very slender. The whole body was clothed by a delicate white skin with a silvery lustre, beneath which there lay alternate smooth and tuberculated bands running the whole length of the body, palpable to the finger through the outer skin, and becoming more perceptible on its removal. Behind the pectoral fin, a few dark bars, which crossed the body obliquely, were very conspicuous when the fish was fresh, and the dorsal had at first an orange tint.

Messrs. Hancock and Embleton's paper mentions that one of the Preventive Service men, in the year 1845, observed a fish of this kind in a pool near Alnmouth. On his approach it bent its body into a circle, and he, ignorantly thinking that it was going to spring upon him, boldly attacked it with his cutlass and cut off its head. It was sixteen feet long, eleven inches deep, and six thick. In the struggles of the dying fish, the sands around were covered with its delicate nacry scales.

On the 26th of March 1849, a fine fish of this genus was captured by the crew of a fishing coble belonging to Cullercoats, consisting of Bartholomew Taylor and his two sons. It was much injured by the captors in their endeavours to secure it, and by subsequent handling during its exhibition at Tynemouth, North and South Shields, and Newcastle. Fortunately it was seen at the last-named town by Albany Hancock, Esq., and Dr. Dennis Embleton, who made drawings of it, and drew up a detailed account of its external appearance and internal structure, which was read at a meeting of the Tyneside Naturalists' Club, and published in the Annals and Magazine of Natural History for July 1849. From that paper the following abridged extracts are taken by permission.

The fish, though much injured and greatly faded, was fresh and had a uniform silvery-grey colour, except a few irregular streaks and dark spots towards the fore-part of the body, and there were remains of a bright iridescence about the pectoral fin and head, a blue tint predominating. The body is excessively compressed, like a double-edged sword-blade, its greatest thickness being below the middle, and the dorsal edge is sharper than the ventral one. The total length when the mouth is retracted is twelve feet three inches, and the depth immediately behind the gills eight inches and a half: two feet farther back the greatest depth of eleven inches and a quarter is attained, and at the end of the dorsal fin it has diminished to three. The skin is covered with a silvery matter in which the scales are invisible to the naked eye, but which is easily detached and adheres to anything it comes in contact with. Submitted to the microscope, this nacre was found to consist of scales like those on the wing of a moth. Round the hind border of the operculum there is a broad dusky patch; a crescentic dark mark exists above the eye, and there are eight or nine narrow oblique streaks on the side, which diminish to mere spots beyond the vent. The lateral line descends gradually from the suprascapula to within two inches of the ventral profile at the vent, and continues descending as it proceeds to the distal end of the fish. Four flattened ridges, each more than an inch in breadth, reach from the head to the tail above the lateral line, the longest and uppermost commencing near the eye. The skin is studded with numerous bony tubercles not regularly arranged, and in the neighbourhood of the head they are replaced by depressed indurations. On the ventral edge the tubercles are numerous and have hooked tips pointing towards the tail.

The head is small, measuring only nine inches to the gill-opening; the orifice of the mouth is circular and capable of being protruded two or three inches by the depression of the mandible. The tongue is small, smooth, and prominent; there are no teeth, and the interior of the mouth is black. Gill-plates proportionally large, preoperculum crescentic, with the lower horn prolonged forwards to the articulation of the mandible. Operculum curved elliptically posteriorly, ending obtusely. Branchiostegals seven. Branchial arches four, with tubercular bristly rakers. Pharyngeal bones above and below furnished with setaceous teeth.

The dorsal fin extends from between the front of the orbits to within three inches of the distal extremity of the fish. The twelve anterior rays were stated by the captors to have been about fourteen inches long, and furnished with a membrane on their posterior edges, which grew wider upwards, somewhat like a peacock's feather. The ends were broken off, but a continuous membrane connected their bases, and their shafts appeared ragged with the remains of the torn membrane. In addition to these there were 268 other rays whose acute points overtopped the connecting membrane, or 280 dorsal rays in all. About the middle of the fish, where the dorsal rays are highest, excepting those on the head, they measure upwards of three inches and a half, and at the termination of the fin their height has decreased to one inch. Behind the termination of the dorsal fin the edge of the back slopes rapidly downwards to within an inch of the line of the belly, and then forms a rounded point which is the distal extremity of the fish. Both the upper and under edges of this extremity are very thin, and the fishermen insisted that when they took the fish this part was entire, and that there was no tail-fin whatever. The edges may be pressed together, and seem to fit. The pectorals are attached low, and contain eleven rays. The ventral fins were represented by a pair of very strong straight spines broken short to the length of four inches, but were said to have been originally twice that length, having even then broken ends; a membranous edge was visible at their bases. The vertebræ, judging from elevations obscurely seen through the muscles, were reckoned at 110. Fin-ray formula—

D. 280: V. 1: P. 11. Vertebræ 110?—*Hancock and Embleton, l.c.*

Messrs. Hancock and Embleton's excellent paper may be consulted for the internal anatomical structure, and several particulars of external form, which have been omitted here from want of space.*

In 1850, another example of this fish, alive but mutilated, was cast ashore on the Yorkshire coast, near Redcar. It measured nearly twelve feet, and weighed sixty-six pounds, as reported in the *Zoologist* (2709) by T. S.

* For a larger cut of the head copied from this paper, see foot of page 35.

VOL. I.　　　　　　　　*(2nd Supp.)*　　　　　　　　D

Rudd, Esq. In the same communication, mention is made of one found on that coast several years previously by a pilot named Slater Potts: if its length was correctly stated at twenty-four feet, it is the largest example of this species that has been recorded. Another was stranded on the 17th of September 1852, near Millar's Stone, in the Bay of Cromarty. This specimen was secured for the museum of Mr. Dunbar at Inverness, and on the dispersion of that collection some years ago, came into the hands of a bird-stuffer of the same place, who kept it hanging in his shop until he tired of looking at it, and no purchaser offering, it was at length consigned by him to the dust-cart.

A northern member of this genus was described by Ascanius under the name of *Ophidium Glesne* in the Copenhagen Memoirs for 1776, the generic name being afterwards changed to *Regalecus*, by which he intended to signify King of the Herrings. Glesne is the name of a village near Bergen, where the fish was taken. This species, which received other names from ichthyologists who came after Ascanius, has been supposed to be the same with the British fish; and the case may be so, but hitherto the Norwegian fish has been described by Ascanius and Brünnich alone, and the one reckons only 126 rays in the dorsal fin, and the other 197, while the figures given by these authors show a greater number. This reckoning, however, is so different from the numbers of the rays in the British fish, that they cannot be considered as the same species until the mistake, if there be one, has been rectified by an accurate examination and comparison of specimens. *Gymnetrus Grillii* of Lindroth, described in 1798, had the large number of 406 dorsal rays, with a total length of eighteen feet, and ventrals measuring five. It is therefore safer for the present to keep Messrs. Hancock and Embleton's fish distinct under the name of *Banksii*, proposed by M. Valenciennes in the *Histoire des Poissons*. The right of priority over the term *Gymnetrus* belongs to *Regalecus*, and this is therefore used here, though M. Valenciennes rejects it for its barbarity.

BANKS'S OAR-FISH (Hancock and Embleton).

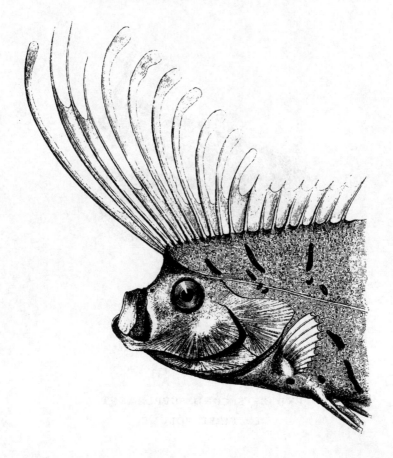

Jaws of Couch's Sea-Bream.   See page 5.

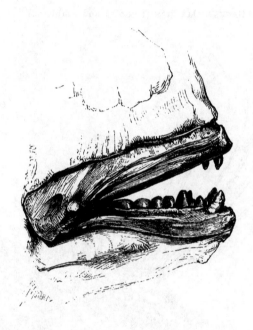

END OF SECOND SUPPLEMENT
TO FIRST VOLUME.

SECOND SUPPLEMENT

TO THE SECOND VOLUME OF

THE HISTORY OF BRITISH FISHES.

ANACANTHINI.
ANISOMERI.

PLEURONECTIDÆ.

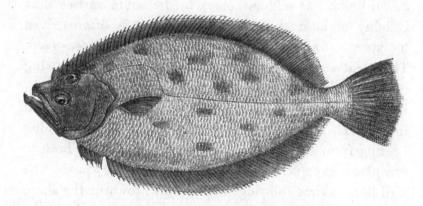

THE SAIL-FLUKE.

Zeugopterus? velivolans.   EDITOR of the Third Ed. of Br. Fishes.

DR. BAIKIE, writing from the banks of the Niger, referred to Dr. Alex. Duguid of Kirkwall for information respecting the very curious habits of this fish, and that gentleman, on being applied to, most kindly sent a sketch made by a friend of his, said to be very correct, from which the above wood-cut has been engraved. He also furnished the following particulars of the history of the fish. This Fluke, he says, is highly prized as an article of food, its flesh being firm and white. It does not take a bait, and he only once saw it caught in a net, but it comes ashore, spontaneously, with its tail erected above the water, like a boat under sail, whence its name. This it does generally in calm weather, and on sandy shores, and the country people residing near such places train

VOL. II.   *(2nd Supp.)*   E their dogs to catch it. The following letter was written by Mr. Robert Scarth, of North Ronaldshay, the northernmost island of the Orkney group, where the Sail-fluke is very common:—" It is never caught by hook or by net, and I have in vain set ground lines for it in the South Bay, baited with Lug-worms, Limpets and Sellocks, neither have Flounder or Skate nets, drawn there, inclosed a Sail-fluke. It seldom comes to the shore earlier than October or later than April, though it is often driven by storms on the beach, entangled among sea-weed. The great supply is, however, obtained in the following manner:—In the winter and early spring a pair of Black-headed Gulls take possession of the Bay, drive away all interlopers, and may be seen at daybreak every morning beating from side to side, on the wing, and never both in one place, except in the act of crossing as they pass. The Sail-fluke skims the ridge of the wave towards the shore with its tail raised over its back, and when the wave recedes is left on the sand, into which it burrows so suddenly and completely, that though I have watched its approach, only once have I succeeded in finding its burrow. The Gull, however, has a surer eye, and casting like a hawk, pounces on the Fluke, from which by one stroke of his bill it extracts the liver. If not disturbed, the Gull no sooner gorges this luscious morsel, than it commences dragging the fish to some outlying rock, where he and his consort may discuss it at leisure. By robbing the Black-backs I have had the house supplied daily with this excellent fish, in weather during which no fishing-boat could put to sea. Close to the beach of South Bay a stone wall has been raised to shelter the crops from the sea-spray. Behind this we posted a smart lad, who kept his eye on the soaring Gulls. The moment one of the birds made its well-known swoop, the boy rushed to the sea-strand, shouting with all his might. He was usually in time to scare the Gull away and secure the Fluke, but in almost every case with the liver torn out. If the Gull by chance succeeded in carrying his prey off to the rock, he and his partner set up a triumphant cackling, as if deriding the disappointed lad. Seals often pursue these Flukes into the bay, and frequently leave serviceable morsels unconsumed. The Sail-fluke exhibits its gambols most frequently before a storm, or when a thaw succeeds a frost. It is the most delicious fish of our seas, but loses its flavour by a day's keeping."—19 Feb. 1849.

Length of a specimen, twenty-one inches. Height, excluding the fins, seven inches and a half, or including them, ten inches. Weight three pounds.

D. 87 : A. 69 : P. 11 : V. 5 : C. 19$\frac{2}{3}$.

The dorsal commences before the upper eye, which is three-quarters of an inch before the lower one. The rough lateral line is much arched over the pectoral fin. Scales large and striated from the centre, roughish on the pale belly, and very small on the fronts of the fin-rays. There is one row of minute sharp teeth on each jaw, and also some teeth on the vomer. Tongue round and conical. Operculum and interoperculum thin and very diaphanous, the bony plates, and the cheeks being as thin as silk paper, so that the smallest type may be read through them. The left or coloured side of the fish is like that of the Common Dab, yellowish-brown, with scattered dark blotches.—Another specimen weighed four pounds, and had ninety-one dorsal rays, with seventy-five anal ones.—(*Dr. Duguid.*)

The oval of the body is wider vertically than that of the *Smähvarf* (Pl. 50) of the *Scandinaviens Fiskar*, but narrower than the ovals of either of our English Top-knots. In the facial profile and forms of the vertical fins there is

E 2

much similarity between the Smähvarf and the Sail-fluke, but none of the four specimens of the former, noticed by Sundevall, have so many dorsal and anal rays as the Orkney Fluke. In the latter, also, the upper eye is more distinctly behind the lower one, and the mandible protrudes farther. Sundevall makes no mention of transparent cheeks ; for the present, therefore, they are described as distinct, such a proceeding being less injurious to science than the junction of two species divers in structure and habits under one name. In the absence of a specimen, the Sail-fluke is placed in the genus *Zeugopterus* interrogatively. A single row of jaw-teeth only is mentioned by Dr. Duguid, but he probably means laterally, where the Smähvarf has only one row, though there is a crowded patch at the symphyses of its jaws.

SUGAR-LOAF, SUTHERLANDSHIRE.

THE CONNEMARA SUCKER.

Lepidogaster cephalus, THOMPS. Nat. Hist. of Irel. iii. p. 214.

THE authority for this species is a specimen, which was taken in Roundstone Bay, Connemara, on the western coast of Ireland, and is preserved in the collection of the late Dr. Ball of Dublin. It has not been observed on the English coasts, nor has any drawing been made of it to which we have access. Mr. Thompson published an account of it in the Annals of Natural History (iii. 34), which has been reprinted in the posthumous edition of his work above quoted. The following passages are extracted from his paper.

" This fish equals *Lepidogaster Cornubiensis* in size, but differs from it in the dorsal and anal fins occupying a considerable portion of its length, and in having a greater breadth of head with a narrower snout: the body likewise is more depressed, and narrows more suddenly behind the ventral disk into the tapering tail. Its specific characters are—a very minute cirrus before each eye; dorsal and anal fins unconnected with the caudal; ventral disk small.

Br. 5 : D. 15 : A. 10 : P. 25 and 4 : C. 15 conspicuous, 20 in all.

" Head very broad posteriorly, forming more than one-third of the entire length; in front of each orbit, and on a line with its upper border, there is a cirrus so minute as to be scarcely visible without the aid of a lens; eyes large, two of their diameters apart: teeth pointed and numerous on the jaws, the outer premaxillary ones being the largest; gill-opening small; ventral disk smaller, and

different from that of *L. Cornubiensis.* Dorsal fin originating behind the middle of the fish, and continued to
near the caudal, with which it does not unite; anal fin
commencing farther back, but reaching as far; the last
rays of the two fins, when laid down, touching the base of
the caudal: rays of all the fins jointed, but not branched:
vent situated midway between the posterior edge of the
ventral disk and the end of the caudal fin: a short anal
tubercle."—*Thompson, l. c.*

The *Lepidogaster Webbianus* of Valenciennes, which inhabits the seas of the Canary Islands, has two cutaneous
filaments at the nostrils on each side, and so has also the
L. zebrinus of Lowe, which inhabits the Madeira coasts,
and is perhaps the same species with *Webbianus.* The
Lepidogaster chupasangue of the same seas, which Mr.
Lowe thinks may be *L. Decandollii* of Risso, has no nasal
cirrus, and the vertical fins are not connected to each
other.

VIEW OF LANCRIGG.

GANOIDEI.   *ACIPENSERIDÆ.*

THE COMMON BRITISH STURGEON.

Acipenser Thompsoni, Ball, Thompson's Nat. Hist. of Ireland, iii. 245.
    ,,   *sturio, Common Sturgeon,* Penn. Brit. Zool. iii. 164, pl. 22.
    ,,   ,,   ,,   ,,   Donov. Brit. Fish. pl. 55.
    ,,   ,,   ,,   ,,   Jenyns, Man. 493.
    ,,   ,, *Sturgeon,*   Flem. Brit. Anim. 173.

Acipenseridæ. *Family Characters.*—Form elongated, pentagonal; the angles of the body formed by the crests of five longitudinal rows of bony shields; mouth on the ventral aspect, protractile, toothless; no branchio-stegals; internal skeleton cartilaginous, except the basal cephalic plate of bone, which extends backwards under the first five dorsal vertebræ; vertical fins supported anteriorly by short bony rays; a long spiral intestinal valve; pancreas glandular.

Acipenser. *Generic Characters.*—Snout tapering, beak-shaped, with four barbels depending from its ventral surface before the mouth; an accessory gill, and, at the upper border of the gill-cover, a spout-hole; trunk of the tail not flattened.—*Heckel.*

The Sturgeons are Ganoid fishes of a lengthened shape, having a cartilaginous skeleton, and the protractile mouth situated under the eyes on the ventral surface considerably behind the tip of the snout. The jaws are much more protractile than those of a Shark, and consist of the premaxillaries going round the upper or anterior border of the mouth, with small maxillaries articulated to them laterally and connected also to the palatines. The mandible is formed principally of a pair of bony limbs, united to each other at the symphysis, ending laterally in a joint furnished with a trochlear cartilage, and moving on the wing-like process of each palatine. A thick fleshy lip, sometimes lobed, covers the premaxillaries; but the mandibular lip is deficient in some groups of species, except at the corners of the mouth; and in other groups the posterior lip crosses the orifice, either in form of a continuous soft roll, or with a mesial depression, or even a mesial interspace The fulness of the lips, in conjunction with the forms of the dorsal crests and a few other characters, have been made to serve for grouping the species. The gills, as in the osseous fishes, consist of five movable arches, and are comb-like, with free tips: a pectinated accessory gill also adheres to the inner surface of the gill-cover, and there is a small spout-hole close behind and above it.

Of the five rows of bony shields on the body, one protects the ridge of the back from the occiput to the dorsal fin; a lateral row extends on each flank from the shoulder to the caudal fin; and a row on each side of the belly ends at the ventrals. Each dorsal shield is more or less distinctly keeled by an acute longitudinal crest, whose apex, in some groups of species, overhangs the posterior edge of the shield, but in other groups is central, the plate sloping off from it both before and behind. The skin intervening between the rows of shields varies also in character, being naked and smooth, or studded with bony grains, either of a granular form, or star-shaped, or with acute points or even hooks. In *Ac. Güldenstädii* of Brandt, which is the *Ac. sturio* of Pallas, the skin of the breast between the coracoid shields is set with elevated star-like or roundish and denticulated ossicles; while in *Ac. nasus* of Heckel the same region is closely covered with flat ganoid scales, like those of *Lepidosteus*. In *Ac. schypa* of Güldenstädt the same part shows stellate ossicles, many of which emit prickles. The *Schypa* is var. β and γ of the *Sturio* of Pallas, who obtained it in the Wolga and Obi.

Age changes the form and size of the body-shields of the Sturgeons, their crests becoming lower and blunter, and their disks smaller, so that in aged fish the sharply-pentagonal form of the body is lost, and the ventral shields often wholly disappear.

The fins, seven in number, are sustained by crowded jointed, and generally flexible rays, finely serrated on the edges; the short graduated rays in front of the dorsal and anal are more or less bony. The anal is situated under the posterior part of the dorsal, which is itself placed far back. A stout, tall, bony first ray supports the pectoral fin.

The skull is cartilaginous throughout, but is supported beneath by an osseous occipito-sphenoidal plate, which extends posteriorly under five cervical vertebræ, and is prolonged anteriorly into a slender vomerine and eth-moidal process; protection is afforded to the skull above by a vaulted crust of ganoid scales or shields, which have received names from Kittary,* Fitzinger and Heckel † and others, accordant with the regions that they cover. ‡ In the views of the upper surface of the head introduced in the subsequent pages, the posterior mesial shield is the first of the dorsal series; anterior to it is the single occipital shield also occupying a mesial place; and whose anterior process enters some way between the coronal or parietal shields which form a pair and come in contact

---

* Dr. Modeste Kittary : Bull. de la Soc. Imp. des Natur. de Moscov. 1850.

† Annalen der Wien. Erster Band.

‡ Professor Owen observes that the attempt to ascertain the homologies of these cranial shields with the true epicranial bones of osseous fishes is difficult and unsatisfactory.

with each other for a part of their length ; before the
coronals and between the eyes lie the frontal shields forming another pair ; in the Frith of Forth Narrow-nosed
Sturgeon the frontals are wholly separated by one or
more interfrontal plates ; the postfrontal and prefrontal
shields are exterior to the main frontals in the positions
that their names indicate ; laterally with respect to the
coronals lie the temporal shields, often coalescent with
a squamosal piece ; and behind them occupying the
posterior lateral angles of the head, and protecting on each
side a styloid process of the cartilaginous skull (which
Owen terms a representation of the par-occipital, but
which Kittary calls the mastoid), lies a shield that articu-
lates with the first dorsal, the occipital, and squamosal,
and the suprascapular : the last-named shield being the
first of the humeral chain that descends behind the gill-
opening, heads the lateral series of body-shields, all of
which partake of its scalene form ; the chevron-shaped
humeral shield gives support to the bony ray of the pectoral ; and the coracoid, the largest piece of the humeral
chain, has wholly a ventral aspect, its crest being on a line
with the crests of the ventral body-shields ; the suprascapular, opercular, and such cranial shields as have a
lateral aspect, are represented in the profiles of the head.

The arterial bulb of the Sturgeons is furnished with
two rows of valves at its commencement, and with one
row at its termination.   The swim-bladder is very large,
and communicates with the gullet by a wide hole.   In
the glandular conglomeration of their pancreatic cæca
the Sturgeons resemble the Sharks.

Heckel and Kner* divide the genus into six groups, three
of which, viz. the *Lionisci*, *Acipenserini*, and *Helopes*, have
the dorsal shields highest at their posterior edges, and the

* Süssw. Fische der Östreich. Mon. 1858.

two first-named have, moreover, fringed barbels, characters which have been attributed to no British Sturgeon. In the other three groups, *Antacei, Sturiones,* and *Husones,* the ridges of the dorsal shields are pointed in the middle, and slope down anteriorly and posteriorly. These groups are further characterized as follows :—4. *Antacei,* having simple and not fringed barbels, a rudimentary mandibular lip, the skin between the rows of body-shields studded with stellate ossicles, and the snout short and broad. 5. *Sturiones,* having a swollen posterior lip, contracted in the middle, simple barbels, and the skin between the rows of body-shields granulated by blunt ossicles. 6. *Husones,* having the mandibular lip divided in the middle; flat, tape-like barbels, and the skin roughened by pointed ossicles. The species are distinguished from each other by the relative positions of the osseous centres of their principal cranial shields. With respect to the value for this purpose of the form of the shields, Dr. Ball says :— " I have collected many specimens, and I do not think that the broadness or sharpness of the nose is a specific distinction, as no two of my specimens can be said to agree in the shape of that member, nor in the arrangement of the scales on it and on the head. A classification of the variations of my numerous examples will reduce the British *Sturio, Thompsoni* and *latirostris* to a single species." Mr. Thompson also observes that " the precise form of the bony plates on the head is of no value as a specific character, neither is the breadth of the snout." These opinions of the Irish naturalists are shared by several English ichthyologists, and the subject requires to be worked out by an investigator who has access to a sufficient number of examples from various British localities, and an opportunity of comparing them with specimens collected from the Continental rivers.

Bloch's figure is worthless from the want of correct details.

Sturgeons named, evidently from their size, *Stör*, *Storje*, *Stoer*, and *Storjer*, by the Scandinavians, inhabit the Baltic, the German Ocean, the English and Irish Channels, and the Mediterranean, Black Sea, Caspian, and Baikal. They abound also in the waters of North America which fall into the Atlantic and Pacific, but they do not appear to frequent rivers which flow into the icy seas. They feed at the bottom, in deep water, beyond the ordinary reach of sea-nets, and are therefore very rarely taken, except in friths, estuaries, or rivers, which they enter for the purpose of spawning. They are more frequently captured in the Scottish waters than on the southern coasts of England, and have been taken, according to Thompson, in the Irish counties of Cork, Derry, Kilkenny, Wexford, Dublin, Down, and Antrim. Examples are by no means uncommon in the fishmongers' shops of London, Edinburgh, Glasgow, Dublin, and other large towns, a few coming into the hands of the principal dealers every season. One caught in a stake-net near Findhorn in Scotland in July 1833, measured eight feet six inches in length, and weighed two hundred and three pounds. Pennant records the capture of one in the Esk which weighed four hundred and sixty pounds; and a head prepared by Mr. Stirling of the Anatomical Museum of the University of Edinburgh, was cut from a Sturgeon caught near Alloa, said to weigh, when entire, fifty stones, or seven hundred pounds; its length was nine feet.

The *débris* of crustaceans and half-digested pieces of fish, mixed with decaying vegetable matters and mud, have been found in the stomachs of Sturgeons, and their food is probably any soft animal or vegetable organisms that they find at the bottom. The flesh of the Sturgeon, like that of most cartilaginous fishes, is more firm and compact than that of osseous fishes ; it generally contains much yellow fat, and is well-flavoured, easy of digestion, and very nutritious. Stewed with rich gravy, it forms a dish in high request for the table. When luxury was at its height in Imperial Rome, a Sturgeon was, according to Athenæus, the most honoured *entrée* in sumptuous repasts ; and Pliny tells us that it was crowned with flowers, and the slaves who bore it into the triclinium were also garlanded. Ovid calls it noble, either because of its costliness or of its excellence.

> Tuque peregrinis Acipenser nobilis undis.

At a later period the price of a Sturgeon had fallen in Rome to four scudi, when a competition among the purveyors of the Catholic dignitaries assembled to elect a pope, in succession to Paul, produced an instantaneous rise in the market, and Cardinal Gualtheri had to pay seventy scudi for his Sturgeon. (*Richter*, Ichth.) In the time of our first Henry, the Sturgeon was reserved for the king's table, and even in the present day, when one is caught in the Thames within the jurisdiction of the Lord Mayor, it is called a Royal Fish, implying that it ought to be sent to the Queen. In Russia and other regions where Sturgeons abound, the roe dried and pressed forms the *Caviare* of commerce ; and the swim-bladder treated in a particular manner furnishes high-priced isinglass.

The editor of the present edition of British Fishes has not had the advantage of personally inspecting the specimens that Mr. Yarrell had before him when he wrote his account of the Common Sturgeon, but he has seen portions of eight specimens caught in the Frith of Forth,

and as such of these as have their cranial plates present a
near agreement with each other in external characters,
and are evidently of one species, he has drawn up an
account of that species in considerable detail, adopting
for it the specific appellation of *Thompsoni,* suggested by
Dr. Ball. It differs in several important characters from
the *Ac. sturio* of Heckel and Kner. These Frith of Forth
examples agree generally with a pencil sketch sent to Mr.
Yarrell by Jonathan Couch, Esq., of the cranial shields
of a Sturgeon caught at Lamorna in Cornwall, in May
1851. The species, therefore, has an extensive range
along the British Coasts, and may be the one to which
the not very appropriate name of " Sharp-nosed " has
been usually applied by English ichthyologists, though a
more comprehensive comparison is needed to establish
that as a fact.

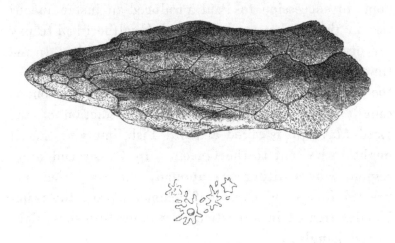

Description of a Sturgeon caught in the *Frith of Forth,*
and preserved in the Museum of the University of Edinburgh. Length nearly six feet. The barbels are rather
nearer to the tip of the snout than to the mouth, and

when laid back do not reach the latter. They are tapering and roundish, but in the dry state show a furrow as if they were composed of two binate cartilaginous rays. Other Frith of Forth specimens do not exhibit this furrow. The lips, having shrunk so that their true form cannot be ascertained, are not described. Shields or bucklers, closely connected by suture, cover the whole dorsal aspect of the head. They are deeply pitted, the pits being only partially disposed in rows so as to form radiating furrows. These well-defined depressions are separated by thin walls, which are crenulated, but do not rise above the general level, so that the character of the surface is not granular. The osseous centres of most of these shields may be made out, but they do not rise into acute crests as in the young fish. Certain lines or ledges are visible rather by their smoothness than by their elevation, and the most remarkable of them present the profile of an obelisk whose apex is in the centre of the occipital shield, whence short lines deflect on each side to the osseous centres of the coronals. Lines proceeding from thence to the centres of the frontals form the sides of the obelisk. Other less conspicuous lines radiate from the same centres of the frontals, namely, backwards over the temporals, and forwards towards the nasal regions, with a convergence coincident with the narrowing of the snout. The polygonal occipital shield receives the anterior point of the first dorsal shield into a sharp mesial notch, and emits anteriorly a salient acute process that enters between the coronals for nearly one half of their length, these plates coming in contact with each other for only about a third of their length, and having the point of an interfrontal plate insinuated between their anterior ends. This oblong interfrontal, and two other intercalary pieces, larger, but otherwise similar to the polygonal shields which

closely cover the whole upper surface of the snout, separate the frontals wholly from each other. The osseous
centres of the temporal shields are somewhat nearer to
the tip of the snout than those of the coronals are. On
the left side of the specimen a small squamosal interposes
between the temporal and mastoid shield, but on the other
side this piece is confluent with the temporal. A moderate inclination of the surfaces of the coronals towards
the mesial line makes a longitudinal furrow, which disap-
pears anteriorly, the interfrontal plates being nearly flat,
and the snout flatly convex transversely. Much of the
gill-flap is occupied by the large opercular shield, which
is marked by pits and furrows, with thin intervening
crenated walls distinctly radiating from a point near the
posterior edge of the plate. This shield being visible
from above merely in profile is not represented in the cut.
Behind and beneath the eye there is a rough rectangular
chevron which, in form and position, represents the preoperculum of osseous fishes. On the under surface of
the snout a raised ledge, narrow at the barbels and widen-
ing gradually in running forwards, as in the *Ac. sturio* of
Heckel, is covered either by a single slightly-rough plate,
or by several coalescent ones. The humeral plate is
deeply pitted, and the coracoid is marked by distinctly-
radiating furrows and pits.

The body-shields are radiately furrowed and pitted,
and have thin longitudinal crests. Eleven saddle-formed
shields occupy the ridge of the back before the dorsal
fin, the fourth or fifth of the series being the largest, and
the ridges of all highest in the middle. The scalene
lateral shields, lying between the suprascapular and
caudal fin, are thirty in number. In other specimens
their number varies from twenty-nine to thirty-two, there
being generally more on one side of the fish than on the

other. Heckel and Kner describe the middle lateral shields of their *Ac. sturio* as having a ·styloid process which proceeds forwards beneath the skin to the preceding shield, and is said to be characteristic of the species; but in the Frith of Forth Sturgeons no such process exists in any one of the whole lateral series, there being merely a notch with a flexible tube, corresponding to the lateral line of osseous fishes. There is, however, a strong and distinct smooth styloid process from the front of all the ventral shields of the Frith of Forth specimens, except the first two of the series. These ventral shields are also unequal in number, on the two sides of the fish, and vary in the specimens from nine to eleven.

The skin between the dorsal and lateral rows of shields is pretty thickly studded with star-like ossicles, intermingled with much more minute angular grains. A cluster of these ossicles is represented under the preceding wood-cut. A pair of small shields intervenes between the dorsal series and the dorsal fin, and on each side of the base of this fin there are about ten star-like ossicles larger than the others. Below the lateral shields the distinctly-stellate ossicles become fewer, and the irregular, crested grains more numerous. Between the limbs of the coracoids, and more especially a little further back below the pectorals, the skin is made rough by extremely irregular ossicles, apparently formed by the confluence of several minute angular grains and acute points, and this roughness continues onwards to the vent. The integument before the opercular shield is studded with small roundish and irregular plates, all with radiating lines from flat centres, and small plates of more oblong forms, but various outlines roughen the surface between the mouth and the coracoids.

The dorsal fin is supported by forty-one rays, the first

VOL. II.   *(2nd Supp.)*   F being a flat, longitudinally-oval plate resembling the dorsal shields in size and texture, but having a small posterior peak, which rises as the first of the rays. About six stumps, seemingly bony and gradually increasing in height, follow it, and are incumbent on each other and on the flexible rays. Behind the dorsal two heart-shaped plates follow one another on the ridge of the tail. The anal, which in this specimen has been injured posteriorly, consists in others of twenty-five rays, the first being very short and incumbent, and in fact the peak of an oblong flat plate, as in the dorsal. Between this plate or fulcrum and the vent there are three pairs of small shields. An upper low caudal fin is composed of a long strap-shaped rough plate with a posterior peak, and of eighteen or nineteen firm, slender, jointless rays lying closely tiled on one another. Underneath these inflexible rays there is a triangular lateral space on each side, which is densely covered by rough, keeled, bony eminences. The under portion and main part of the caudal is lobed anteriorly, and contains numerous jointed rays. In young individuals the anterior under lobe is said not to be developed. The pectoral contains thirty-eight rays, which are prickly on the edges, and the first one is stout and bony, seemingly formed by the coalescence of about ten rays, whose number is shown by the prickly ridges which rib its surface.

Dr. James McBain, of Leith, possesses the head of a Sturgeon that was caught near Stirling, in which the cranial plates correspond almost exactly with those of the specimen described above, except that the squamosals on both sides are coalescent with the mastoidal shields. In this preparation the thin vertical plate of bone which descends from the mastoidal shield into the cranial cartilage is well shown. Dr. McBain's fish seems to have been nearly one-third larger than the one in the Edinburgh University Museum described above.

Another perfect specimen of smaller size, being only three feet eight inches and a half long, preserved in the Museum of the Free Kirk College of Edinburgh, presents also a close similarity in the cranial plates to the two preceding, but the squamosals are both united to the temporals, and the mastoids have consequently smaller disks. In this younger fish the pits and furrows in the shields are deeper and more distinctly radiated from osseous centres. The crests of the dorsal shields are higher, and the styloid anterior processes of the ventral shields are very distinctly perceptible through the skin. The ossicles which stud the skin of the body are more generally and perfectly star-like, more of the rays being acute. There are twenty-nine lateral shields on one side, and thirty-two on the other, and the ventral shields on the right side number ten, but there are only nine on the left side. The fin-rays are—

D. 41: A. 25: P. 1 | 37 : V. 27 or 28.

The osseous centres of the temporals are equidistant from the tip of the snout with those of the coronals, instead of being a little nearer, as in the other two examples, the difference being probably due to the squamosals having in this example a common centre with the temporals. The snout is also narrower, and the shields covering it are closely pressed together so as to seem confluent. This probably arises from the cartilage not having been so fully cleared out in preparation, and shrinking much in drying. To this cause also may perhaps be attributed the very slender snouts of some of the younger Sturgeons preserved in English Museums. The exact place of capture of this individual is not mentioned on the

F 2

label attached to it, but it has a special interest as
belonging to the Museum formed by the late Professor
Fleming, and, therefore, representing the *Acipenser sturio*
of his British Animals.

In the Museum of the College of Surgeons of Edinburgh, there is a stuffed Sturgeon in excellent order,
which measures six feet and a half in length, and does
not differ materially in the form and arrangement of the
cranial shields from that of the University Museum.
The shields both on the head and body are, however,
more deeply pitted and furrowed, their radiation is more
complete, and the intervening walls of the furrows are
more granulated; more of the imbedded ossicles also are
star-shaped. There are eleven dorsal shields; thirty-two
lateral ones on the left side, thirty on the right side; and
the ventral shields are ten and eleven on the right and
left sides respectively. The label to this Sturgeon does
not indicate its place of capture.

In the Anatomical Museum of Edinburgh University,
there are preparations of several Sturgeons caught near
Alloa, and in other parts of the Frith of Forth, made to
exhibit the structure of the cartilaginous cranium and
other internal parts. One of these shows that the
occipital spine of the cartilaginous cranium is acute,
and that it does not project so far back as the mastoid
or par-occipital processes which are also acute. Kittary's
figure of the cartilaginous cranium of his *Ac. sturio,* an
inhabitant of the Caspian, represents the occiput and
snout as being both widely rounded (*l. c.* Pl. vi. f. 5).

Taking Heckel and Kner as the best authorities for
the continental *Ac. sturio,* and more especially for the
fish of that name in the Danube, we find that, though
their figures and description present many characters in
common with the Frith of Forth Sturgeon, there are

some points of difference which prevent us from pronouncing on their identity without further investigation. The specific marks they assign to their *sturio* are, " premaxillary lip with an incurvature, short barbels, osseous centres of the temporals nearer to the point of the snout than those of the coronals; the process of the occipital shield that interposes between the ends of the coronals broad and chisel-shaped or truncated, and the coracoid bucklers roughly granulated, not rayed." Supposing these characters to be constant, the last-mentioned one and the truncation of the salient process of the occipital do not correspond with those parts in the British fish. The skin also of the Austrian *Sturio* is described as being studded with rough, blunt ossicles, mostly uniform in size, being merely a little larger near the head, but nowhere either radiated or stellate. In the form of the dermal ossicles the Frith of Forth Sturgeon agrees with the *Antacei* rather than with the *Sturiones*, but not with any of the six *Antacei* figured in Heckel and Kner's book. The inflexion of the upper lip belongs to all these *Antacei* except *A. schypa*.

Respecting the young, the Austrian authors so often referred to say that *Ac. sturio*, when not exceeding ten inches in length, has a stiletto-shaped snout bent upwards, the occipital enters further between the coronals; in place of the interfrontal shields there is a fontenelle, and the under anterior caudal lobe is not developed. In the Museum of the Free Kirk College at Edinburgh there is a Sturgeon, about eighteen inches long, which may probably be the young of the Frith of Forth species, described at such length in the preceding pages. It has a slender, elongated snout, evidently greatly shrunk in drying, but the arrangement of the cranial shields has much resemblance to that which exists in the older fish. The inter-

frontal plate is composed of six pieces, and the squamosals
are not united to either the mastoids or temporals. The
reason for entertaining a doubt of the identity of the
species is the difference of character of the surface of
both cranial and body-shields. In the small specimen
the roughness of the plates is produced by round grains of
various sizes disposed in radiating lines with furrows between, while in the old the crenulated edges of the thin
walls that bound the depressions do not rise above the
general level. The osseous centres form in the young
thin crests, which on the dorsal shields have a hooked
apex. A continuous crest runs from the centre of the
temporal along the side of the head to that of the mastoid; and owing to the greater elevation of the centres
of the coronals the mesial trough is deeper than in the
larger fish, but the lines which in the latter form the profile of an obelisk are not evident. The suprascapular is
pitted with radiating grooves towards its edges, and the
coracoid is also decidedly radiated. The skin between
the rows of body-shields is studded with roundish, irregular, very small osseous grains, the larger ones being
radiated on the edges. In the recent state this radiation
would be concealed by the epidermis. The barbels are
short, tapering, and more remote from the tip of the
snout than in larger fish. An under caudal lobe is already formed. The body-shields number fourteen on the
dorsal row; thirty-eight on the right side in a line with
the suprascapular, and forty on the left side; eleven
ventral ones on the right side, and ten on the left. The
fin-rays are—

D. 37: V. 27: A. 26.

The osseous centres of the coronals and temporals are
equidistant from the tip of the snout.

The following wood-cut, reproduced from the first

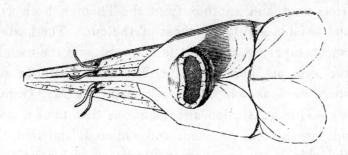

edition of British Fishes, resembles the young Sturgeon of the Free Kirk Museum in the middle ledge of the ventral aspect of the snout, not being dilated gradually towards the point, as in the larger examples. The scale is too small to give a correct idea of the form of the lips, and the figure was probably taken from a small and dried specimen.

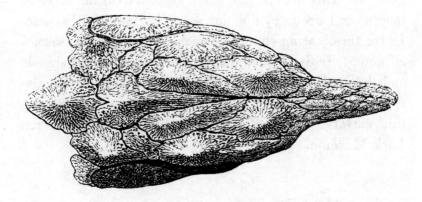

This wood-cut, which was introduced into the second edition of British Fishes, represents an arrangement of the cranial shields, differing from all the Frith of Forth examples in the want of the interfrontal plates, and in the longer tapering prolongation of the salient process of the occipital shield : the squamosals, it will be observed, are distinct pieces. The British Museum possesses a Sturgeon from Teignmouth which is four feet and three-quarters long, and another from the Thames, both with cranial shields similar to those of the cut. The barbels are short, tapering, and a little flattened, and the ossicles in the skin are partly stellate, but mostly minute and angular, as a sketch obligingly made by Mr. Gerard shows. This gentleman also mentions that the coracoid shields are netted, grooved, and radiated, and that the cranial shields are grooved and radiated with a series of ridges connecting the centres of the principal pairs of shields. If this be not a species distinct from the Frith of Forth Sturgeon it is at least a notable variety, but to be ranked equally with it among the *Antacei*, if the form of the mandibular lip will allow, and not with the *Sturiones* of Heckel and Kner. There is still needed a good description of the recent fish in various stages of its growth. This article has been extended to an unusual length, but accuracy did not seem attainable otherwise. In the terminating vignette, which appeared in the second edition of British Fishes, the intervals between the shields and the smallness of the opercular plate denote that the original was a young fish, though it does not show the thin elevated crests of the small specimen in the Free Kirk Museum.

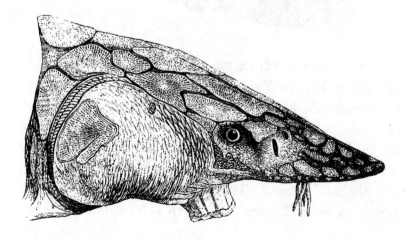

THE GRAY NOTIDANUS.

Notidanus griseus, JONATHAN COUCH, Zool. for 1846, p. 1337, fig.
Hexanchus   ,,   MÜLL. und HENLE, Plagiost. p. 80.
    ,,   ,,   GRAY, Cat. of Chondropt. Brit. Mus. p. 67.

NOTIDANIDÆ. *Family Characters.*—Sharks with a single dorsal and an anal fin. Head flat. A small three-cornered nasal lappet. Upper fold at the corner of the mouth very large, the under one small. Nictitating membrane wanting. Tongue adherent. Spout-holes small, perpendicular. Six or seven stigmata, diminishing successively in length, and all before the pectoral fin. A mesial tooth on the mandible : the next five or six under-teeth form a saw, by the projection of their conical cusps ; the fore or inner borders of the mandibular teeth are either smooth or wholly and finely serrated ; and the distal teeth of that jaw are small and flat. In the upper jaw the teeth are longer, more slender and more pointed, and their first denticle is much longer than the rest : the outer border of the upper teeth is thick, the inner one finely serrated towards its base : the foremost are hook-shaped, on a broad base, and are clustered : the next in succession have exteriorly one or two lateral denticles ; and towards the corner of the mouth, the upper teeth resemble the under ones. Lateral line distinct. The single dorsal stands behind the ventrals, and partly before, partly over the anal. The caudal has small under lobes, with a notch towards the end, which is obliquely or directly docked. No caudal pits. Intestinal valve screw-shaped.

NOTIDANUS.—The only genus, subdivided by Raffinesque into *Hexanchus* and *Heptanchus,* according as the gill-openings are six or seven.

IN the year 1846 a specimen of this fish, caught by a fisherman at Polperro, was brought to Jonathan Couch, Esq., who immediately recognised it as the grey sexbranchial Notidanus, and he soon afterwards published an account of it with a figure in the Zoologist (1337).

A specimen had been also taken in the preceding year by Captain Swinburne, and presented to the British Museum, which likewise possesses a portion of the jaw from Dr. Mantell's collection. The origin of the latter is not stated, and the other two are the only instances known of this Shark being taken in the British seas.

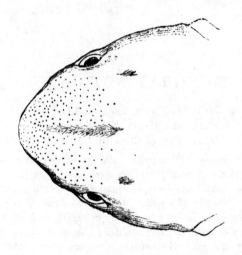

The nostril is nearer to the tip of the rounded snout than to the angle of the mouth, and the spout-hole nearer to the first gill-opening than to the eye. The hooked upper front teeth rise from a broad base, and the succeeding seven or eight large teeth are serrated on the distal edge, the first denticle being decidedly the tallest. The front mandibular tooth has lateral serratures, but no middle cusp, and the following five or six broad teeth on each side are equally serrated on the exterior and longer border by from nine to eleven denticles: their thinner inner borders are finely serrated. Dorsal notched on the edge, so placed that the anal commences before the middle of its base, and half-way between the vent and the caudal fin. Pectoral quadrangular, with rounded corners. Anal rounded anteriorly, pointed posteriorly. A distinct under lobe to the

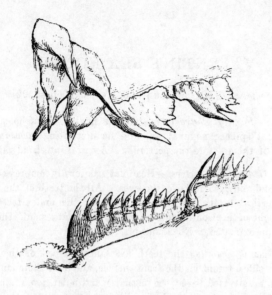

caudal with an obliquely-cut end.   Scales entire and very small, leaf-shaped, with a medial keel which reaches to the acute point.—*Müller and Henle.*

Mr. Couch's specimen was two feet two inches and a half long, and was a male with small claspers.   Captain Swinburne's fish was about eleven feet in length, and was captured off Ventnor in November 1845.   The cuts were all drawn from this specimen.

GRAY NOTIDANUS.

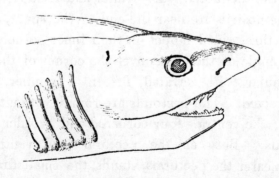

VALENTIN'S SEA-HOUND.

Scymnus lichia, MÜLLER und HENLE, Plagiost. 92.

SCYMNIDÆ. *Family Characters.*—Sharks having spout-holes and two dorsals without spines: no anal fin, and no nictitating membrane. Five stigmata, all of them before the pectorals. A spiral intestinal valve.

SCYMNUS. *Generic Characters.*—Head flat or laterally compressed. Spout-holes far behind and rather above the eyes. All the teeth of the upper jaw straight or vertical to the jaw, slender, hooked: the under teeth broader, with an upright or horizontal cutting edge. Stigmata small, the last two moderately approximated. No caudal pit.

In the sub-genus *Scymnus* the teeth are lancet-shaped on the mandible, sharp on the sides, tumid on the front surface, and have their lancet-shaped tips raised on an elevated base: the mesial mandibular tooth is not smaller than its neighbours, and has the basis alike on both sides, with a notch at the origin of the root. The rest of the mandibular teeth have an impress on the inner side formed by the overlying root of the next tooth. Their roots are bilobate, with a furrow. There is no prickle in the claspers.

THIS Shark occurs in Mr. Yarrell's list of new British Fishes intended for his third edition, but without any intimation of the time or place of its capture. As it is a species which inhabits the Mediterranean and the Bay of Biscay, and may be expected to enter the British Channel occasionally at least, a notice of it is given for the benefit of practical ichthyologists. The following description is quoted from Müller and Henle:—

" The nostrils are near the end of the snout, and have a small three-sided lappet on their inner border, and the hinder angle of the eye is over the corner of the mouth. Mandibular teeth serrated, fifteen in number, with two rows erected. The pectorals are round, without a hinder corner ; the ventrals four-cornered and broader than the pectorals. Between the pectorals and ventrals, and rather nearer the pectorals, stands the small first dorsal, rounded, and without a hinder angle. The second dorsal is larger and four-cornered, and is situated immediately behind the ventrals; it is notched in its upper border, blunt at its fore corner, and pointed behind. Caudal fin destitute of an under lobe, three-cornered. Scales having three or more points on a four-cornered base, with three keels on the fore part. Colour violet-blackish or brownish, uniform, but with some black clouds posteriorly."

LOCH LONG, NEAR THE ENTRANCE OF LOCH GOIL.

There floated Haco's banner trim,
Above Norweyan warriors grim,
Savage of heart, and large of limb;
Threatening both continent and isle,
Bute, Arran, Cunningham, and Kyle.
SCOTT.

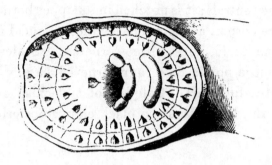

TEETH OF *Petromyzon Juræ.*

The *Petromyzon Juræ*, figured in Mac Culloch's Western Isles (vol. ii. pp. 186-7, t. 29, f. 1) is probably of this
genus. If the teeth are correctly represented, the fish
has no relation to *Lampetra fluvialis* (*vide* GRAY, Cat.
Fish, B. Mus. p. 139). It was found adhering to a
Gurnard on the coast of the island of Jura, one of the
Hebrides.

END OF SECOND SUPPLEMENT
TO SECOND VOLUME.

NOTICE TO THE BINDER.

In binding the First and Second Supplements with the First Edition, place in the First Volume, after page 408 :—

> The Title and Preface to the First Supplement;
> The Title and Preface to the Second Supplement;
> Pages 1 to 48 of the First Supplement;
> Pages 1 to 36 of the Second Supplement;
> The Portrait to face the Title;
> The Memoir and List of Writings to follow the Title of the Work.

And in the Second Volume, after page 472, place :—

> The First Supplement, pages 1 to 72;
> The Second Supplement, pages 1 to 30.

In binding the Second Supplement with the Second Edition :—

> The Portrait should face the Title-page;
> The Memoir and List of Writings should follow the Title-page of Vol. 1;
> The Title-page and Preface, and pages 1 to 36, should follow page 464 in Vol. 1.

In the Second Volume :—

> Pages 1 to 30, Supplement, should follow page 628 of the Work.

N.B. The First Edition requires the two Supplements; the Second Edition requires the Second Supplement only; the third Title-page is for the convenience of those who may have already bound the First Edition in 2 vols. The two Supplements would, in this case, form the third vol.